AF411331

Adipose Tissue Inflammation 2022

Adipose Tissue Inflammation 2022

Editor

Javier Gómez-Ambrosi

Basel • Beijing • Wuhan • Barcelona • Belgrade • Novi Sad • Cluj • Manchester

Editor
Javier Gómez-Ambrosi
Metabolic Research Lab
Department of Endocrinology and Nutrition
Clínica Universidad de Navarra
Pamplona
Spain

Editorial Office
MDPI
St. Alban-Anlage 66
4052 Basel, Switzerland

This is a reprint of articles from the Special Issue published online in the open access journal *Cells* (ISSN 2073-4409) (available at: www.mdpi.com/journal/cells/special_issues/Adipose_Tissue_Inflammation).

For citation purposes, cite each article independently as indicated on the article page online and as indicated below:

Lastname, A.A.; Lastname, B.B. Article Title. *Journal Name* **Year**, *Volume Number*, Page Range.

ISBN 978-3-0365-8675-5 (Hbk)
ISBN 978-3-0365-8674-8 (PDF)
doi.org/10.3390/books978-3-0365-8674-8

Contents

About the Editor . . . . **vii**

Preface . . . . **ix**

Javier Gómez-Ambrosi
Adipose Tissue Inflammation
Reprinted from: *Cells* **2023**, *12*, 1484, doi:10.3390/cells12111484 . . . . **1**

Nir Goldstein, Yarden Kezerle, Yftach Gepner, Yulia Haim, Tal Pecht and Roi Gazit et al.
Higher Mast Cell Accumulation in Human Adipose Tissues Defines Clinically Favorable
Obesity Sub-Phenotypes
Reprinted from: *Cells* **2020**, *9*, 1508, doi:10.3390/cells9061508 . . . . **4**

**Ioannis G. Lempesis, Nicole Hoebers, Yvonne Essers, Johan W. E. Jocken, Kasper M. A.
Rouschop and Ellen E. Blaak et al.**
Physiological Oxygen Levels Differentially Regulate Adipokine Production in Abdominal and
Femoral Adipocytes from Individuals with Obesity Versus Normal Weight
Reprinted from: *Cells* **2022**, *11*, 3532, doi:10.3390/cells11223532 . . . . **22**

**Shihab Kochumon, Hossein Arefanian, Sardar Sindhu, Reeby Thomas, Texy Jacob and
Amnah Al-Sayyar et al.**
Expression of Steroid Receptor RNA Activator 1 (SRA1) in the Adipose Tissue Is Associated
with TLRs and IRFs in Diabesity
Reprinted from: *Cells* **2022**, *11*, 4007, doi:10.3390/cells11244007 . . . . **30**

**Leonardo Sandrini, Alessandro Ieraci, Patrizia Amadio, Marta Zarà, Nico Mitro and Francis
S. Lee et al.**
Physical Exercise Affects Adipose Tissue Profile and Prevents Arterial Thrombosis in *BDNF*
Val66Met Mice
Reprinted from: *Cells* **2019**, *8*, 875, doi:10.3390/cells8080875 . . . . **48**

**Talita da Silva Mendes de Farias, Regislane Ino da Paixao, Maysa Mariana Cruz, Roberta
Dourado Cavalcante da Cunha de Sa, Jussara de Jesus Simão and Vitor Jaco Antraco et al.**
Melatonin Supplementation Attenuates the Pro-Inflammatory Adipokines Expression in
Visceral Fat from Obese Mice Induced by A High-Fat Diet
Reprinted from: *Cells* **2019**, *8*, 1041, doi:10.3390/cells8091041 . . . . **66**

**Tassiane C. Morais, Luiz C. de Abreu, Ocilma B. de Quental, Rafael S. Pessoa, Mahmi
Fujimori and Blanca E. G. Daboin et al.**
Obesity as an Inflammatory Agent Can Cause Cellular Changes in Human Milk due to the
Actions of the Adipokines Leptin and Adiponectin
Reprinted from: *Cells* **2019**, *8*, 519, doi:10.3390/cells8060519 . . . . **79**

**Marcin   Trojnar,   Jolanta   Patro-Małysza,   Żaneta   Kimber-Trojnar,   Bożena
Leszczyńska-Gorzelak and Jerzy Mosiewicz**
Associations between Fatty Acid-Binding Protein 4–A Proinflammatory Adipokine and Insulin
Resistance, Gestational and Type 2 Diabetes Mellitus
Reprinted from: *Cells* **2019**, *8*, 227, doi:10.3390/cells8030227 . . . . **91**

Hui-Hua Chang and Guido Eibl
Obesity-Induced Adipose Tissue Inflammation as a Strong Promotional Factor for Pancreatic
Ductal Adenocarcinoma
Reprinted from: *Cells* **2019**, *8*, 673, doi:10.3390/cells8070673 . . . . **106**

María Eugenia Cornide-Petronio, Mónica B. Jiménez-Castro, Jordi Gracia-Sancho and Carmen Peralta
New Insights into the Liver–Visceral Adipose Axis During Hepatic Resection and Liver Transplantation
Reprinted from: *Cells* **2019**, *8*, 1100, doi:10.3390/cells8091100 . . . . **122**

Fien Demeulemeester, Karin de Punder, Marloes van Heijningen and Femke van Doesburg
Obesity as a Risk Factor for Severe COVID-19 and Complications: A Review
Reprinted from: *Cells* **2021**, *10*, 933, doi:10.3390/cells10040933 . . . . **137**

About the Editor

Javier Gómez-Ambrosi

Javier Gomez-Ambrosi is a Researcher at the Metabolic Research Laboratory of the Clínica Universidad de Navarra and Associate Professor at the School of Medicine in the University of Navarra, Pamplona, Spain. His main area of research is obesity and related morbidities, from a clinical and molecular point of view. His research combines basic research in experimental animals and cells with the clinical setting, trying to disentangle the pathophysiological mechanisms responsible of the impact of excess adiposity on the development of comorbidities. He has published more than 200 articles (h-index 58) and has been the PI in more than 20 research projects.

Preface

Over the last few decades, obesity has become one of the most prevalent metabolic disorders. Excess adiposity favors the development of cardiometabolic alterations, such as type 2 diabetes (T2D), cardiovascular disease, dyslipidemia, non-alcoholic fatty liver disease, and cancer. In the last years, adipose tissue inflammation represents one of the major mechanisms underlying adipose tissue dysfunction, contributing to the development of metabolic derangements in other organs. The contribution of the different adipose tissue depots, the discovery of the function of new adipokines, the involvement of the inflammasome, or the dual effect of macrophage polarization have greatly contributed to the improved understating produced in previous years, of the role played by adipose tissue inflammation in the development of metabolic derangements. This reprint presents recent advances in understanding the molecular processes that take place in adipose tissue inflammation; moreover, it discusses the impact of adipose tissue inflammation on systemic metabolic alterations associated with excess adiposity, as well as its repercussion in several pathological conditions.

Javier Gómez-Ambrosi
Editor

Editorial

Adipose Tissue Inflammation

Javier Gómez-Ambrosi [1,2,3]

1 Metabolic Research Laboratory, Clínica Universidad de Navarra, 31008 Pamplona, Spain; jagomez@unav.es
2 Centro de Investigación Biomédica en Red-Fisiopatología de la Obesidad y Nutrición (CIBEROBN), Instituto de Salud Carlos III, 31008 Pamplona, Spain
3 Obesity and Adipobiology Group, Instituto de Investigación Sanitaria de Navarra (IdiSNA), 31008 Pamplona, Spain

Citation: Gómez-Ambrosi, J. Adipose Tissue Inflammation. *Cells* **2023**, *12*, 1484. https://doi.org/10.3390/cells12111484

Received: 22 May 2023
Accepted: 24 May 2023
Published: 26 May 2023

In recent decades, obesity has become one of the most common metabolic diseases. Excess adiposity increases the risk of developing type 2 diabetes (T2D), cardiometabolic diseases, dyslipidemia, fatty liver, and several types of cancer [1]. Much progress has been made in understanding the major regulatory pathways underlying adipose tissue inflammation, which represent one the main drivers of adipose tissue dysfunction and, consequently, of obesity-associated metabolic alterations [2].

This Special Issue presents recent advances in understanding the molecular processes that take place in adipose tissue inflammation; moreover, it discusses the impact of adipose tissue inflammation on systemic metabolic alterations associated with excess adiposity, as well as its repercussion in several pathological conditions.

Although obesity has traditionally been considered a single medical entity, in recent years, greater importance has been placed on phenotyping the different obesities in order to improve their clinical management [3,4]. In their cross-sectional and prospective study, Goldstein et al. analyze the usefulness of determining mast cell accumulation in human adipose tissue as a proxy of metabolic phenotyping [5]. They suggest that patients with obesity with high expression of mast cell genes exhibit a healthier metabolic phenotype than individuals with low expression. The authors also find that higher mast cell accumulation in adipose tissue in patients undergoing bariatric surgery is a predictor of greater weight loss after surgery. They conclude that a high number of mast cells defines a clinically favorable obesity phenotype [5].

Several studies discuss the molecular mechanisms that regulate the impact of adipose tissue inflammation. Lempesis et al. show that low physiological oxygen tension decreases the expression and secretion of proinflammatory adipokines in adipocytes obtained from patients with obesity, an effect that is not found in cells derived from donors of normal weight [6]. Kochumon and colleagues report that the adipose tissue expression of steroid receptor RNA activator 1 (SRA1) may represent a novel surrogate marker of metabolic inflammation through its association with Toll-like receptors (TLRs) [7].

Other studies provide interesting information regarding the regulation of inflammation in adipose tissue in mouse models. In one such study, Sandrini et al. study the effects of physical exercise on *BDNF* Val66Met mice, a model of increased adiposity associated with a proinflammatory and prothrombotic profile [8]. They find that four weeks of voluntary wheel running changes epididymal adipose tissue morphology and the expression of proinflammatory genes, inducing reversion of the prothrombotic phenotype; this suggests that a reduction in adipose tissue inflammation is important in promoting the positive effects of physical activity [8]. In another study, Mendes de Farias and colleagues find that daily melatonin supplementation for 10 weeks in mice on a high-fat diet reduces fat accumulation, adipocyte size, and the expression of proinflammatory adipokines in adipose tissue and the circulation; this suggests that melatonin could be considered as a therapeutic molecule for the treatment of obesity [9].

Adipokines and adipose tissue inflammation have been shown to play a role in several physiologic and pathophysiologic conditions [10,11], as stated in several studies included in this Special Issue. Morais et al. show that an adequate balance between adiponectin and leptin concentrations in human milk may regulate colostrum mononuclear cell activity, eliciting a more effective response against neonatal infection in breastfeeding infants [12]. Another adipokine, fatty acid-binding protein 4 (FABP4), implicated in the control of cellular lipid metabolism, is also involved in inflammation and the development of insulin resistance. In an exhaustive review, Trojnar et al. detail the different mechanisms by which FABP4 is involved in inflammation and insulin resistance and the potential role of this adipokine in T2D, gestational diabetes, and fatty liver, among other conditions [13]. Chang and Eibl describe the relevance of adipose tissue inflammation as an important driver of obesity-associated pancreatic ductal adenocarcinoma, and consider strategies aimed at reducing inflammation in this tissue as a weapon against this type of cancer [14]. In another interesting review, Cornide-Petronio and colleagues reinforce current knowledge regarding the interaction between the liver and adipose tissue during liver surgery. The scientific and clinical controversies in this area are reviewed, as are potential therapeutic approaches. The information provided could help to develop protective measures focused on manipulating the liver–visceral adipose tissue axis to enhance the postoperative results of hepatic surgery [15]. Finally, Demeulemeester et al. summarize scientific research on the link between obesity and COVID-19 severity, and analyze probable mechanisms that could help to understand why patients living with obesity exhibit an increased risk of serious consequences during COVID-19 [16].

In recent years, there have been significant advancements in the understanding of the cellular and molecular mechanisms involved in adipose tissue inflammation. Thanks to this progress, more tools and approaches are available for the treatment of obesity and T2D. However, in order to optimize the management of patients with obesity, more research needs to be conducted.

Funding: This paper was supported by ISCIII–FEDER (PI20/00080) and CIBEROBN, ISCIII, Spain, and by PC098-099 MEPERTROBE and the Department of Health 58/2021, Gobierno de Navarra-FEDER, Spain.

Conflicts of Interest: The author declares no conflict of interest.

References

1. Catalán, V.; Avilés-Olmos, I.; Rodríguez, A.; Becerril, S.; Fernández-Formoso, J.A.; Kiortsis, D.; Portincasa, P.; Gómez-Ambrosi, J.; Frühbeck, G. Time to consider the "Exposome Hypothesis" in the development of the obesity pandemic. *Nutrients* **2022**, *14*, 1597. [CrossRef] [PubMed]
2. Frühbeck, G.; Balaguer, I.; Méndez-Giménez, L.; Valentí, V.; Becerril, S.; Catalán, V.; Gómez-Ambrosi, J.; Silva, C.; Salvador, J.; Calamita, G.; et al. Aquaporin-11 contributes to TGF-b1-induced endoplasmic reticulum stress in human visceral adipocytes: Role in obesity-associated inflammation. *Cells* **2020**, *9*, 1403. [CrossRef] [PubMed]
3. Salmón-Gómez, L.; Catalán, V.; Frühbeck, G.; Gómez-Ambrosi, J. Relevance of body composition in phenotyping the obesities. *Rev. Endocr. Metab. Disord.* **2023**, 1–15. [CrossRef] [PubMed]
4. Perdomo, C.M.; Avilés-Olmos, I.; Dicker, D.; Frühbeck, G. Towards an adiposity-related disease framework for the diagnosis and management of obesities. *Rev. Endocr. Metab. Disord.* **2023**, 1–13. [CrossRef] [PubMed]
5. Goldstein, N.; Kezerle, Y.; Gepner, Y.; Haim, Y.; Pecht, T.; Gazit, R.; Polischuk, V.; Liberty, I.F.; Kirshtein, B.; Shaco-Levy, R.; et al. Higher mast cell accumulation in human adipose tissues defines clinically favorable obesity sub-phenotypes. *Cells* **2020**, *9*, 1508. [CrossRef] [PubMed]
6. Lempesis, I.G.; Hoebers, N.; Essers, Y.; Jocken, J.W.E.; Rouschop, K.M.A.; Blaak, E.E.; Manolopoulos, K.N.; Goossens, G.H. Physiological oxygen levels differentially regulate adipokine production in abdominal and femoral adipocytes from individuals with obesity versus normal weight. *Cells* **2022**, *11*, 3532. [CrossRef] [PubMed]
7. Kochumon, S.; Arefanian, H.; Sindhu, S.; Thomas, R.; Jacob, T.; Al-Sayyar, A.; Shenouda, S.; Al-Rashed, F.; Koistinen, H.A.; Al-Mulla, F.; et al. Expression of steroid receptor RNA activator 1 (SRA1) in the adipose tissue is associated with TLRs and IRFs in diabesity. *Cells* **2022**, *11*, 4007. [CrossRef] [PubMed]
8. Sandrini, L.; Ieraci, A.; Amadio, P.; Zara, M.; Mitro, N.; Lee, F.S.; Tremoli, E.; Barbieri, S.S. Physical exercise affects adipose tissue profile and prevents arterial thrombosis in BDNF Val66Met mice. *Cells* **2019**, *8*, 875. [CrossRef] [PubMed]

9. Farias, T.; Paixao, R.I.D.; Cruz, M.M.; de Sa, R.; Simao, J.J.; Antraco, V.J.; Alonso-Vale, M.I.C. Melatonin supplementation attenuates the pro-inflammatory adipokines expression in visceral fat from obese mice induced by a high-fat diet. *Cells* **2019**, *8*, 1041. [CrossRef] [PubMed]
10. Frühbeck, G.; Catalán, V.; Rodríguez, A.; Gómez-Ambrosi, J. Adiponectin-leptin ratio: A promising index to estimate adipose tissue dysfunction. Relation with obesity-associated cardiometabolic risk. *Adipocyte* **2018**, *7*, 57–62. [CrossRef] [PubMed]
11. Frühbeck, G.; Catalán, V.; Rodríguez, A.; Ramírez, B.; Becerril, S.; Salvador, J.; Colina, I.; Gómez-Ambrosi, J. Adiponectin-leptin ratio is a functional biomarker of adipose tissue inflammation. *Nutrients* **2019**, *11*, 454. [CrossRef] [PubMed]
12. Morais, T.C.; de Abreu, L.C.; de Quental, O.B.; Pessoa, R.S.; Fujimori, M.; Daboin, B.E.G.; Franca, E.L.; Honorio-Franca, A.C. Obesity as an inflammatory agent can cause cellular changes in human milk due to the actions of the adipokines leptin and adiponectin. *Cells* **2019**, *8*, 519. [CrossRef] [PubMed]
13. Trojnar, M.; Patro-Malysza, J.; Kimber-Trojnar, Z.; Leszczynska-Gorzelak, B.; Mosiewicz, J. Associations between fatty acid-binding protein 4. A proinflammatory adipokine and insulin resistance, gestational and type 2 diabetes mellitus. *Cells* **2019**, *8*, 227. [CrossRef] [PubMed]
14. Chang, H.H.; Eibl, G. Obesity-induced adipose tissue inflammation as a strong promotional factor for pancreatic ductal adenocarcinoma. *Cells* **2019**, *8*, 673. [CrossRef] [PubMed]
15. Cornide-Petronio, M.E.; Jiménez-Castro, M.B.; Gracia-Sancho, J.; Peralta, C. New insights into the liver-visceral adipose axis during hepatic resection and liver transplantation. *Cells* **2019**, *8*, 1100. [CrossRef] [PubMed]
16. Demeulemeester, F.; de Punder, K.; van Heijningen, M.; van Doesburg, F. Obesity as a risk factor for severe COVID-19 and complications: A review. *Cells* **2021**, *10*, 933. [CrossRef] [PubMed]

 *cells*

Article

Higher Mast Cell Accumulation in Human Adipose Tissues Defines Clinically Favorable Obesity Sub-Phenotypes

Nir Goldstein [1,2], Yarden Kezerle [3], Yftach Gepner [2], Yulia Haim [1], Tal Pecht [1], Roi Gazit [4,5], Vera Polischuk [6], Idit F. Liberty [6], Boris Kirshtein [6], Ruthy Shaco-Levy [3], Matthias Blüher [7,8] and Assaf Rudich [1,5,*]

[1] Department of Clinical Biochemistry and Pharmacology, Ben-Gurion University of the Negev, Beer-Sheva 8410501, Israel; goldsnir@tauex.tau.ac.il (N.G.); beckyu@bgu.ac.il (Y.H.); pacht@post.bgu.ac.il (T.P.)

[2] Department of Epidemiology and Preventive Medicine, School of Public Health, Sackler Faculty of Medicine and Sylvan Adams Sports Institute Tel Aviv University, Tel-Aviv 6997801, Israel; gepner@tauex.tau.ac.il

[3] Institute of Pathology, Soroka University Medical Center Ben-Gurion University of the Negev, Beer-Sheva 84101, Israel; kezerley@bgu.ac.il (Y.K.); rsl@bgu.ac.il (R.S.-L.)

[4] The Shraga Segal Department of Microbiology Immunology and Genetics Faculty of Health Sciences, Ben-Gurion University of the Negev, Beer-Sheva 8410501, Israel; gazitroi@bgu.ac.il

[5] The National Institute of Biotechnology in the Negev, Ben-Gurion University of the Negev, Beer-Sheva 8410501, Israel

[6] Soroka University Medical Center and Faculty of Health Sciences, Ben-Gurion University of the Negev, Beer-Sheva 84101, Israel; VeraP@clalit.org.il (V.P.); iliberty@bgu.ac.il (I.F.L.); borkirsh@bgu.ac.il (B.K.)

[7] Department of Medicine, University of Leipzig, 04103 Leipzig, Germany; bluma@medizin.uni-leipzig.de

[8] Helmholtz Institute for Metabolic, Obesity and Vascular Research (HI-MAG) of the Helmholtz Zentrum München at the University of Leipzig and University Hospital Leipzig, 04103 Leipzig, Germany

* Correspondence: rudich@bgu.ac.il

Received: 4 April 2020; Accepted: 18 June 2020; Published: 20 June 2020

Abstract: The identification of human obesity sub-types may improve the clinical management of patients with obesity and uncover previously unrecognized obesity mechanisms. Here, we hypothesized that adipose tissue (AT) mast cells (MC) estimation could be a mark for human obesity sub-phenotyping beyond current clinical-based stratifications, both cross-sectionally and prospectively. We estimated MC accumulation using immunohistochemistry and gene expression in abdominal visceral AT (VAT) and subcutaneous (SAT) in a human cohort of 65 persons with obesity who underwent elective abdominal (mainly bariatric) surgery, and we validated key results in two clinically similar, independent cohorts ($n = 33$, $n = 56$). AT-MC were readily detectable by immunostaining for either c-kit or tryptase and by assessing the gene expression of KIT (KIT Proto-Oncogene, Receptor Tyrosine Kinase), TPSB2 (tryptase beta 2), and CMA1 (chymase 1). Participants were characterized as VAT-MC^{low} if the expression of both CMA1 and TPSB2 was below the median. Higher expressers of MC genes (MC^{high}) were metabolically healthier (lower fasting glucose and glycated hemoglobin, with higher pancreatic beta cell reserve (HOMA-β), and lower triglycerides and alkaline-phosphatase) than people with low expression (MC^{low}). Prospectively, higher MC accumulation in VAT or SAT obtained during surgery predicted greater postoperative weight-loss response to bariatric surgery. Jointly, high AT-MC accumulation may be used to clinically define obesity sub-phenotypes, which are associated with a "healthier" cardiometabolic risk profile and a better weight-loss response to bariatric surgery.

Keywords: obesity; type 2 diabetes; bariatric surgery; adipose tissue; mast cells

1. Introduction

The inflammation of adipose tissue (AT) may link obesity to its cardiometabolic comorbidities. Although macrophages (CD68+ cells) were the first immune cell type realized to infiltrate adipose tissue (AT) in obesity and to associate with obesity-related metabolic dysfunction [1], the current view in the field engages virtually all immune cell types in obesity-associated AT inflammation, including T and B lymphocytes (and their sub-classes), dendritic cells, neutrophils, and natural killer (NK) cells [2]. Within the changing environment of adipose tissues in obesity, these cell types undergo complex phenotypic alterations, rendering them highly diverse. This challenges the efforts to establish clear contributions of specific cell types to obesity-associated metabolic deterioration, even for adipose tissue macrophages [3] (also recently reviewed in [4,5]). In 2009, elevated numbers of mast cells (MC) were demonstrated in obese AT, located mainly near micro-vessels, both in humans and in mouse models [6]. Mast cells are hematopoietic, bone-marrow-derived immune cells, which mature and differentiate in the tissue in which they eventually reside. Although mainly found in tissues with greater interaction with the outer environment such as the reticular layers of the skin, gastrointestinal system, and the airways, MC can be found in all organs and tissues. In AT of lean mice, MC are more abundant in subcutaneous AT (SAT) than in epididymal, mesenteric, or perirenal fat pads. Interestingly, obesity is associated with increases in MC numbers in all white AT depots besides SAT [7].

The physiological/functional impact of increased numbers of AT-MC in obesity currently remains controversial: Two independent groups demonstrated that MC-deficient mice by mutation in the growth factor receptor KIT (KitW-sh/W-sh), which is required for MC development, are protected from diet-induced obesity and its co-morbidities [6,8]. In contrast, two different, none KIT-dependent MC-deficient mouse models (Cpa3cre/+ (Cre recombinase in the carboxypeptidase A locus) and Mcpt5-Cre+ R-DTA+ (MC specific expression of diphtheria toxin A)) exhibited no phenotypic impact on the development of obesity or on its metabolic consequences, including insulin resistance, hepatic steatosis, or inflammation [9,10].

In humans, the AT of people with obesity exhibited higher MC numbers compared to lean patients, both in SAT [6,11] and VAT [11]. Importantly, this finding was particularly evident among patients with obesity and type 2 diabetes [11], suggesting a link between high AT-MC infiltration and the severity of obesity-related metabolic disturbance. Consistently, a sub-analysis among 20 persons with obesity suggested that higher MC numbers associate with higher fasting glucose and HbA1c [11]. Interestingly, in that study, two sub-populations of AT-MC were assessed based on the expression of their proteases—tryptase+ and tryptase+/chymase+ MC (MCT and MCTC, respectively), rendering tryptase, the gene product of tryptase beta-2 (TPSB2), a common marker for MC. Yet, the ratio between these two sub-populations remained similar in different depots and in leanness versus obesity. Isolated MC from the SAT of patients with obesity and type 2 diabetes were more activated, releasing more inflammatory cytokines and proteases such as tryptase [11]. Correspondingly, serum tryptase concentrations were higher in people with obesity compared to lean [6]. In a different study examining a mixed cohort of persons without or with obesity, the expression of MC-tryptase (TPSB2) in VAT did not associate with parameters of glucose homeostasis or insulin sensitivity [12]. Collectively, obesity seems to associate with increased numbers of MC in AT. Mouse models so far yielded conflicting results. Human studies suggest that higher AT-MC numbers (i.e., accumulation), and perhaps their activation, may associate with poor glycemic control, but assessment of whether and how AT-MC associate within cohorts of patients with obesity, and whether it corresponds to obesity sub-phenotypes, is limited.

In the present study, we aimed to address this current gap of knowledge by hypothesizing that among persons with obesity, higher numbers of AT-MC associate with an obesity phenotype characterized by a poorer metabolic profile. We sought to assess the relevance of such MC-based obesity sup-phenotyping both in cross-sectional analyses of patients with obesity, with and without type 2 diabetes, and for predicting patients' response to bariatric surgery.

2. Materials and Methods

2.1. Human Cohorts

We recruited persons with obesity (Body mass index (BMI) $\geq$ 30 kg/m^2) undergoing elective abdominal surgery (mainly bariatric surgery or elective cholecystectomy) as part of the coordinated human adipose tissue bio-banks in Beer-Sheva, Israel (n = 65, main cohort) and in Leipzig, Germany (n = 32 and n = 56, validation cohorts 1 and 2, respectively) (Table 1). Prior to operation, under overnight fasting conditions, body weight and blood samples were obtained. Both visceral (omental) and superficial-subcutaneous adipose tissues biopsies were obtained during the surgery and processed for histology and gene expression using coordinated methodologies, as we previously described in detail [13,14]. Persons were identified as normoglycemic if fasting plasma glucose (FPG) levels were lower than 5.6 mmol/L, HbA1c $\leq$ 38 mmol/mol (5.6%), and with no anti-diabetic medications on the day of operation. Prediabetes was defined as FPG 5.6–6.9 mmol/L and HbA1c 39–46 mmol/mol (5.7–6.4%), and type 2 diabetes if glucose $\geq$ 7.0 mmol/L or HbA1c $\geq$ 48 mmol/mol (6.5%). In 13 patients (20%), in whom glycemic status was ambivalent, medical records were screened up to 4 months pre-operation for additional FPG and HbA1c measurements, and final categorization was made by co-author IFL, who is a Diabetologist. A homeostatic model assessment of insulin resistance (HOMA-IR) and homeostatic model assessment of beta cells reserve (HOMA-β) were calculated as follows: HOMA-IR: (FPG (mmol/L) $\times$ Insulin (μIU/mL)/22.5) and HOMA-β: (20 $\times$ Insulin (μIU/mL)/ FPG (mmol/L)-3.5) [15]. For post-operation follow-up sub-study, only people undergoing bariatric surgery for the first time and for whom postoperative information was available were included. In addition to the main Beer-Sheva cohort, we included two independent cohorts—validation cohorts 1 and 2, with n = 32 and n = 56 individuals, respectively, all with obesity (BMI range: 30–75 kg/m^2, Table 1), from the University of Leipzig Obesity Treatment Center. Paired abdominal subcutaneous and omental adipose tissue biopsies were taken during elective sleeve gastrectomy, Roux-en-Y gastric bypass, hernia, or cholecystectomy surgeries and processed as previously described [16]. As for validation cohort 2, we included data from 56 patients who underwent a two-step bariatric surgery strategy with laparoscopic gastric sleeve resection as the first step and a Roux-en-Y gastric bypass as the second step 12 ± 2 months later (Table 1). At both time points, serum/plasma samples, omental, and abdominal subcutaneous adipose tissue biopsies were obtained. All patients provided before the study a written informed consent to participate, and all procedures were approved in advance by the local ethical committees and conducted in accordance with the declaration of Helsinki guidelines (0348-15-SOR; for Leipzig cohorts:, 017-12-23012012, and Reg. No. 031-2006).

Table 1. Baseline characteristics of obese patient.

	Normal Values/Units	Beer-Sheva (Main Cohort)			Leipzig-1 (Validation Cohort 1)			Leipzig 2 (Validation Cohort 2)		
		Total	VAT-Low	VAT-High	Total	VAT-Low	VAT-High	Total	VAT-Low	VAT-High
N		**65**	19	46	**32**	7	25	**56**	18	38
Age	(year)	**46.7 ± 14.2**	50.8 ± 13.7	45.0 ± 14.3	**47.7 ± 15.3**	44.1 ± 8.5	48.9 ± 11.7	**43.9 ± 10.3**	45.4 ± 6.9	43.2 ± 11.6
Sex	M/F	**25/40**	8/11	17/29	**9/23**	1/6	8/17	**13/43**	3/15	10/28
Weight	(kg)	**112.5 ± 23.6**	111.7 ± 22.6	112.8 ± 24.2	**134.9 ± 32.6**	120.4 ± 24.3	139.0 ± 33.8	**144.2 ± 30.3**	132.2 ± 24.8	149.9 ± 31.2 *
BMI	<25 (kg/m^2)	**40.7 ± 6.0**	40.2 ± 5.8	40.9 ± 6.2	**47.9 ± 11.2**	44.1 ± 8.5	48.9 ± 11.7	**50.5 ± 8.6**	47.9 ± 7.8	51.7 ± 8.8
FPG	3.9–5.4 mmol/L	**6.5 ± 2.7**	7.7 ± 3.9	6.0 ± 1.8	**6.7 ± 2.2**	6.8 ± 1.8	6.7 ± 2.3	**6.6 ± 1.7**	7.0 ± 1.8	6.4 ± 1.6
Insulin	<174 pmol/L	**105.1 ± 72.5**	89.9 ± 49.8	111.4 ± 79.8	**135.1 ± 158.9**	74.3 ± 49.5	147.2 ± 171.1	**206.1 ± 163.4**	211.6 ± 215.2	203.5 ± 135.1
HbA1c	<39 mmol/mol	**45.9 ± 16.7**	49.2 ± 22.6	43.9 ± 13.9	**44.3 ± 9.1**	47.7 ± 10.7	43.2 ± 8.5	**46.4 ± 11.8**	46.9 ± 9.6	46.2 ± 12.9
	<5.7%	**6.3 ± 1.6**	6.8 ± 2.1	6.2 ± 1.3	**6.2 ± 0.8**	6.5 ± 1.0	6.1 ± 0.8	**6.4 ± 1.1**	6.4 ± 0.9	6.4 ± 1.2
HOMA-IR	<2.5	**5.2 ± 3.8**	5.3 ± 3.8	5.1 ± 3.9	**6.7 ± 11.4**	3.0 ± 1.9	7.4 ± 12.3	**8.9 ± 7.4**	9.0 ± 9.3	8.8 ± 6.5
HOMA-β %		**173.4 ± 159.6**	112.4 ± 80.6	198.6 ± 177.4 *	**130.6 ± 92.3**	82.6 ± 66.7	140.3 ± 95.1	**219.3 ± 166.8**	210.6 ± 208.1	223.6 ± 145.9
Total cholesterol	<5.2 mmol/L	**4.8 ± 1.1**	5.1 ± 1.3	4.6 ± 1.0	**4.7 ± 1.1**	5.2 ± 1.2	4.4 ± 1.0	**5.3 ± 1.1**	5.2 ± 1.0	5.3 ± 1.2
LDL-c	<1.8 mmol/L	**2.8 ± 0.9**	2.7 ± 0.9	2.8 ± 0.9	**3.2 ± 1.3**	3.3 ± 0.9	3.2 ± 1.5	**3.3 ± 1.0**	3.1 ± 0.5	3.4 ± 1.2
Triglycerides	<1.7 mmol/L	**2.0 ± 1.2**	2.7 ± 1.6	1.7 ± 0.8 *	**1.9 ± 1.0**	1.7 ± 0.8	1.9 ± 1.2	**2.0 ± 1.0**	1.9 ± 0.7	2.1 ± 1.1
HDL: male	>1.03 mmol/L	**1.0 ± 0.2**	1.0 ± 0.2	1.0 ± 0.2	**1.1 ± 0.2**	1.2 ± 0.2	1.1 ± 0.2	**1.1 ± 0.3**	1.0 ± 0.1	1.1 ± 0.4
female	>1.29 mmol/L	**1.2 ± 0.4**	1.2 ± 0.4	1.2 ± 0.4				**1.2 ± 0.3**	1.2 ± 0.4	1.2 ± 0.3
TG/HDL ratio: male		**2.6 ± 1.5**	3.1 ± 1.8	2.3 ± 1.4	**1.8 ± 1.2**	1.3 ± 0.6	2.0 ± 1.3	**2.6 ± 1.9**	2.0 ± 0.4	2.7 ± 2.1
female		**1.6 ± 1.3**	2.2 ± 2.1	1.4 ± 0.8 *				**1.7 ± 1.1**	1.6 ± 0.9	1.7 ± 1.2
CRP	<47.6 nmol/L	**15.7 ± 25.2**	10.5 ± 9.7	17.9 ± 29.2	**17.6 ± 19.1**	10.1 ± 8.6	19.4 ± 20.6	**59.4 ± 65.3**	46.5 ± 63.8	65.4 ± 65.6
AST	<0.68 μkat/L	**0.5 ± 0.4**	0.6 ± 0.4	0.5 ± 0.4	**0.7 ± 0.7**	0.5 ± 0.1	0.8 ± 0.8	**0.6 ± 0.3**	0.6 ± 0.3	0.6 ± 0.3
ALT	<0.68 μkat/L	**0.6 ± 0.4**	0.7 ± 0.4	0.5 ± 0.4	**0.7 ± 0.5**	0.5 ± 0.1	0.8 ± 0.6	**0.7 ± 0.4**	0.7 ± 0.5	0.6 ± 0.4
AP	<2 μkat/L	**1.5 ± 0.5**	1.8 ± 0.6	1.4 ± 0.4 *						
Diastolic BP	<85 mmHg	**81.0 ± 17.2**	88.2 ± 27.7	78.4 ± 10.6						
Systolic BP	<130 mmHg	**138.5 ± 16.1**	139.1 ± 19.7	138.3 ± 14.8						
Visceral Adipocyte area	(μm^2)	**4213.6 ± 1323.8**	4455.8 ± 1138.7	4097.8 ± 1412.8						
Subcutaneous Adipocyte area	(μm^2)	**5720.9 ± 1284.1 †**	5771.7 ± 1337.8	5696.7 ± 1290.7						
SAT-KIT		**1.0 ± 0.7**	0.8 ± 0.3	1.1 ± 0.8	**5.9 ± 0.3**	5.9 ± 0.2	5.9 ± 0.4	**1.2 ± 0.7**	1.0 ± 0.5	1.3 ± 0.8
SAT-TPSB2		**1.2 ± 0.8**	1.23 ± 0.8	1.2 ± 0.8	**6.3 ± 0.5**	6.1 ± 0.3	6.2 ± 0.5	**1.2 ± 0.9**	0.8 ± 0.5	1.3 ± 0.9 *
SAT-CMA1		**2.1 ± 1.0**	1.6 ± 0.8	2.3 ± 1.1	**5.3 ± 0.1**	5.2 ± 0.1	5.3 ± 0.5	**1.3 ± 0.8**	1.1 ± 1.1	1.3 ± 0.6
VAT fibrosis grade		**1.8 ± 0.7**	1.8 ± 0.6	1.8 ± 0.7						

Clinical characteristics of participants with obesity from the Beer–Sheva and Leipzig cohorts included in this study. VAT-MC[low/high] were defined based on VAT expression of CMA1 and TPSB2, as detailed in methods (see also Table S2 for MC[low/high] definition based on VAT expression of KIT). M, male; F, female; FPG, fasting plasma glucose; HOMA-IR, homeostatic model assessment of insulin resistance; HOMA-β, homeostatic model assessment of beta cells reserve; LDL-c, low-density lipoprotein cholesterol; TG, triglycerides; HDL-c, high-density lipoprotein cholesterol; CRP, c-reactive protein; AST; aspartate aminotransferase; ALT, alanine transaminase; AP, alkaline phosphatase; VAT-MC: visceral adipose tissue with mast cells; Values are mean ± standard deviation. * $p < 0.05$ different from VAT-MC[low] by independent *t*-test. † $p < 0.05$ compared visceral adipocyte area, by paired *t*-test. # C- reactive protein (CRP) values are higher compared with validation cohort 1 and the main cohort since inclusion criterion of patients with CRP < 47.6 nmol/L (5.0 mg/L) was not applied in this cohort.

2.2. RNA Extraction and Quantification

mRNA was extracted as described previously [13]. Briefly, 300 mg of tissue were minced and extraction was done with an RNeasy lipid tissue minikit (Qiagen. Hilden Germany). cDNA was produced using the reverse transcriptase kit (Applied Biosystems. Beverly, MA, USA). mRNA was quantified using the Taqman system, where expression levels of selected genes were calculated as a fold from AT-specific endogenous control genes (PGK1 and PPIA) [17] and calculated as $2^{-\Delta\Delta ct}$ [18]. Probes assay IDs are in Table S1. For the Leipzig validation cohort 1, mRNA expression of candidate genes (cKIT, CMA1, TPSB2) was analyzed using Illumina human HT-12 expression chips. RNA integrity and concentration were examined using an Agilent 2100 Bioanalyzer (Agilent Technologies, USA). In validation cohort 2, additional measurements of cytokine serum concentrations and adipose tissue expression (Interleukin -6 (IL-6), Interleukin -1beta (IL-1beta), Tumor necrosis factor alpha (TNFalpha)) were performed as described previously [3].

2.3. Histology and Immunostaining

Immunostaining for C-Kit (Dako (Agilent), Santa Clara, CA USA) or Tryptase (ThermoFisher. Waltham, MA USA), and CD68 (Dako (Agilent), performed in a sub-cohort of $n = 30$ from the Beer-Sheva cohort), is performed routinely in clinical pathology to stain tissue MC and macrophages, respectively. Immunostaining was performed in 5 microns-thick sections from paraffin embedded visceral AT (VAT) and subcutaneous (SAT) samples as described before [14]. Cell count/field was performed, blindly and independently, by two pathologists (Y.K and R.S-L) using an Olympus BX43 light microscope, in 10 consecutive high-power fields (X400), and the number of C-Kit+ and tryptase+ per 100 adipocytes was calculated.

2.4. Statistical Analyses

Baseline clinical characteristics values are presented as mean ± SD. An independent t-test was used to compare between groups with Levene's test for equality of variances. In cases of non-normal distribution, ln-transformation was made. Comparison between percentages of medications usage was calculated with a chi-square test. An ANOVA test was used to compare between the four groups when sex, age, and diabetes stratification was used in combination with MC high/low stratifications. In order to detect differences between the four groups, Least Significant Difference (LSD) post-hoc analysis was used. A 2×2 ANOVA was performed to test the relationship between non- and type 2 diabetes groups and the MC group (MClow/high) using an age-adjusted model. Extreme outliers were excluded using the interquartile range $3 \times$ (Q3-Q1). We also used Spearman's correlation to test the association between c-Kit+, tryptase+, and CD68$^+$ cells, clinical parameters, and MC-related genes. A linear regression model was used to assess the association between VAT-CMA1 high/low expression and percentage of weight loss. The model was adjusted for baseline BMI or surgery type. Analyses were performed using SPSS Version 20.0 (SPSS, Inc., Chicago, IL, USA) and GraphPad prism (8.4.2., San-Diego, CA, USA). All tests were two-tailed and $p < 0.05$ was considered statistically significant.

3. Results

3.1. Assessing AT-MC Histologically and by Gene Expression

To test if the degree of AT-MC accumulation may reflect obesity sub-phenotypes, we first investigated a cohort of persons with obesity (BMI $\geq$ 30 kg/m^2, $n = 65$) (Table 1, Beer–Sheva cohort). Among the persons with obesity, 38 persons (58.5%) were morbidly obese (Class III, BMI $\geq$ 40 kg/m^2). Compared to those with 30 $\leq$ BMI < 40 kg/m^2, persons with morbid obesity differed only in weight, BMI, waist circumference, and low-density lipoprotein (LDL).

We contemplated both histological and gene expression approaches to estimate AT-MC accumulation, utilizing several genes/proteins considered as MC-specific, including KIT/c-Kit (c-KIT Proto-Oncogene Receptor Tyrosine Kinase), TPSB2/Tryptase (mast cell tryptase beta II), and CMA1 (mast cell chymase 1 [19]). The specificity of selected genes to MC is presented in Supplemental Figures S1–S3. By immunohistochemistry, c-Kit+ or tryptase+ cells were readily discernable in human visceral AT from obese patients (Figure 1A,B). When assessing serial sections, the number of c-Kit+ cells per 100 adipocytes (percentage of c-Kit+) highly correlated with the percentage of tryptase+ cells (Figure 1C). AT-MC could be detected both in fibrotic areas within AT and dispersed between the adipocytes, being more prevalent in VAT sections rated by Clinical Pathologists (co-authors YK and RSL) as exhibiting more severe fibrosis (Figure 1D, for SAT see Figure S4). Interestingly, AT-MC correlated with CD68$^+$ macrophages only within fibrotic areas of the tissues (Figure 1E), but not in parenchymal areas between adipocytes ($r(\varrho) = 0.264$, $p = 0.159$, $n = 30$). Using a gene expression approach, the expression of all three MC genes was significantly intercorrelated in both VAT (Figure 1F) and SAT (not shown). Moreover, VAT-KIT gene expression levels correlated with visceral percentage of c-Kit+ cells (log-transformed, $r = 0.363$, $p = 0.038$), and VAT-CMA1 expression correlated with both percentage of c-Kit+ and percentage of tryptase+ ($r(\varrho) = 0.348$, $p = 0.044$; $r(\varrho) = 0.561$, $p = 0.008$, respectively). Consistent with the proposed association between MC and fibrosis, we found significant associations between AT-MC genes and the expression of collagens 1A1, 3A1, and 6A1 in VAT (Table 2), but less so in SAT that exhibited only correlations between SAT-MC genes with collagen 6A1 (Table S3).

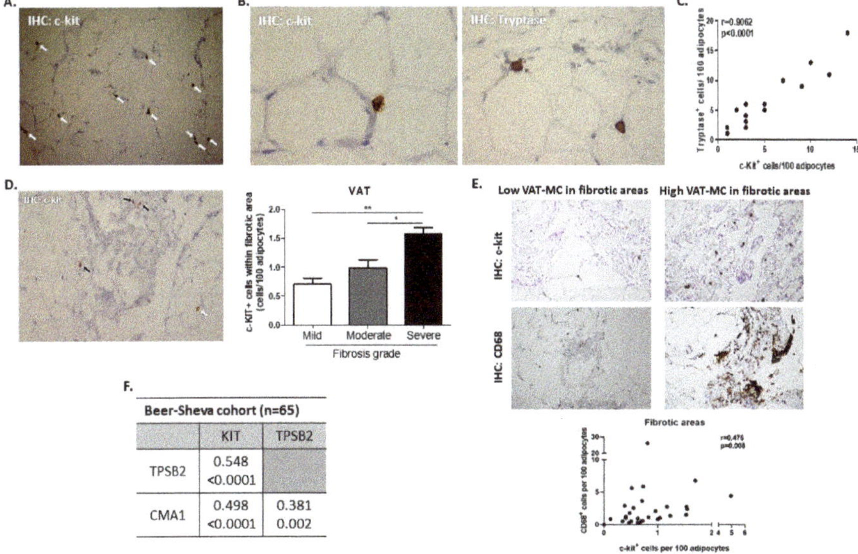

Figure 1. Identification of human visceral (omental) adipose tissue mast cells (AT MC). AT stained for c-KIT Proto-Oncogene Receptor Tyrosine Kinase (C-Kit+, white arrows) with (**A**) ×100 and (**B**) ×400 magnification (left), or stained for tryptase+ cells ×400 (**B**, right). (**C**) Spearman's correlation between %C-Kit+ and %tryptase+ cells (percentage being per 100 adipocytes) in serial AT sections ($n = 14$). (**D**) C-Kit+ is discernable both within fibrotic areas (black arrows) and around adipocytes (white arrow), and bar graph—the number of C-Kit+ cells in sections rated by clinical pathologists (co-authors YD and RSL) as exhibiting mild, moderate, or severe degree of fibrosis. (**E**) Representative histological sections from two representative patients—one with low and the second with high MC in fibrotic areas within VAT, stained for either C-Kit (MC) or CD68 (macrophages). Graph below depicts Spearman's correlation between %C-Kit+ and %CD68+ cells (i.e., macrophages, percentage being per 100 adipocytes) in serial VAT sections within the fibrotic areas, $n = 30$. (**F**) Spearman's intercorrelations ($n = 65$) between VAT MC-related genes (KIT, TPSB2, and CMA1). In each correlation, the upper line indicates $r(\varrho)$ value and lower line—p-value.

Table 2. Intercorrelation between VAT-MC and collagens' gene expression in the Beer-Sheva cohort ($n = 65$). n.s, not significant. Values are r and p value (upper and lower line respectively) of Spearman's correlations. CMA1: mast cell chymase 1, TPSB2: mast cell tryptase beta II, COLA1A1: collagen type I alpha 1 chain, COLA3A1: collagen type III alpha 1 chain, COLA6A1: collagen type VI alpha 1 chain.

	KIT	TPSB2	CMA1	COLA3A1	COLA6A1
COLA1A1	0.342 0.007	n.s.	0.236 0.065	0.895 <0.001	0.305 0.015
COLA3A1	0.333 0.009	n.s.	0.275 0.032		0.303 0.017
COLA6A1	n.s.	0.435 <0.001	0.26 0.038		

In the independent validation, cohort 1 of $n = 32$ patients with obesity whose VAT gene expression was assessed by microarrays [16], all MC genes probe sets (KIT, TPSB2, CMA1) exhibited significant negative correlations with macrophage gene probes (CD68, Mac2, IGTAX(CD11c)). Most significant were correlations between higher VAT-KIT and lower expression of CD68 ($r(\varrho) = -0.559$, $p = 0.001$, $n = 32$), and between higher VAT-CMA1 and lower expression of Mac2 ($r(\varrho) = -0.448$, $p = 0.010$, $n = 32$). Consistently, AT-MC genes' expression in VAT negatively associated with the number of macrophage crown-like structures (CLS) in validation cohort 2 ($r(\varrho) = -0.294$, $p = 0.047$, $n = 56$). Jointly, these analyses demonstrate (1) that AT-MC accumulation can be estimated confidently by histology and gene expression, (2) that histologically, AT-MC may correlate with the abundance of macrophages only within fibrotic regions. By gene expression, AT-MC do not positively correlate, and may even exhibit a negative association with the total abundance of AT macrophages.

3.2. Cross-Sectional Analyses

Next, we wished to challenge our hypothesis that higher AT-MC accumulation would signify an obese sub-phenotype characterized by worse cardiometabolic risk. For this purpose, we used the larger Beer-Sheva cohort of $n = 65$ patients with obesity to compare those with obesity and high versus low VAT-MC accumulation. Given the suspicion that KIT may not be sufficiently MC-specific [9], but that c-Kit and TPSB2 largely identify the same cells in the tissue (Figure 1D), and that AT-MC express TPSB2–/+ CMA1 [11], we based MC gene expression stratification on the combined VAT expression levels of TPSB2 and CMA1: Patients were categorized as VAT-MC^{low} only if the AT expression of both TPSB2 and CMA1 genes was below the median level of expression in the entire cohort, whereas VAT-MC^{high} were all those with above-median gene expression in either or both genes.

Contrary to our hypothesis, patients with obesity and low VAT-MC gene expression (below median expression of both TPSB2 and CMA1) were either not significantly different, or in some parameters trended to, or exhibited significantly worse clinical parameters (Table 1, Figure 2A). Significant differences between patients with obesity and VAT-MC^{low} and VAT-MC^{high} were observed in triglyceride (TG) levels and HDL (high-density lipoprotein) ratio, in circulating alkaline phosphatase and in HOMA-β. Sensitivity analysis in which outliers were excluded rendered the difference in HOMA-β between those with VAT-MC^{low} versus VAT-MC^{high} insignificant. We could not attribute the differences to the different use of medications (Table S4). Validation cohorts 1 and 2 exhibited similar trends to those observed in the main cohort (Table 1), and such trends were less robust when defining VAT-MC accumulation based on the expression of KIT alone (Table S2). To further explore if this unexpected association between higher VAT-MC expression and seemingly improved clinical markers of cardiometabolic risk are contributed by specific sub-groups of patients, the analysis was repeated after stratification of the main cohort by either sex, age (above/below median of 45 years old), T2DM status, and obesity class (Figure 2 and Figure S5, respectively). Remarkably, significant differences exhibiting improved parameters in VAT-MC^{high} were more readily observed in females (Figure S5A), in participants whose age was above median (Figure 2B), in patients with type 2 diabetes mellitus

(T2DM) (Figure 2C), and in those with BMI 30.0–39.9 kg/m^2 (Figure S5B). Intriguingly, a significant p interaction was observed between diabetes status and VAT-MC in FPG and in HbA1c ($p = 0.016$ and $p = 0.041$, respectively). This could not be attributed to different diabetes duration or different use of medications (Tables S4 and S5).

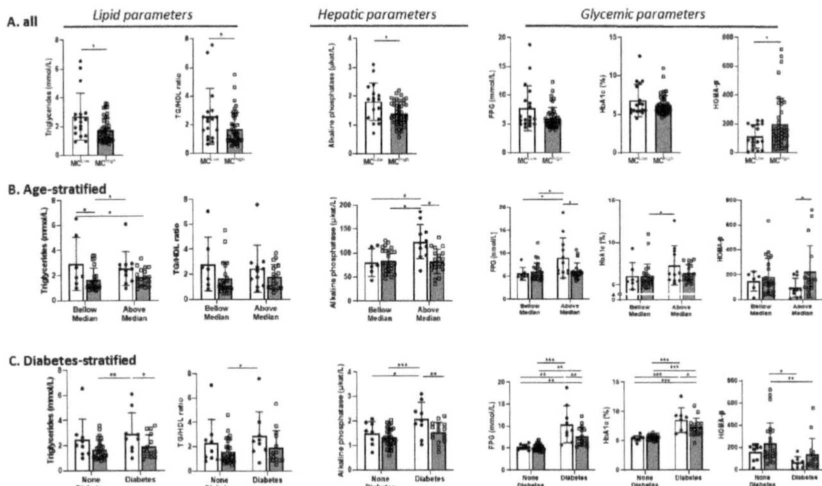

Figure 2. Comparison of clinical parameters between participants with high versus low MC accumulation in VAT. MC^{low} (white bars) were defined as people in whom the expression of both *TPSB2* and *CMA1* were below the median value of the cohort for each of the genes; all others were defined as MC^{high} (i.e., with either one or two of the genes expressed above median, gray bars). (**A**) Differences in triglycerides (TG)/HDL ratio, alkaline phosphatase, FPG, Hemoglobin A1c (HbA1c), and HOMA-β between VAT- MC^{low} versus VAT- MC^{high}. (**B**,**C**) Same analysis as in **A**, but the cohort was stratified by age (below/above median = 45 years, B), or type 2 diabetes status (**C**). Age-adjusted 2 by 2 ANOVA was used to test for variance. Age-adjusted Least Significant Difference (LSD) post-hoc test was used to compare between groups. Values are expressed as mean ± SD. * $p < 0.05$, ** $p < 0.01$, *** $p < 0.001$.

Indeed, by correlation (rather than group statistics) analysis, a significant negative association was observed between VAT-CMA1 expression level and FPG or HbA1c among the sub-group of patients with obesity and T2DM (association with TPSB2 exhibited a similar trend) (Figure 3A,B, respectively). Importantly, this association was also evident in the two independent Leipzig cohorts (Figure 3C,D). In SAT, associations between the expression of MC genes and clinical parameters in the 3 cohorts is presented in Table S6, displaying less consistent cross-sectional associations than those observed with VAT-MC. Jointly, these cross-sectional analyses disproved our initial hypothesis that increased VAT-MC accumulation would signify a worse obesity sub-phenotype. Rather. they provided evidence that the reverse association may in fact hold true. This was particularly apparent in certain sub-groups of patients with obesity, such as females, older patients, in those whose obesity was also complicated with T2DM, and in patients with obesity class I + II.

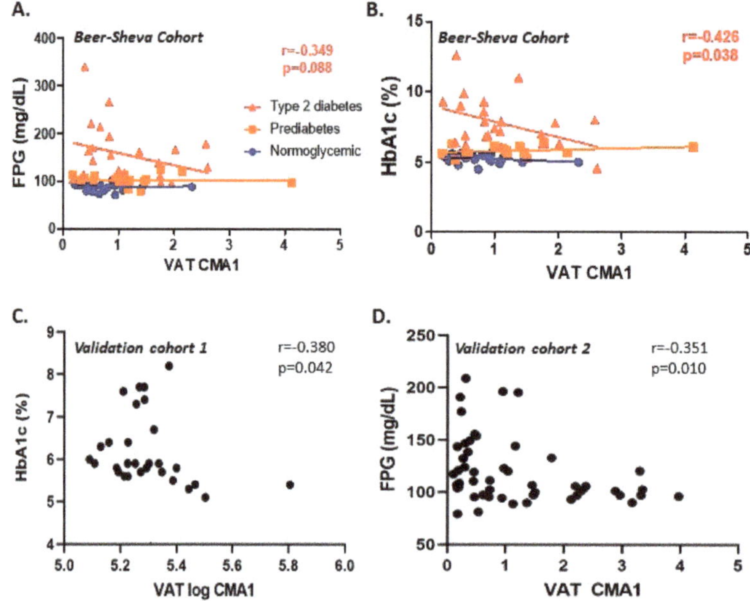

Figure 3. Associations between VAT CMA1 gene expression and glycemic parameters in the Beer–Sheva and Leipzig independent cohorts. Spearman's associations between VAT CMA1 gene expression and FPG (**A**) or HbA1c (**B**) among the Beer–Sheva cohort ($n = 65$, all with body mass index ≥ 30 kg/m^2), stratified into those with normal glucose homeostasis (normoglycemic, blue circle), prediabetes (orange square), or type 2 diabetes (red triangle). *r* denotes Spearman's rank correlation coefficient (Rho, ϱ) among participants with type 2 diabetes. (**C**) Spearman's rank correlation between log-transformed CMA1 expression in VAT assessed by microarray and HbA1c in the Leipzig validation cohort 1 ($n = 32$, all with BMI ≥ 30 kg/m^2). (**D**) Spearman's rank correlation between CMA1 expression in VAT and FPG in Leipzig validation cohort 2 ($n = 56$, all with BMI ≥ 30 kg/m^2).

3.3. *In Vitro Analyses*

The results described so far suggest that increased VAT-MC infiltration may cross-sectionally associate with a more favorable obese sub-phenotype, but the underlying mechanism remains obscure. To this end, given the centrality of the VAT–liver access in metabolic health/dysfunction in obesity, particularly when complicated with diabetes, we hypothesized that secreted factors from VAT-MC^{high} AT mediate a more metabolically favorable communication with liver cells compared to VAT-MC^{low} AT. To test this hypothesis, we treated human hepatocyte-derived cells with conditioned media (CM) obtained from the VAT of obese people with either high or low MC gene expression and assessed the ensuing acute signaling response to insulin stimulation (Figure 4A). HepG2 cells treated with CM from VAT-MC^{high} were more insulin-responsive compared to cells treated with CM from VAT-MC^{low}, as determined by the insulin-induced phosphorylation level of GSK3 (Figure 4B,C).

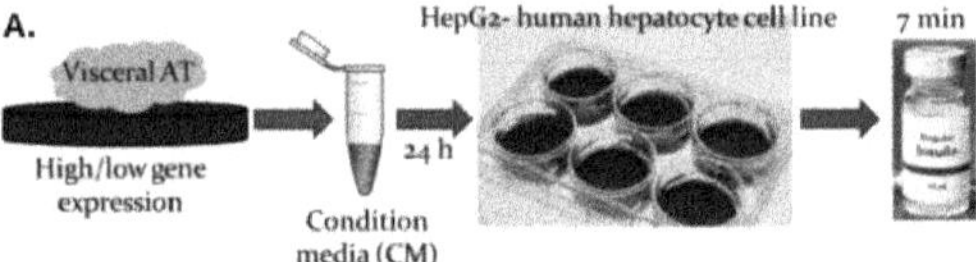

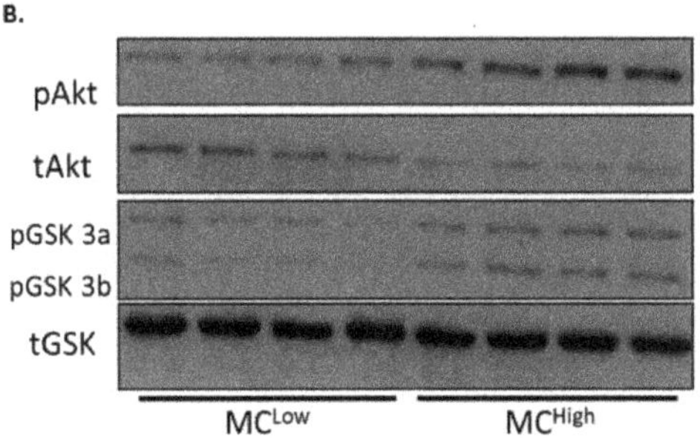

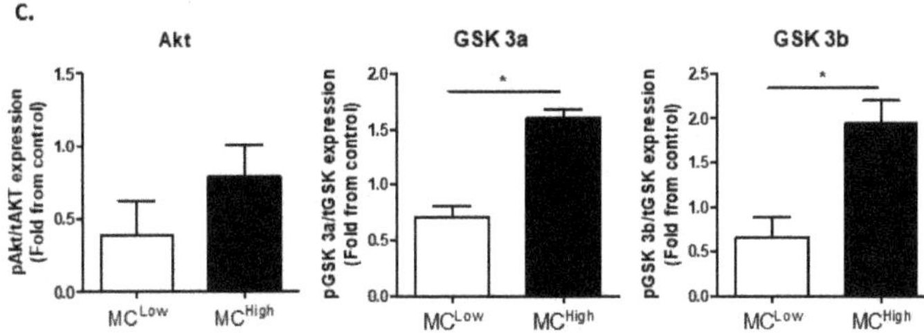

Figure 4. Insulin signaling in HepG2 cells after treatment with condition media from the VAT of obese MC$^{low/high}$. (**A**) Schematic experimental design: VAT explants were incubated in media for 24 h for the preparation of conditioned media (CM). CM was used to expose human hepatocyte cell line (HepG2) for 24 h, followed by insulin (100 nM) stimulation for 7 min. (**B**) Representative blots of Western blot analysis for the following antibodies: p(Ser473)Akt, tAkt, p(Ser21/9)GSK3, and tGSK3. (**C**) Expression quantification of the proteins above (**B**) from two independent experiments, $n = 4$ for each group. Mann–Whitney non-parametric test was performed in order to compare means. Values are expressed as mean ± standard error of the mean (SEM), * $p < 0.05$.

3.4. Prospective Analyses

Beyond the potential to utilize the estimation of MC accumulation of VAT for sub-typing obese patients in a cross-sectional analysis, we aimed to determine in a subset of patients from the main cohort, for whom post-bariatric surgery data was already available, whether the expression level of MC genes in VAT can predict clinically meaningful outcomes of intervention. Using the same approach to assess the VAT-MC accumulation level based on both TPSB2 and CMA1, patients with VAT-MC^{high} (above median value of the entire cohort) lost significantly more weight 6 months after bariatric surgery compared to VAT-MC^{low} (31.7 ± 1.6% versus 13.4 ± 6.5% of baseline body weight, corresponding to 79.1 ± 5.7% and 30.7 ± 14.5% loss of excess body weight, respectively, $p = 0.002$). For this analysis, we found a yet stronger association when using VAT-CMA1 expression only (Figure 5A,B): People with obesity and high VAT-CMA1 expression achieved 1.74-fold greater weight reduction compared to obese people with low VAT-CMA1 expression 6 months post-surgery ($p = 0.016$). This difference remained significant ($p = 0.038$) even when the two patients that showed minimal weight loss were excluded. Different weight losses corresponded to a 1.9-fold greater loss of excess body weight among the VAT-CMA1^{High} compared to the VAT-CMA1^{Low}, $p = 0.008$, Figure 5C). The difference in weight reduction between the two groups remained significant even when adjusting to either baseline BMI or surgery type (Figure 5D). Intriguingly, people with obesity and low VAT-CMA1 had higher levels of FPG (Figure 5E), HbA1c (Figure 5F), and triglycerides (Figure 5G) pre-operation, and despite lower extent of weight loss, exhibited a greater reduction, particularly in triglycerides, 6 months postoperatively. Yet, unlike with the weight change, these differences could be explained by their higher baseline values (data not shown). We did not observe any additional difference in the other clinical parameters that reached statistical significance (data not shown).

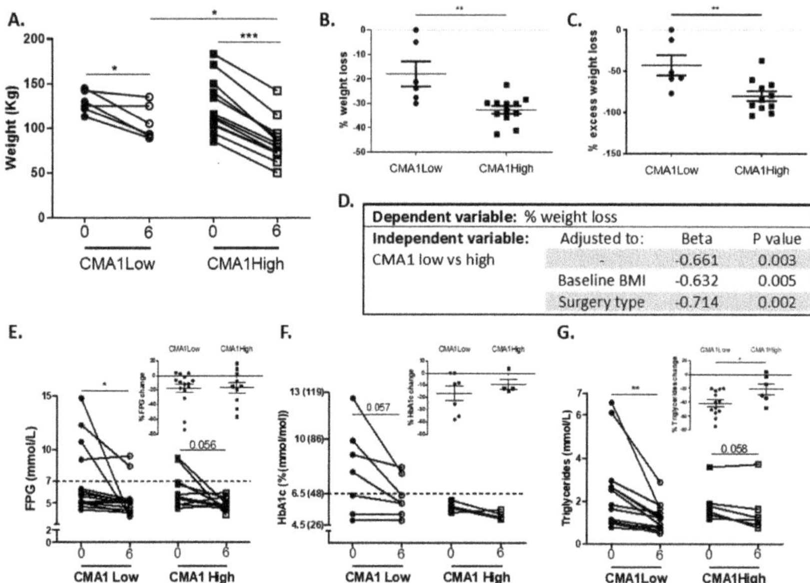

Figure 5. Six months postoperative weight reduction and metabolic changes comparison between people with VAT CMA1 high or CMA1 low. (**A**) Body weight change between operation day (0) and 6 months post-surgery in persons with low expression (i.e., below median) of CMA1 (n = 6, circles) or high CMA1 mRNA levels (n = 12, squares). Percentage of weight loss (**B**) from preoperative weight and of excess weight loss six months following bariatric surgery (**C**). A paired t-test was used to compare between pre-operation (0) and 6 months post-surgery (6). A Mann–Whitney non-parametric test was performed in order to compare means between CMA1 low and CMA1 high. * $p < 0.05$, ** $p < 0.01$. (**D**) Multi-variant model for association between VAT CMA1 groups and 6 months weight changes as dependent variable, adjusted to baseline BMI and surgery type. Change in FPG (**E**), HbA1c (**F**), and triglycerides (**G**) between operation day (0) and 6 months post-surgery in persons with low expression CMA1 or high CMA1 mRNA levels. The percentage of change in FPG, HbA1c, and triglycerides (insets). Values are expressed as mean ± standard error of the mean (SEM).

To strengthen the limited observation achievable with our main cohort, we utilized validation cohort 2 of the Leipzig bio-bank, in which 1 year post-laparoscopic sleeve gastrectomy data was available (see details of this 2-step bariatric surgery cohort in the Methods section). As expected, there were significant declines in weight and anthropometric measures, in glycemic and insulin resistance indices, and in lipids (increase in HDL) (Table 3).

Table 3. Changes in clinical and inflammatory markers and AT genes, 12 months following bariatric surgery in validation cohort 2. Values are mean ± SD. CLS—crown-like structures, VAT—visceral adipose tissue, SAT—subcutaneous adipose tissue. ns—not significant ($p > 0.05$).

Characteristics	Baseline ($n = 56$)	After 12 Months ($n = 56$)	p-Value
Weight (kg)	144.2 ± 30.3	105.6 ± 23.8	<0.0001
BMI (kg/m^2)	50.5 ± 8.6	37.1 ± 8.0	<0.0001
Body fat (%)	45.1 ± 7.6	35.5 ± 8.2	<0.0001
FPG (mmol/L)	6.6 ± 1.7	5.6 ± 1.5	<0.0001
HbA1c (%)	6.4 ± 1.1	5.4 ± 0.8	<0.0001
Insulin (pmol/L)	206.1 ± 163.4	88.8 ± 102.4	<0.0001
HOMA-IR	8.9 ± 7.4	3.8 ± 5.3	<0.0001
Total cholesterol (mmol/L)	5.3 ± 1.1	4.9 ± 0.9	0.067
LDL-c (mmol/L)	3.3 ± 1.0	3.1 ± 1.0	ns
HDL (mmol/L)	1.2 ± 0.3	1.4 ± 0.4	<0.0001
Triglycerides (mmol/L)	2.0 ± 1.0	1.3 ± 0.6	<0.0001
CRP (nmol/L)	59.4 ± 65.0	32.9 ± 42.8	0.024
Interleukin 6 (pg/mL)	5.5 ± 3.4	2.6 ± 2.0	<0.0001
Interleukin 1 beta (pg/mL)	9.6 ± 6.7	9.1 ± 8.0	ns
Tumor necrosis factor alpha (pg/mL)	8.3 ± 3.1	8.0 ± 3.7	ns
VAT			
CLS (per 100 adipocytes)	9.0 ± 2.9	7.7 ± 3.3	<0.0001
TNF (AU)	2.3 ± 1.9	1.7 ± 1.9	ns
IL1b (AU)	1.9 ± 1.2	1.2 ± 0.9	0.003
TGFb (AU)	1.5 ± 1.4	1.4 ± 1.1	ns
KIT (AU)	1.1 ± 0.7	1.1 ± 0.7	ns
TPSB2 (AU)	1.0 ± 1.0	1.1 ± 1.0	ns
CMA1 (AU)	1.2 ± 1.1	0.8 ± 1.0	0.044
SAT			
CLS (per 100 adipocytes)	4.3 ± 2.1	3.6 ± 2.3	0.005
TNF (AU)	1.9 ± 1.8	1.5 ± 1.2	ns
IL1b (AU)	1.9 ± 1.8	1.3 ± 0.9	0.035
TGFb (AU)	1.5 ± 0.9	1.2 ± 0.7	0.024
KIT (AU)	1.3 ± 0.7	1.1 ± 0.8	ns
TPSB2 (AU)	1.2 ± 0.9	1.1 ± 0.7	ns
CMA1 (AU)	1.3 ± 0.8	1.4 ± 1.0	0.146

Among circulating cytokines tested, IL-6 robustly declined ($p < 0.0001$), and a significant reduction was observed in CRP levels ($p = 0.024$). Several inflammatory markers were measured in both VAT and SAT. In VAT, both macrophage CLS number and IL1b expression decreased significantly ($p < 0.0001$ and $p = 0.003$, respectively). Among MC genes, only CMA1 expression exhibited a significant expression reduction (delta of 0.3 ± 1.0, $p = 0.044$). Similar results were achieved in SAT, with significant reductions in macrophages CLS and IL1b ($p = 0.005$, $p = 0.035$, respectively), but with no significant change in CMA1 expression but with a reduction in TGFbeta expression ($p = 0.024$). Next, we looked for associations between baseline VAT and SAT MC genes expression and selected inflammatory markers and changes in anthropometric parameters 1 year following bariatric surgery (Figure 6). In this cohort of extremely obese patients (average: BMI > 50 kg/m^2, Table 1), the baseline expression of KIT and TPSB2 in VAT seemed to be better predictors of greater weight loss 1 y postoperatively, reaching statistically significant associations ($r(\varrho) = 0.295$, $p = 0.044$, $r(\varrho) = 0.313$, $p = 0.03$, respectively). Baseline VAT-KIT was also associated with BMI loss ($p = 0.04$) and trended with percentage of weight loss ($r(\varrho) = 0.283$, $p = 0.054$). This is somewhat consistent with the results of the main cohort of less obese patients (Table 1); baseline VAT-CMA1 in validation cohort 2 trended to associate with a greater loss in BMI 1 y after surgery ($r(\varrho) = 0.229$, $p = 0.099$). In addition, baseline VAT-TGFbeta was also associated with a greater reduction in WC 1 year postoperatively ($r(\varrho) = 0.284$, $p = 0.048$).

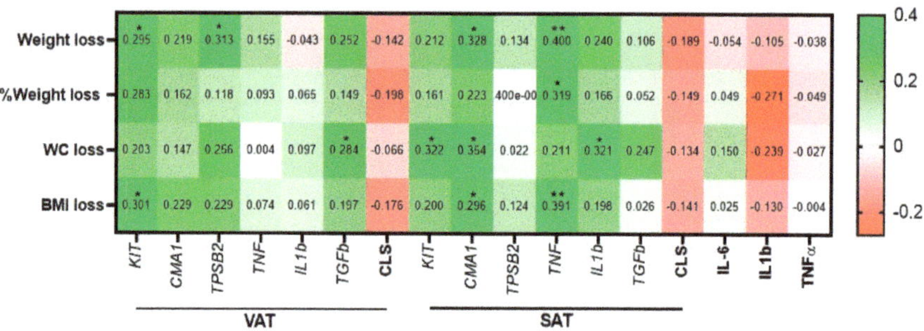

Figure 6. Correlation between baseline parameters and 1 year postoperative weight loss response in validation cohort 2. Values are $r(\varrho)$ of Spearman's correlations. WC: waist circumference. * $p < 0.05$; ** $p < 0.01$.

Interestingly, unlike the cross-sectional analyses (Section 3.2), in which VAT-MC exhibited more robust associations with patients' clinical characteristics than SAT-MC (Figure 3 and Figure S5 and Table S6), SAT-MC genes and inflammatory markers exhibited more associations with 1 year postoperative outcomes (Figure 6). Higher SAT-CMA1 associated with greater weight, waist circumference (WC), and BMI loss ($r(\varrho) = 0.328$, $p = 0.015$, $r(\varrho) = 0.354$, $p = 0.015$ and $r(\varrho) = 0.296$, $p = 0.030$, respectively). Baseline SAT-KIT similarly associated with greater WC loss ($p = 0.045$). Intriguingly, higher SAT-TNFalpha expression highly associated with greater weight, percentage weight, and BMI losses ($r(\varrho) = 0.400$, $p = 0.004$, $r(\varrho) = 0.319$, $p = 0.024$ and $r(\varrho) = 0.391$, $p = 0.005$, respectively), as did a higher baseline expression of SAT-IL1beta (association with WC decline—$r(\varrho) = 0.321$, $p = 0.028$). Baseline macrophages CLS numbers, in both VAT and SAT, and circulating cytokines were not significantly associated with a reduction in any anthropometric parameters. Moreover, although non-significantly, they exhibit negative $r(\varrho)$ coefficients with 1 year postoperative weight loss measures, which is opposite to the associations (both statistically significant or non-significant) between higher baseline VAT or SAT-MC gene expression and greater weight loss following surgery.

Jointly, these results provide proof-of-principle for VAT and SAT MC and inflammatory gene expression as putative molecular predictors of weight loss and metabolic improvement six months to 1 y following bariatric surgery.

4. Discussion

In this study, we hypothesized that AT-MC accumulation may aid in characterizing obesity sub-phenotypes, and we addressed this hypothesis by cross-sectional and prospective/predictive analyses of three independent cohorts of patients, all with obesity. Our results suggest the following: (1) We anticipated that greater AT-MC accumulation of VAT would characterize a more adverse obese phenotype as determined by cardiometabolic risk parameters. This hypothesis could be clearly rejected by the data. Moreover, contrasting our initial hypothesis, we found that MC numbers and MC-related gene expression in omental fat (VAT) associated in 3 independent cohorts with a better cardiometabolic risk profile, associations that are not consistently observed with SAT. Conducting post-hoc exploratory analyses, this association was strengthened when stratifying the larger cohort by basic clinical sub-groups, such as sex, age, and type 2 diabetes status. In the latter group, high VAT-MC gene expression associated with a metabolically healthier phenotype compared to those with low MC gene expression pattern. In some parameters, patients with obesity and diabetes, and with high MC gene expression in VAT were indistinguishable from patients with obesity without diabetes. (2) In vitro studies suggest that VAT with high MC gene expression communicates more favorably with liver-derived cells, rendering them more insulin-responsive compared to cells exposed to VAT with low MC gene expression. (3) We observed that analyzing VAT-MC gene expression in samples obtained during bariatric surgery can provide predictive information on the degree of weight loss and metabolic response to bariatric surgery: high pre-operative VAT CMA1 or KIT expression predicted significantly greater weight loss 6 months or 1 year postoperatively, respectively. Unlike in the cross-sectional analyses, higher SAT-MC gene expression also correlated with greater weight-loss response to bariatric surgery.

Whether AT MC accumulation plays a role in obesity is still debatable. In mice, results are contradictory depending on the models used to generate MC deficiency [6,8–10]. Moreover, MC depletion via c-kit prevented obesity development [6,8–10], and therefore it may not contribute to uncovering MC's role when obesity is established, and how it relates to the development of obesity-associated cardiometabolic morbidity. In humans, particular interest in obesity sub-typing/sub-phenotyping is emerging [20], since with its high prevalence, there is a pressing need to increase the identification of obesity subtypes that may require more intensive treatment. Compared to lean people, persons with obesity exhibit higher numbers of MC both in VAT and SAT [6,11], especially if obesity was accompanied with type 2 diabetes mellitus [11]. Compared to the study by Divoux et al. that examined persons with obesity with/without type 2 diabetes (albeit with a limited $n = 10$ in each group) [11], we could not detect a clear increase in MC parameters in VAT from persons with diabetes compared to those without ($n = 40$ and 25, respectively). Our findings suggest that a larger cohort than that previously analyzed may have been required to uncover heterogeneity within the group of patients with obesity and type 2 diabetes. Indeed, associations between VAT-MC and metabolic parameters were mainly discernable among those with diabetes and higher expression levels of MC genes in VAT. In those patients, VAT-MC gene expression was associated with a healthier metabolic phenotype, suggesting a possible compensatory increase in VAT-MC in response to metabolic impairment, which, differently than previously suspected, exerts protective effects on metabolic health.

This work has some limitations. Our main cohort's size ($n = 65$), though the largest (to our knowledge) used to examine in humans AT-MC, is still relatively small, and the sub-analysis for predicting response to surgery includes an even smaller sub-cohort ($n = 18$). In addition, the high HOMA-β values may indicate the possibility that we investigated a specific obesity sub-group with relatively high beta cell reserve. Yet, we confirmed the key findings in 2 independent cohorts, using different approaches to estimate AT-MC gene expression. Findings were opposite to our pre-defined hypothesis that increased MC accumulation would associate with a worse metabolic outcome, and subsequent post-hoc analyses are more prone to biases and type 2 error and should best be viewed as hypothesis-generating observations. Yet, the original hypothesis was clearly rejected, so MC accumulation, at least in VAT, is not associated with a worse metabolic phenotype. Regarding the estimation of MC accumulation, we did not use flow-cytometry methods that would have enabled characterizing MC sub-types, and potentially their activation state. Rather, we critically assessed the use of several, nicely intercorrelated, "MC-specific genes", defining the AT-MC^{low} sub-group stringently, using combined thresholds for two genes (TPSB2 and CMA1) that would capture the two MC sub-populations. We also verified gene expression agreement with immunohistochemical assessment of AT-MC accumulation. Therefore, we cannot draw conclusions about MC activation state, including a possible association with LDL particles—which are known regulators of MC activity [21], although most of the associations we observed were between AT-MC accumulation and triglycerides, not with LDL or total cholesterol levels. Limiting the clinical implementation of our results is our use of VAT measures as a biomarker, since this AT is not accessible for sampling in the regular clinical setting. People with obesity have been shown to have higher levels of serum tryptase compared to lean people [6], and a positive association has been noted with BMI [22]. Yet, the source for serum tryptase is not restricted to MC in VAT, and therefore it is likely a poor correlate of visceral AT-MC accumulation. Therefore, future studies are required to assess whether MC-derived, circulating blood biomarkers, such as serum tryptase and/or chymase levels, could be clinically used to estimate visceral AT-MC accumulation and obesity sub-types. Our results suggest that clinically meaningful information may be obtained for those undergoing abdominal surgery for obesity management, by assessing tissues obtained during bariatric surgery molecularly and/or histopathologically. Although such analyses could be predictive of post-operative endpoints and may guide post-operative care, before obtaining more supportive evidence by other groups, it may still be premature to propose adipose tissue MC accumulation as a clinical predictor for the outcome of a bariatric surgery.

Although speculative, the data suggest putative mechanisms to explain the seemingly-positive effect of VAT-MC. Although our study was not focused on providing a systematic analysis of adipose tissue fibrosis, we report that in VAT, MC also populate fibrotic areas, and their gene expression correlates with several AT collagens (in SAT, correlations were only with COL6A1). Yet, recent studies suggest that in certain contexts, MC can exert anti-fibrotic/collagen-degrading effects that may result in a protective, rather than a pathogenic, role [23,24]. In addition, while AT fibrosis in mice is largely thought to contribute to tissue dysfunction [25,26], in humans, the pathological role of AT fibrosis may be more complex: higher SAT fibrosis may associate with poor metabolic and weight loss response to bariatric surgery [27], but VAT fibrosis associates with smaller adipocytes, potentially by limiting visceral adipocyte hypertrophy, thereby mediating better metabolic profile [28,29]. Additionally, MC secrete prostaglandins including 15-deoxy-delta PGJ2, which is an endogenous ligand for PPARγ, particularly in response to high-glucose conditions, resulting in increased adipogenesis [8]. Such MC-mediated PPARγ activation may support "healthy" AT expansion [30]. Indeed, the expression of CPA3, an MC marker, is associated with Uncoupling Protein 1 (UCP1) in the SAT of lean people, and MC-related histamine and IL-4 induce UCP1 expression in adipocytes, promoting AT beiging [31,32]. MC also contribute to angiogenesis, as demonstrated in cancer [33,34]. Thus, AT-MC adjacent to micro-vessels may imply their involvement in increased angiogenesis, possibly further supporting healthy AT expansion by limiting hypoxia. Finally, we show that AT-MC correlate with macrophages only in fibrotic areas, and at the whole-tissue level, MC genes may in fact inversely correlate with macrophage-specific genes or with macrophage CLS numbers. This suggests that higher MC accumulation in VAT might in fact associate with lower macrophage infiltration/activity, which in turn is indicative of a lower inflammatory burden-related adipose tissue dysfunction. Furthermore, MC^high exhibited greater postoperative weight loss. A putative explanation might be high preoperative fasting insulin that enabled greater postoperative decline in insulin levels, thereby contributing to more substantial weight loss.

5. Conclusions

Clinically used methods for better phenotyping people with obesity are still insufficient and greatly limit stratified or more precision obesity management. Our study suggests that VAT-MC accumulation estimates could be used as a tool for obesity sub-phenotyping. This can rely on relatively available laboratory procedures, such as molecular (real-time PCR) and/or histopathological assays using clinically common, known cell markers. Moreover, we propose that molecular and/or histological examinations of tissues obtained during surgery can uncover clinically important information, such as the prediction of post bariatric-surgery outcomes. This could aid in the post-operative management to optimize patient care in an individualized manner, similar to the common practice in the management of other diseases, such as cancer. Jointly, we propose that adipose tissue composition holds clinically relevant information that should be better exploited for improving the care provided to people with obesity.

Supplementary Materials: The following are available online at http://www.mdpi.com/2073-4409/9/6/1508/s1, Figure S1: The expression of KIT in cells that can be found in human AT, Figure S2: The expression of TPSB2 in cells that can be found in human AT, Figure S3: The expression of CMA1 in cells that can be found in human AT, Figure S4: SAT C-Kit+ cells within fibrotic area and fibrosis grading. Figure S5: Comparison of clinical parameters between participants with high versus low MC accumulation in VAT, stratified by sex or obesity category. Table S1: PCR probe ID. Table S2: Clinical characteristics of participants with obesity from the Beer–Sheva and Leipzig cohorts, each also stratified by VAT-MC low/high based on expression of KIT. Table S3: Inter-correlation between SAT-MC gene expression and SAT-collagens. Table S4: Usage of medications in persons with VAT MC^high and MC^low among the entire main cohort. Table S5: Diabetes duration and medication among persons with type

2 diabetes stratified to VAT MC^{high} and MC^{low}, main cohort. Table S6: Association between SAT-MC genes expression and clinical parameters in the three cohorts.

Author Contributions: N.G. conceived and designed the study, obtained the data, proposed and performed the statistical analyses, conducted the literature search, drafted the report, and reviewed/edited the manuscript. Y.K. and R.S.-L. obtained the histopathological data and provided technical and professional support. Y.G. proposed and performed the statistical analyses and reviewed the manuscript. Y.H. contributed to the literature search and reviewed the manuscript. T.P. contributed to the literature search and study design. R.G. contributed bioinformatics analyses and central intellectual content. V.P., I.F.L., and B.K. recruited patients, obtained clinical data and samples, and reviewed the manuscript. M.B. provided the clinical data of the Leipzig cohorts, reviewed/edited the manuscript and revised the report for important intellectual content. A.R. conceived the study, reviewed and analyzed the data, and wrote the manuscript. All authors have read and agreed to the published version of the manuscript.

Funding: This research was funded by grants from the DFG (German Research Foundation)—Projektnummer 209933838—SFB 1052 (specific projects: B2 to A.R., project B1 to M.B.), and the Israel Science Foundation (ISF 928/14 and 2176/19 to A.R.).

Conflicts of Interest: The authors declare no conflict of interest

References

1. Weisberg, S.P.; McCann, D.; Desai, M.; Rosenbaum, M.; Leibel, R.L.; Ferrante, A.W., Jr. Obesity is associated with macrophage accumulation in adipose tissue. *J. Clin. Investig.* **2003**, *112*, 1796–1808. [CrossRef] [PubMed]

2. Mraz, M.; Haluzik, M. The role of adipose tissue immune cells in obesity and low-grade inflammation. *J. Endocrinol.* **2014**, *222*, R113–R127. [CrossRef] [PubMed]

3. Schmitz, J.; Evers, N.; Awazawa, M.; Nicholls, H.T.; Brönneke, H.S.; Dietrich, A.; Mauer, J.; Blüher, M.; Brüning, J.C. Obesogenic memory can confer long-term increases in adipose tissue but not liver inflammation and insulin resistance after weight loss. *Mol. Metab.* **2016**, *5*, 328–339. [CrossRef] [PubMed]

4. Arivazhagan, L.; Ruiz, H.H.; Wilson, R.A.; Manigrasso, M.B.; Gugger, P.F.; Fisher, E.A.; Moore, K.J.; Ramasamy, R.; Schmidt, A.M. An eclectic cast of cellular actors orchestrates innate immune responses in the mechanisms driving obesity and metabolic perturbation. *Circ. Res.* **2020**, *126*, 1565–1589. [CrossRef] [PubMed]

5. Weinstock, A.; Moura Silva, H.; Moore, K.J.; Schmidt, A.M.; Fisher, E.A. Leukocyte heterogeneity in adipose tissue, including in obesity. *Circ. Res.* **2020**, *126*, 1590–1612. [CrossRef] [PubMed]

6. Liu, J.; Divoux, A.; Sun, J.; Zhang, J.; Clément, K.; Glickman, J.N.; Sukhova, G.K.; Wolters, P.J.; Du, J.; Gorgun, C.Z.; et al. Genetic deficiency and pharmacological stabilization of mast cells reduce diet-induced obesity and diabetes in mice. *Nat. Med.* **2009**, *15*, 940–945. [CrossRef] [PubMed]

7. Altintas, M.M.; Azad, A.; Nayer, B.; Contreras, G.; Zaias, J.; Faul, C.; Reiser, J.; Nayer, A. Mast cells, macrophages, and crown-like structures distinguish subcutaneous from visceral fat in mice. *J. Lipid Res.* **2011**, *52*, 480–488. [CrossRef]

8. Tanaka, A.; Nomura, Y.; Matsuda, A.; Ohmori, K.; Matsuda, H. Mast cells function as an alternative modulator of adipogenesis through 15-deoxy-delta-12, 14-prostaglandin J2. *Am. J. Physiol. Cell Physiol.* **2011**, *301*, C1360–C1367. [CrossRef]

9. Gutierrez, D.A.; Muralidhar, S.; Feyerabend, T.B.; Herzig, S.; Rodewald, H.R. Hematopoietic kit deficiency, rather than lack of mast cells, protects mice from obesity and insulin resistance. *Cell Metab.* **2015**, *21*, 678–691. [CrossRef]

10. Chmelař, J.; Chatzigeorgiou, A.; Chung, K.J.; Prucnal, M.; Voehringer, D.; Roers, A.; Chavakis, T. No role for mast cells in obesity-related metabolic dysregulation. *Front. Immunol.* **2016**, *7*, 524. [CrossRef]

11. Divoux, A.; Moutel, S.; Poitou, C.; Lacasa, D.; Veyrie, N.; Aissat, A.; Arock, M.; Guerre-Millo, M.; Clement, K. Mast cells in human adipose tissue: Link with morbid obesity, inflammatory status, and diabetes. *J. Clin. Endocrinol. Metab.* **2012**, *97*, E1677–E1685. [CrossRef] [PubMed]

12. Einwallner, E.; Kiefer, F.W.; Di Caro, G.; Orthofer, M.; Witzeneder, N.; Hörmann, G.; Itariu, B.; Zeyda, M.; Penninger, J.M.; Stulnig, T.M.; et al. Mast cells are not associated with systemic insulin resistance. *Eur. J. Clin. Investig.* **2016**, *46*, 911–919. [CrossRef] [PubMed]

13. Haim, Y.; Blüher, M.; Slutsky, N.; Goldstein, N.; Klöting, N.; Harman-Boehm, I.; Kirshtein, B.; Ginsberg, D.; Gericke, M.; Guiu Jurado, E.; et al. Elevated autophagy gene expression in adipose tissue of obese humans: A potential non-cell-cycle-dependent function of E2F1. *Autophagy* **2015**, *11*, 2074–2088. [CrossRef] [PubMed]

14. Shapiro, H.; Pecht, T.; Shaco-Levy, R.; Harman-Boehm, I.; Kirshtein, B.; Kuperman, Y.; Chen, A.; Blüher, M.; Shai, I.; Rudich, A. Adipose tissue foam cells are present in human obesity. *J. Clin. Endocrinol. Metab.* **2013**, *98*, 1173–1181. [CrossRef]

15. Wallace, T.M.; Levy, J.C.; Matthews, D.R. Use and abuse of HOMA modeling. *Diabetes Care* **2004**, *27*, 1487–1495. [CrossRef]

16. Keller, M.; Hopp, L.; Liu, X.; Wohland, T.; Rohde, K.; Cancello, R.; Klös, M.; Bacos, K.; Kern, M.; Eichelmann, F.; et al. Genome-wide DNA promoter methylation and transcriptome analysis in human adipose tissue unravels novel candidate genes for obesity. *Mol. Metab.* **2017**, *6*, 86–100. [CrossRef]

17. Neville, M.J.; Collins, J.M.; Gloyn, A.L.; McCarthy, M.I.; Karpe, F. Comprehensive human adipose tissue mRNA and microRNA endogenous control selection for quantitative real-time-PCR normalization. *Obesity* **2011**, *19*, 888–892. [CrossRef]

18. Livak, K.J.; Schmittgen, T.D. Analysis of relative gene expression data using real-time quantitative PCR and the 2(-Delta Delta C(T)). *Methods* **2001**, *25*, 402–408. [CrossRef]

19. Saito, H.; Matsumoto, K.; Okumura, S.; Kashiwakura, J.I.; Oboki, K.; Yokoi, H.; Kambe, N.; Ohta, K.; Okayama, Y. Gene expression profiling of human mast cell subtypes: An in silico study. *Allergol. Int.* **2006**, *55*, 173–179. [CrossRef]

20. Piche, M.E.; Tchernof, A.; Despres, J.P. Obesity Phenotypes, Diabetes, and Cardiovascular Diseases. *Circ. Res.* **2020**, *126*, 1477–1500. [CrossRef]

21. Kovanen, P.T. Mast Cells as Potential Accelerators of Human Atherosclerosis-From Early to Late Lesions. *Int. J. Mol. Sci.* **2019**, *20*, 4479. [CrossRef] [PubMed]

22. Fenger, R.V.; Linneberg, A.; Vidal, C.; Vizcaino, L.; Husemoen, L.L.; Aadahl, M.; Gonzalez-Quintela, A. Determinants of serum tryptase in a general population: The relationship of serum tryptase to obesity and asthma. *Int. Arch. Allergy. Immunol.* **2012**, *157*, 151–158. [CrossRef] [PubMed]

23. Amiot, L.; Vu, N.; Drenou, B.; Scrofani, M.; Chalin, A.; Devisme, C.; Samson, M. The anti-fibrotic role of mast cells in the liver is mediated by HLA-G and interaction with hepatic stellate cells. *Cytokine* **2019**, *117*, 50–58. [CrossRef] [PubMed]

24. Bradding, P.; Pejler, G. The controversial role of mast cells in fibrosis. *Immunol. Rev.* **2018**, *282*, 198–231. [CrossRef]

25. Sun, K.; Tordjman, J.; Clement, K.; Scherer, P.E. Fibrosis and adipose tissue dysfunction. *Cell Metab.* **2013**, *18*, 470–477. [CrossRef]

26. Crewe, C.; An, Y.A.; Scherer, P.E. The ominous triad of adipose tissue dysfunction: Inflammation, fibrosis, and impaired angiogenesis. *J. Clin. Investig.* **2017**, *127*, 74–82. [CrossRef]

27. Bel Lassen, P.; Charlotte, F.; Liu, Y.; Bedossa, P.; Le Naour, G.; Tordjman, J.; Poitou, C.; Bouillot, J.L.; Genser, L.; Zucker, J.D.; et al. The FAT Score, a Fibrosis Score of Adipose Tissue: Predicting Weight-Loss Outcome After Gastric Bypass. *J. Clin. Endocrinol. Metab.* **2017**, *102*, 2443–2453. [CrossRef]

28. Muir, L.A.; Neeley, C.K.; Meyer, K.A.; Baker, N.A.; Brosius, A.M.; Washabaugh, A.R.; Varban, O.A.; Finks, J.F.; Zamarron, B.F.; Flesher, C.G.; et al. Adipose tissue fibrosis, hypertrophy, and hyperplasia: Correlations with diabetes in human obesity. *Obesity* **2016**, *24*, 597–605. [CrossRef]

29. Divoux, A.; Tordjman, J.; Lacasa, D.; Veyrie, N.; Hugol, D.; Aissat, A.; Basdevant, A.; Guerre-Millo, M.; Poitou, C.; Zucker, J.D.; et al. Fibrosis in human adipose tissue: Composition, distribution, and link with lipid metabolism and fat mass loss. *Diabetes* **2010**, *59*, 2817–2825. [CrossRef]

30. Smith, U.; Kahn, B.B. Adipose tissue regulates insulin sensitivity: Role of adipogenesis, de novo lipogenesis and novel lipids. *J. Intern. Med.* **2016**, *280*, 465–475. [CrossRef]

31. Finlin, B.S.; Zhu, B.; Confides, A.L.; Westgate, P.M.; Harfmann, B.D.; Dupont-Versteegden, E.E.; Kern, P.A. Mast Cells Promote Seasonal White Adipose Beiging in Humans. *Diabetes* **2017**, *66*, 1237–1246. [CrossRef] [PubMed]

32. Finlin, B.S.; Confides, A.L.; Zhu, B.; Boulanger, M.C.; Memetimin, H.; Taylor, K.W.; Johnson, Z.R.; Westgate, P.M.; Dupont-Versteegden, E.E.; Kern, P.A. Adipose Tissue Mast Cells Promote Human Adipose Beiging in Response to Cold. *Sci. Rep.* **2019**, *9*, 8658. [CrossRef] [PubMed]

33. Norrby, K. Mast cells and angiogenesis. *APMIS* **2002**, *110*, 355–371. [CrossRef]

34. Ribatti, D.; Crivellato, E. Mast cells, angiogenesis, and tumour growth. *Biochim. Biophys. Acta* **2012**, *1822*, 2–8. [CrossRef] [PubMed]

Article

Physiological Oxygen Levels Differentially Regulate Adipokine Production in Abdominal and Femoral Adipocytes from Individuals with Obesity Versus Normal Weight

Ioannis G. Lempesis [1,2,3,*], Nicole Hoebers [2], Yvonne Essers [2], Johan W. E. Jocken [2], Kasper M. A. Rouschop [4], Ellen E. Blaak [2], Konstantinos N. Manolopoulos [1,3,†] and Gijs H. Goossens [2,*,†]

1   Institute of Metabolism and Systems Research (IMSR), College of Medical and Dental Sciences, University of Birmingham, Birmingham B15 2TT, UK
2   Department of Human Biology, NUTRIM School of Nutrition and Translational Research in Metabolism, Maastricht University Medical Centre+, 6229 ER Maastricht, The Netherlands
3   Centre for Endocrinology, Diabetes and Metabolism, Birmingham Health Partners, Birmingham B15 2TT, UK
4   Radiotherapy, GROW School for Oncology & Reproduction, Maastricht University Medical Centre+, 6229 ER Maastricht, The Netherlands
*   Correspondence: lemp.ioan@gmail.com (I.G.L.); g.goossens@maastrichtuniversity.nl (G.H.G.)
†   These authors contributed equally to this work.

Citation: Lempesis, I.G.; Hoebers, N.; Essers, Y.; Jocken, J.W.E.; Rouschop, K.M.A.; Blaak, E.E.; Manolopoulos, K.N.; Goossens, G.H. Physiological Oxygen Levels Differentially Regulate Adipokine Production in Abdominal and Femoral Adipocytes from Individuals with Obesity Versus Normal Weight. *Cells* **2022**, *11*, 3532. https://doi.org/10.3390/cells11223532

Academic Editor: Javier Gómez-Ambrosi

Received: 4 October 2022
Accepted: 6 November 2022
Published: 8 November 2022

Abstract: Adipose tissue (AT) inflammation may increase obesity-related cardiometabolic complications. Altered AT oxygen partial pressure (pO_2) may impact the adipocyte inflammatory phenotype. Here, we investigated the effects of *physiological* pO_2 levels on the inflammatory phenotype of abdominal (ABD) and femoral (FEM) adipocytes derived from postmenopausal women with normal weight (NW) or obesity (OB). Biopsies were collected from ABD and FEM subcutaneous AT in eighteen postmenopausal women (aged 50–65 years) with NW (BMI 18–25 kg/m^2, $n = 9$) or OB (BMI 30–40 kg/m^2, $n = 9$). We compared the effects of prolonged exposure to different *physiological* pO_2 levels on adipokine expression and secretion in differentiated human multipotent adipose-derived stem cells. Low *physiological* pO_2 (5% O_2) significantly increased leptin gene expression/secretion in ABD and FEM adipocytes derived from individuals with NW and OB compared with high *physiological* pO_2 (10% O_2) and standard laboratory conditions (21% O_2). Gene expression/secretion of IL-6, DPP-4, and MCP-1 was reduced in differentiated ABD and FEM adipocytes from individuals with OB but not NW following exposure to low compared with high *physiological* pO_2 levels. Low *physiological* pO_2 decreases gene expression and secretion of several proinflammatory factors in ABD and FEM adipocytes derived from individuals with OB but not NW.

Keywords: adipose tissue; adipokines; inflammation; body fat distribution; obesity pathophysiology; hypoxia

1. Introduction

Excess fat mass in obesity poses a major health risk [1]. The research in the past decades has clearly demonstrated that body fat distribution is a better predictor of cardiometabolic complications than total fat mass, with abdominal obesity increasing and lower-body (gluteofemoral) fat accumulation conferring relative protection against chronic cardiometabolic diseases [2–6]. This seems related to the distinct functional properties of these different AT depots. Many studies in rodents and humans have shown that AT dysfunction in obesity is characterised by adipocyte hypertrophy, mitochondrial dysfunction, reactive oxygen species (ROS) production, impaired lipid metabolism, reduced blood flow, and inflammation, together contributing to an increased risk of developing cardiometabolic diseases and cancer [6–11].

The AT microenvironment impacts metabolic and inflammatory processes [8,9]. We, and others, have previously demonstrated that AT oxygen partial pressure (pO_2), which is determined by the balance between local oxygen supply (determined by adipose tissue blood flow) and consumption (primarily mitochondrial oxygen consumption), may be an important determinant of the AT phenotype and whole-body insulin sensitivity [9,12–14]. Interestingly, differences in adipose tissue blood flow and/or adipose tissue oxygen consumption between individuals with normal weight and obesity, and between upper and lower AT depots, have previously been demonstrated [9,10,12,15,16]. Although AT pO_2 is reduced in rodent models of obesity [17–19], conflicting findings on AT pO_2 have been reported in humans [9,20–24]. We have previously shown that AT pO_2 was higher in individuals with obesity and was positively associated with AT gene expression of proinflammatory markers and whole-body insulin resistance [22,25]. Moreover, we found that AT pO_2 was lower in femoral than in abdominal subcutaneous AT in women with obesity [16].

The normal *physiological* range of AT pO_2 in human AT is ~3–11% O_2 (~23–84 mmHg) [9,21–23,25]. Therefore, the outcomes of experiments comparing the effects of pO_2 below and well above these *physiological* levels should be interpreted with caution, because the results may not directly translate to the human in vivo situation [9]. Several in vitro studies have demonstrated that the expression and secretion of many adipokines are sensitive to changes in pO_2 levels, as extensively reviewed [9,26]. Most of these studies have shown that acute exposure to severe, non-physiological hypoxia (1% O_2 for 1–24 h) induces a proinflammatory expression and secretion profile in (pre)adipocytes, while prolonged exposure to mild *physiological* hypoxia (5% O_2 for 14 days) seems to elicit a different adipokine expression/secretion profile [9,16,27]. Recently, we found that prolonged exposure to low *physiological* hypoxia decreased proinflammatory gene expression in abdominal and femoral adipocytes derived from women with obesity [16]. The metabolic and inflammatory responses to changes in the AT microenvironment may differ between individuals and AT depots. Thus, oxygen levels might exert distinct effects on AT function in people with different adiposity and in different AT depots. Importantly, however, studies investigating the impact of altered pO_2 levels on the inflammatory phenotype of adipocytes derived from people with normal weight and obesity are lacking.

Therefore, the aim of the present study was to investigate the impact of prolonged exposure to various *physiological* oxygen levels on gene expression and secretion of inflammatory factors within upper and lower body differentiated human multipotent adipose-derived stem (hMADS) cells derived from women with normal weight or obesity.

2. Materials and Methods

2.1. Upper and Lower Body Adipose Tissue Biopsies

Paired abdominal (ABD) and femoral (FEM) subcutaneous AT needle biopsies were obtained from eighteen postmenopausal women (aged 50–65 years) with normal weight (NW: BMI 18–25 kg/m^2, $n = 9$) or obesity (OB: BMI 30–40 kg/m^2, $n = 9$) (Table 1). The U.K. Health Research Authority National Health System Research Ethics Committee approved the present study (approval no. 18/NW/0392). Briefly, the biopsy specimens (up to ~1 g) were collected 6 to 8 cm lateral from the umbilicus (ABD AT) and from the anterior aspect of the upper leg (FEM AT) under local anaesthesia (1% lidocaine) after an overnight fast. Samples were immediately rinsed with sterile saline, and visible blood vessels were removed with sterile tweezers. Isolation of hMADS cells followed, as described before [16].

2.2. Human Primary Adipocyte Experiments

Human multipotent abdominal (ABD) and femoral (FEM) adipose-derived stem cells, an established human white adipocyte model [28], were seeded at a density of 2000 cells/cm^2 and kept in proliferation medium for seven days. Thereafter, these cells were differentiated under different *physiological* O_2 levels (10% O_2, high *physiological* pO_2; 5% O_2, low *physiological* pO_2) [9,16,22,29] as well as standard laboratory conditions (room air, 21% O_2) for 14 days. Gas mixtures were refreshed every 8 h (to maintain variation <0.1% O_2), whereas the medium was refreshed three times per week.

Table 1. Subjects' characteristics.

Parameter	Normal Weight ($n = 9$)	Obesity ($n = 9$)	p Value
Age (years)	56.7 ± 1.8	56 ± 1.3	0.566
BMI (kg/m^2)	22.8 ± 0.4	34.8 ± 1.3	<0.001
Waist circumference (cm)	79.4 ± 3.1	105.2 ± 3.8	<0.001
Hip circumference (cm)	94.4 ± 2.8	119.9 ± 4.8	<0.001
Waist-to-hip ratio	0.84 ± 0.02	0.88 ± 0.04	0.127
Visceral fat mass (g)	402.5 ± 118	$1,325 \pm 153.3$	0.003
Abdominal fat mass (kg)	10.01 ± 1.48	24.4 ± 2.37	<0.001
Leg fat mass (kg)	7.67 ± 0.86	16.03 ± 1.43	0.001
Fasting glucose (mmol/L)	4.91 ± 0.10	5.10 ± 0.23	0.416
2-hour glucose (mmol/L)	4.90 ± 0.34	4.70 ± 0.33	0.684
Fasting insulin (pmol/L)	28.40 ± 5.80	43.30 ± 10.20	0.202
HOMA2 IR	0.46 ± 0.10	0.72 ± 0.20	0.187
SBP (mmHg)	119.6 ± 4.4	133.0 ± 3.1	0.039
DBP (mmHg)	73.6 ± 4.3	81.7 ± 1.9	0.153

BMI, body mass index; DBP, diastolic blood pressure; HOMA2 IR, Homeostasis Model Assessment 2 Insulin Resistance; SBP, systolic blood pressure. Data are mean $\pm$ SEM.

2.3. Adipocyte Gene Expression

Total RNA was extracted from hMADS cells using TRIzol reagent (Invitrogen, Breda, The Netherlands), and SYBR-Green-based real-time PCRs were performed to assess the gene expression of leptin, dipeptidyl-peptidase (DPP)-4, interleukin (IL)-6, plasminogen activator inhibitor (PAI)-1, adiponectin, tumour necrosis factor (TNF)α, and monocyte chemoattractant protein (MCP)-1; the adipocyte differentiation markers peroxisome proliferator-activated receptor γ (PPARγ), CCAAT-enhancer binding protein α (C/EBPα), fatty acid synthase (FAS), and perilipin 1 (PLIN1); as well as the hypoxia markers glucose transporter 1 (GLUT1), Bcl-2 interacting protein 3 (BNIP3), and vascular endothelial growth factor A (VEGFA) using an iCycler (Bio-Rad, Veenendaal, The Netherlands). Results were normalised to 18S ribosomal RNA.

2.4. Adipokine Secretion

The medium of the hMADS cells was collected over 24 h, from day 13 (after replacement of medium) to day 14 of differentiation, to determine the secretion of adipokines using high-sensitive ELISAs (leptin and DPP-4 from R&D Systems, Inc., Minneapolis, MN, USA; IL-6 and MCP-1 from Diaclone SAS, Besancon Cedex, France; adiponectin and PAI-1 from BioVendor–Laboratorni medicina a.s., Brno, Czech Republic). If necessary, samples were diluted with the dilution buffer provided by the manufacturer prior to the assay, which was performed in duplicates according to the manufacturer's instructions.

2.5. Statistical Analyses

Data are presented as mean $\pm$ SEM. The effects of exposure to different oxygen levels on adipocyte gene expression and adipokine secretion were analysed using one-way ANOVA or the Friedman test when data were not normally distributed, followed by post hoc comparison using Student's paired t-tests or the Wilcoxon signed-rank test in case of skewed data. GraphPad Prism version 8 for Windows (GraphPad Software, San Diego, CA, USA) was used to perform statistical analyses. $p < 0.05$ was considered statistically significant.

3. Results

3.1. The Effects of Oxygen Partial Pressure on Adipocyte Gene Expression

The exposure of differentiated hMADS cells derived from ABD and FEM AT to different pO$_2$ levels induced distinct gene expression patterns. Specifically, exposure to low *physiological* pO$_2$ (5% O$_2$) increased *leptin* expression compared with exposure to high *physiological* pO$_2$ (10% O$_2$) or room air (21% O$_2$) in differentiated ABD and FEM hMADS derived from individuals with NW as well as OB (all $p < 0.01$, Figure 1A). Furthermore, low *physiological* pO_2 markedly reduced the gene expression of the proinflammatory factors *DPP-4* and *IL-6* in both ABD and FEM differentiated hMADS derived from donors with OB (all $p < 0.01$) but not NW compared with high *physiological* pO_2 (Figure 1B,C). Low *physiological* pO_2 levels did not significantly alter the gene expression of *PAI-1*, *TNFα*, or *MCP-1* in differentiated ABD and FEM hMADS derived from NW and OB individuals (Figure 1D–G), except for a modest but significant ($p = 0.041$) increase in adiponectin gene expression in FEM differentiated hMADS derived from individuals with obesity (Figure 1E). In addition, high *physiological* AT pO_2 (10% O_2) increased the *PAI-1* ($p = 0.005$) and reduced the *adiponectin* expression ($p = 0.010$) in FEM differentiated hMADS derived from individuals with OB compared with those at 21% O_2 exposure. As expected, exposure to *physiological* oxygen levels, i.e., lower oxygen levels compared with standard laboratory conditions, increased the gene expression of the classical hypoxia markers *GLUT1* and *VEGFA*, and, to a lesser extent, increased that of *BNIP3* (Figure S1A–C). Furthermore, exposure to low *physiological* oxygen levels (5% O_2) did not alter the gene expression of adipocyte differentiation markers compared with room air (21% O_2) in differentiated hMADS derived from individuals with NW as well as OB (Supplemental Figure S1D–G). In the differentiated hMADS derived from individuals with OB, the gene expression of *PPARγ*, *C/EBPα*, and FAS was lower, and expression of *PLIN1* higher, following exposure to 5% compared with 10% O_2.

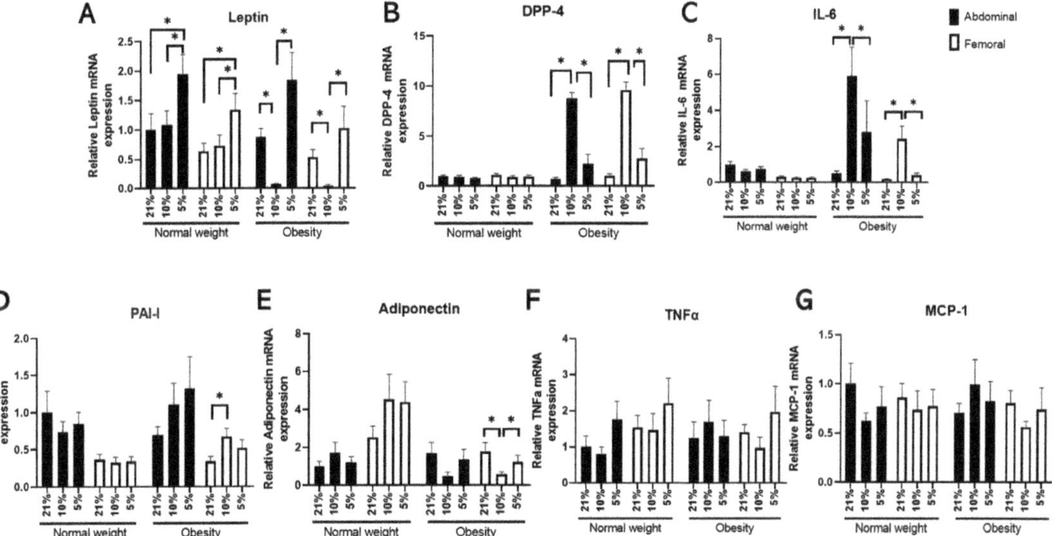

Figure 1. Adipokine and inflammatory markers' gene expression in hMADS cells following differentiation under different pO_2s (21% vs. 10% vs. 5% O_2) ($n = 9$ paired samples): (**A**) *leptin*, (**B**) *dipeptidylpeptidase (DPP)-4*, (**C**) *interleukin (IL)-6*, (**D**) *plasminogen activator inhibitor (PAI)-1*, (**E**) *adiponectin*, (**F**) *tumour necrosis factor (TNF)α*, and (**G**) *monocyte chemoattractant protein (MCP)-1*. Data are expressed as mean ± SEM. * $p < 0.05$.

3.2. The Effects of Oxygen Partial Pressure on Adipokine Secretion

Next, we investigated whether exposure to different pO_2 levels elicited functional changes in adipokine secretion from differentiated ABD and FEM hMADS. We found that adipokine secretion from both differentiated ABD and FEM hMADS was significantly affected by changes in oxygen availability (Figure 2). Specifically, low *physiological* pO_2 (5% O_2) exposure increased leptin secretion in differentiated ABD and FEM hMADS derived from individuals with OB compared with exposure to high *physiological* pO_2 (10% O_2: ABD,

$p = 0.009$; FEM, $p = 0.021$), and in differentiated ABD and FEM hMADS derived from individuals with NW compared with exposure to room air (21% O_2: ABD, $p = 0.014$; FEM, $p = 0.006$) (Figure 2A). Furthermore, DPP-4 secretion was significantly lower following exposure to low (5% O_2) compared with high (10% O_2) *physiological* pO_2 in differentiated ABD ($p = 0.027$) and FEM hMADS ($p = 0.004$) and IL-6 secretion in differentiated FEM hMADS only ($p = 0.007$), derived from donors with OB but not NW (Figure 2B,C). Moreover, low *physiological* pO_2 (5% O_2) reduced MCP-1 secretion ($p = 0.030$) but did not alter PAI-1 secretion from differentiated ABD hMADS derived from individuals with OB compared with 10% O_2 (Figure 2D,E). Finally, low *physiological* pO_2 (5% O_2) reduced both MCP-1 ($p = 0.028$) and PAI-1 ($p = 0.003$) secretion from differentiated FEM hMADS derived from donors with NW compared with 21% O_2 (Figure 2D,E). Adiponectin secretion was not detectable, and these data are therefore not reported.

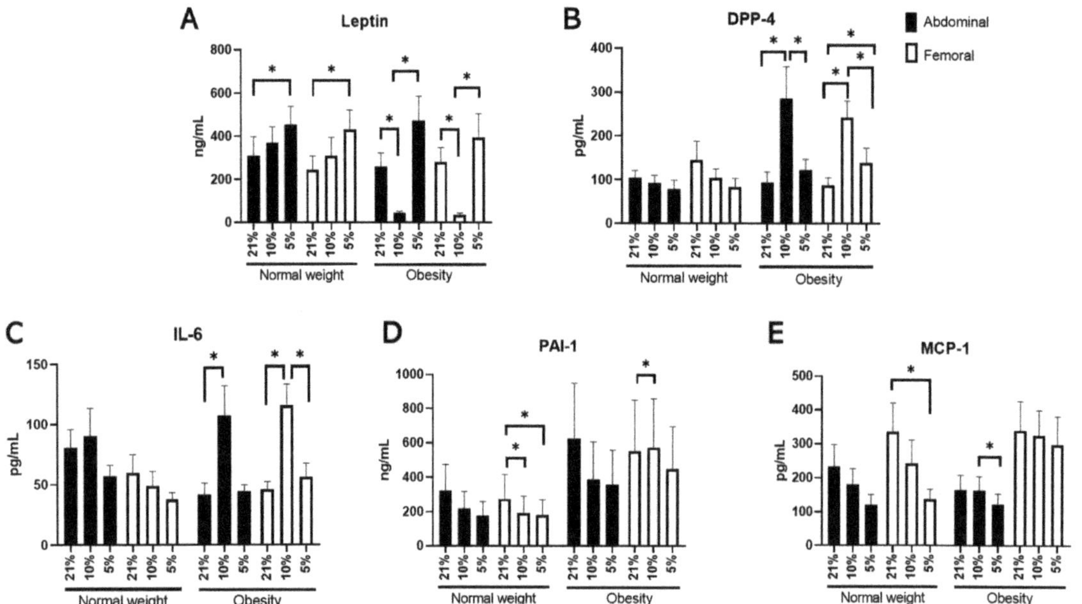

Figure 2. Adipokine and inflammatory markers' secretion in hMADS cells following differentiation under different pO_2s (21% vs. 10% vs. 5% O_2) ($n = 9$ paired samples). (**A**) Leptin, (**B**) dipeptidyl-peptidase (DPP)-4, (**C**) interleukin (IL)-6, (**D**) plasminogen activator inhibitor (PAI)-1, and (**E**) monocyte chemoattractant protein (MCP)-1. Data are expressed as mean $\pm$ SEM. * $p < 0.05$.

4. Discussion

In the present study, we investigated the impact of oxygen tension on adipokine gene expression and secretion in differentiated human multipotent ABD and FEM adipose-derived stem cells from women with NW or OB. Here, we demonstrate that low *physiological* pO_2 decreased gene expression and secretion of the proinflammatory factors *DDP-4* and *IL-6* in both differentiated ABD and FEM hMADS derived from individuals with OB, while these responses were not present in differentiated hMADS cells from NW individuals. Our findings highlight that the changes in pO_2 within the human *physiological* range in the adipocyte microenvironment contribute to alterations in the AT inflammatory phenotype and that these effects may differ between individuals with normal weight and obesity.

To determine whether the amount of oxygen present in the AT microenvironment affects the gene expression of adipokines, we exposed differentiating hMADS cells from ABD and FEM AT to low (5%) and high (10%) *physiological* pO_2 levels in human AT [9,16,21–24]. As expected, low *physiological* pO_2 levels increased the gene expression of several hypoxia markers. Strikingly, we show for the first time that low *physiological* pO$_2$ during adipogenesis consistently decreased the expression and secretion of the proinflammatory markers IL-6 and DPP-4 in both differentiated FEM and ABD hMADS derived from individuals with OB, but not NW. Moreover, the present data suggest that these cells maintain a memory of origin (i.e., a normal-weight or obese microenvironment) in vitro, even after 14 days of exposure to the same experimental conditions. In agreement with our findings, we have previously reported that in vivo ABD AT pO$_2$ was positively associated with the AT gene expression of several proinflammatory markers [22] and that low *physiological* pO$_2$ exposure reduced the gene expression of IL-6 and DPP-4 in adipocytes derived from women with obesity [16]. In addition, the present results show that low *physiological* pO$_2$ levels consistently increased leptin gene expression and secretion in differentiated ABD and FEM hMADS derived from donors with NW or OB. Leptin is an important regulator of appetite and energy expenditure, providing important feedback in relation to energy storage in the body through the hypothalamus, and is involved in multiple *physiological* processes such as the regulation of immunity [9,30–32]. Changes in leptin secretion due to altered oxygen tension in the AT microenvironment may thus affect these processes. Notably, pO$_2$-induced alterations in adipokine gene expression were paralleled by comparable changes in adipokine secretion. Importantly, the modest effects of pO$_2$ levels on adipocyte differentiation, if present at all, do not seem to explain the observed changes in adipokine expression and secretion, exemplified by the opposing effects of low pO$_2$ on the expression and secretion of leptin and the proinflammatory markers Il-6 and DPP-4. Famulla et al. [27] previously showed increased DPP-4, adiponectin, and IL-6 secretion following prolonged exposure to high *physiological* pO$_2$ (10% O$_2$), while low *physiological* pO$_2$ (5% O$_2$) tended to reduce the secretion of adiponectin. These differences between studies may at least partly be explained by differences in donor characteristics.

A strength of the present study is the paired comparisons between differentiated adipose-derived multipotent stem cells derived from ABD and FEM AT of individuals with NW and OB. Previous studies examining the effects of pO$_2$ levels on adipocyte inflammation have either used cell lines, adipose-derived multipotent stem cells from a single donor, or a pool of stem cells obtained from different donors. Because our findings demonstrate that the impact of changes in the AT microenvironment (i.e., different *physiological* pO$_2$ levels) on adipokine expression and secretion depends on the characteristics of the donors, future studies in the field of AT biology should take this '*memory-of-origin effect*' into account. Secondly, in contrast to many studies showing that acute exposure to severe (non-physiological) hypoxia evokes a proinflammatory response in murine and human (pre)adipocytes [13,14], we aimed to mimic *physiological* in vivo conditions in terms of pO$_2$ levels as well as the prolonged exposure duration in the present study. This study also has some limitations. We examined the effects of various oxygen levels in cells derived from postmenopausal women. Therefore, our findings cannot be translated to other subgroups of the population such as men or individuals of different age. Furthermore, we used a targeted approach to examine the gene expression and secretion of several adipokines. Future studies using an untargeted approach (e.g., microarray analysis, RNA sequencing, or proteomics) are warranted.

5. Conclusions

In conclusion, the present findings demonstrate that *physiological* oxygen levels regulate adipokine expression and secretion in differentiated ABD and FEM hMADS. Differentiated hMADS cells derived from women with OB display lower expression and secretion of several (proinflammatory) adipokines at low (5% O$_2$) compared with high (10% O$_2$) *physiological* oxygen tension. Except for the effects on leptin expression, no significant effects of low compared with high *physiological* oxygen levels were observed in differentiated hMADS cells derived from individuals with NW. Our findings thus indicate that pO$_2$ levels alter the expression and secretion of several adipokines in differentiated human ABD and FEM hMADS, and that donor characteristics determine experimental outcomes.

This has important implications for future mechanistic in vitro studies in the field of AT biology. For example, the outcomes of studies in which the effects of certain interventions on adipocyte inflammation and related biological mechanisms are investigated may depend on the microenvironmental oxygen tension. Furthermore, our findings highlight that it is important to report detailed the characteristics of the cell donor(s) in studies examining human adipocyte biology.

Supplementary Materials: The following supporting information can be downloaded at: https://www.mdpi.com/article/10.3390/cells11223532/s1, Figure S1: Adipocyte differentiation and hypoxia markers' gene expression in adipose tissue-derived mesenchymal stem cells following differentiation under different pO$_2$s (21% vs. 10% vs. 5% O$_2$).

Author Contributions: K.N.M. and G.H.G. acquired funding, conceived and designed research, interpreted data, and revised the manuscript; I.G.L. performed experiments, analysed data, interpreted data, prepared figures, and drafted the manuscript; N.H., Y.E., and J.W.E.J. performed experiments and analysed data. K.M.A.R. and E.E.B. interpreted data and revised the manuscript. All authors approved the final version of the manuscript.

Funding: This work was supported by the European Foundation for the Study of Diabetes (EFSD) under an EFSD/Lilly European Diabetes Research Program grant to Gijs H. Goossens and Konstantinos N. Manolopoulos, and Maastricht University (The Netherlands) and the University of Birmingham (U.K.) under a joint Ph.D. scholarship grant to Gijs H. Goossens and Konstantinos N. Manolopoulos.

Institutional Review Board Statement: The study was conducted in accordance with the Declaration of Helsinki, and the U.K. Health Research Authority National Health System Research Ethics Committee approved the present study (approval no. 18/NW/0392).

Informed Consent Statement: Informed consent was obtained from all subjects involved in the study.

Data Availability Statement: Data presented in this manuscript are available from the corresponding author on reasonable request.

Acknowledgments: The authors would like to express their gratitude to the study participants, the teams of the NIHa/Wellcome Trust Clinical Research Facility at Queen Elizabeth Hospital Birmingham and the NIHR CRN West Midlands (U.K.), as well as the Department of Human Biology at Maastricht University Medical Center+ (The Netherlands) for excellent practical support.

Conflicts of Interest: The authors declare no conflict of interest. The funders had no role in the design of the study; in the collection, analyses, or interpretation of data; in the writing of the manuscript; or in the decision to publish the results.

References

1. Kopelman, P.G. Obesity as a medical problem. *Nature* **2000**, *404*, 635–643. [CrossRef] [PubMed]
2. Karpe, F.; Pinnick, K.E. Biology of upper-body and lower-body adipose tissue—Link to whole-body phenotypes. *Nat. Rev. Endocrinol.* **2014**, *11*, 90–100. [CrossRef] [PubMed]
3. Goossens, G.H. The Metabolic Phenotype in Obesity: Fat Mass, Body Fat Distribution, and Adipose Tissue Function. *Obes. Facts* **2017**, *10*, 207–215. [CrossRef] [PubMed]
4. Yusuf, S.; Hawken, S.; Ôunpuu, S.; Bautista, L.; Franzosi, M.G.; Commerford, P.; Lang, C.C.; Rumboldt, Z.; Onen, C.L.; Lisheng, L.; et al. Obesity and the risk of myocardial infarction in 27 000 participants from 52 countries: A case-control study. *Lancet* **2005**, *366*, 1640–1649. [CrossRef]
5. Seidell, J.; Perusse, L.; Després, J.-P.; Bouchard, C. Waist and hip circumferences have independent and opposite effects on cardiovascular disease risk factors: The Quebec Family Study. *Am. J. Clin. Nutr.* **2001**, *74*, 315–321. [CrossRef]
6. Goossens, G.H.; Jocken, J.W.; Blaak, E.E. Sexual Dimorphism in Cardiometabolic Health: The Role of Adipose Tissue, Muscle and Liver. *Nat. Rev. Endocrinol.* **2021**, *17*, 47–66. [CrossRef]
7. Goossens, G.H. The role of adipose tissue dysfunction in the pathogenesis of obesity-related insulin resistance. *Physiol. Behav.* **2008**, *94*, 206–218. [CrossRef]
8. Rosen, E.D.; Spiegelman, B.M. What We Talk About When We Talk About Fat. *Cell* **2014**, *156*, 20–44. [CrossRef]
9. Lempesis, I.G.; Van Meijel, R.L.J.; Manolopoulos, K.N.; Goossens, G.H. Oxygenation of adipose tissue: A human perspective. *Acta Physiol.* **2019**, *228*, e13298. [CrossRef]
10. Heinonen, S.; Jokinen, R.; Rissanen, A.; Pietiläinen, K.H. White adipose tissue mitochondrial metabolism in health and in obesity. *Obes. Rev.* **2019**, *21*, e12958. [CrossRef]

11. Chen, K.; Zhang, J.; Beeraka, N.M.; Tang, C.; Babayeva, Y.V.; Sinelnikov, M.Y.; Zhang, X.; Zhang, J.; Liu, J.; Reshetov, I.V.; et al. Advances in the Prevention and Treatment of Obesity-Driven Effects in Breast Cancers. *Front. Oncol.* **2022**, *12*, 2663. [CrossRef] [PubMed]

12. Goossens, G.H.; Blaak, E.E. Adipose Tissue Dysfunction and Impaired Metabolic Health in Human Obesity: A Matter of Oxygen? *Front. Endocrinol.* **2015**, *6*, 55. [CrossRef] [PubMed]

13. Goossens, G.H.; Blaak, E.E. Adipose Tissue Oxygen Tension: Implications for Chronic Metabolic and Inflammatory Diseases. *Curr. Opin. Clin. Nutr. Metab. Care* **2012**, *15*, 539–546. [CrossRef]

14. Trayhurn, P. Hypoxia and Adipose Tissue Function and Dysfunction in Obesity. *Physiol. Rev.* **2013**, *93*, 1–21. [CrossRef] [PubMed]

15. Lempesis, I.G.; Goossens, G.H.; Manolopoulos, K.N. Measurement of human abdominal and femoral intravascular adipose tissue blood flow using percutaneous Doppler ultrasound. *Adipocyte* **2021**, *10*, 119–123. [CrossRef]

16. Vogel, M.A.; Jocken, J.W.; Sell, H.; Hoebers, N.; Essers, Y.; Rouschop, K.M.; Čajlaković, M.; Blaak, E.E.; Goossens, G.H. Differences in Upper and Lower-Body Adipose Tissue Oxygen Tension Contribute to the Adipose Tissue Phenotype in Humans. *J. Clin. Endocrinol. Metab.* **2018**, *103*, 3688–3697. [CrossRef]

17. Rausch, M.E.; Weisberg, S.; Vardhana, P.; Tortoriello, D.V. Obesity in C57BL/6J mice is characterized by adipose tissue hypoxia and cytotoxic T-cell infiltration. *Int. J. Obes.* **2007**, *32*, 451–463. [CrossRef]

18. Ye, J.; Gao, Z.; Yin, J.; He, Q. Hypoxia Is a Potential Risk Factor for Chronic Inflammation and Adiponectin Reduction in Adipose Tissue of Ob/Ob and Dietary Obese Mice. *Am. J. Physiol. Endocrinol. Metab.* **2007**, *293*, E1118–E1128. [CrossRef]

19. Yin, J.; Gao, Z.; He, Q.; Zhou, D.; Guo, Z.; Ye, J. Role of hypoxia in obesity-induced disorders of glucose and lipid metabolism in adipose tissue. *Am. J. Physiol. Metab.* **2009**, *296*, E333–E342. [CrossRef]

20. Hodson, L.; Humphreys, S.M.; Karpe, F.; Frayn, K.N. Metabolic Signatures of Human Adipose Tissue Hypoxia in Obesity. *Diabetes* **2013**, *62*, 1417–1425. [CrossRef]

21. Pasarica, M.; Sereda, O.R.; Redman, L.M.; Albarado, D.C.; Hymel, D.T.; Roan, L.E.; Rood, J.C.; Burk, D.H.; Smith, S.R. Reduced Adipose Tissue Oxygenation in Human Obesity: Evidence for Rarefaction, Macrophage Chemotaxis, and Inflammation without an Angiogenic Response. *Diabetes* **2009**, *58*, 718–725. [CrossRef] [PubMed]

22. Goossens, G.H.; Bizzarri, A.; Venteclef, N.; Essers, Y.; Cleutjens, J.P.; Konings, E.; Jocken, J.W.E.; Čajlaković, M.; Ribitsch, V.; Clément, K.; et al. Increased Adipose Tissue Oxygen Tension in Obese Compared with Lean Men Is Accompanied by Insulin Resistance, Impaired Adipose Tissue Capillarization, and Inflammation. *Circulation* **2011**, *124*, 67–76. [CrossRef] [PubMed]

23. Lawler, H.M.; Underkofler, C.M.; Kern, P.A.; Erickson, C.; Bredbeck, B.; Rasouli, N. Adipose Tissue Hypoxia, Inflammation, and Fibrosis in Obese Insulin-Sensitive and Obese Insulin-Resistant Subjects. *J. Clin. Endocrinol. Metab.* **2016**, *101*, 1422–1428. [CrossRef]

24. Cifarelli, V.; Beeman, S.C.; Smith, G.I.; Yoshino, J.; Morozov, D.; Beals, J.W.; Kayser, B.D.; Watrous, J.D.; Jain, M.; Patterson, B.W.; et al. Decreased adipose tissue oxygenation associates with insulin resistance in individuals with obesity. *J. Clin. Investig.* **2020**, *130*, 6688–6699. [CrossRef] [PubMed]

25. Goossens, G.H.; Vogel, M.A.A.; Vink, R.G.; Mariman, E.C.; Van Baak, M.A.; Blaak, E.E. Adipose tissue oxygenation is associated with insulin sensitivity independently of adiposity in obese men and women. *Diabetes Obes. Metab.* **2018**, *20*, 2286–2290. [CrossRef] [PubMed]

26. Trayhurn, P. Hypoxia and Adipocyte Physiology: Implications for Adipose Tissue Dysfunction in Obesity. *Annu. Rev. Nutr.* **2014**, *34*, 207–236. [CrossRef] [PubMed]

27. Famulla, S.; Schlich, R.; Sell, H.; Eckel, J. Differentiation of Human Adipocytes at Physiological Oxygen Levels Results in Increased Adiponectin Secretion and Isoproterenol-Stimulated Lipolysis. *Adipocyte* **2012**, *1*, 132–181. [CrossRef] [PubMed]

28. Jocken, J.W.E.; Goossens, G.H.; Popeijus, H.; Essers, Y.; Hoebers, N.; Blaak, E.E. Contribution of lipase deficiency to mitochondrial dysfunction and insulin resistance in hMADS adipocytes. *Int. J. Obes.* **2015**, *40*, 507–513. [CrossRef]

29. Rouschop, K.M.; Ramaekers, C.H.; Schaaf, M.B.; Keulers, T.G.; Savelkouls, K.G.; Lambin, P.; Koritzinsky, M.; Wouters, B. Autophagy is required during cycling hypoxia to lower production of reactive oxygen species. *Radiother. Oncol.* **2009**, *92*, 411–416. [CrossRef]

30. Dardeno, T.A.; Chou, S.H.; Moon, H.-S.; Chamberland, J.P.; Fiorenza, C.G.; Mantzoros, C.S. Leptin in human physiology and therapeutics. *Front. Neuroendocr.* **2010**, *31*, 377–393. [CrossRef]

31. Fasshauer, M.; Blüher, M. Adipokines in health and disease. *Trends Pharmacol. Sci.* **2015**, *36*, 461–470. [CrossRef] [PubMed]

32. Celik, O.; Aydin, S.; Celik, N.; Yilmaz, M. Peptides: Basic determinants of reproductive functions. *Peptides* **2015**, *72*, 34–43. [CrossRef] [PubMed]

Article

Expression of Steroid Receptor RNA Activator 1 (SRA1) in the Adipose Tissue Is Associated with TLRs and IRFs in Diabesity

Shihab Kochumon [1], Hossein Arefanian [1], Sardar Sindhu [1,2], Reeby Thomas [1], Texy Jacob [1], Amnah Al-Sayyar [1], Steve Shenouda [1], Fatema Al-Rashed [1], Heikki A. Koistinen [3,4,5], Fahd Al-Mulla [6], Jaakko Tuomilehto [4,7,8] and Rasheed Ahmad [1,*]

1   Department of Immunology & Microbiology, Dasman Diabetes Institute, Dasman 15462, Kuwait
2   Animal and Imaging Core Facilities, Dasman Diabetes Institute, Dasman 15462, Kuwait
3   Department of Medicine, University of Helsinki and Helsinki University Hospital, 00029 Helsinki, Finland
4   Department of Public Health and Welfare, Finnish Institute for Health and Welfare, 00271 Helsinki, Finland
5   Minerva Foundation Institute for Medical Research, 00290 Helsinki, Finland
6   Department of Genetics and Bioinformatics, Dasman Diabetes Institute, Dasman 15462, Kuwait
7   Department of Public Health, University of Helsinki, 00100 Helsinki, Finland
8   Saudi Diabetes Research Group, King Abdulaziz University, Jeddah 21589, Saudi Arabia
* Correspondence: rasheed.ahmad@dasmaninstitute.org; Tel.: +965-2224-2999 (ext. 4311)

Citation: Kochumon, S.; Arefanian, H.; Sindhu, S.; Thomas, R.; Jacob, T.; Al-Sayyar, A.; Shenouda, S.; Al-Rashed, F.; Koistinen, H.A.; Al-Mulla, F.; et al. Expression of Steroid Receptor RNA Activator 1 (SRA1) in the Adipose Tissue Is Associated with TLRs and IRFs in Diabesity. *Cells* **2022**, *11*, 4007. https://doi.org/10.3390/cells11244007

Academic Editor: Javier Gómez-Ambrosi

Received: 5 November 2022
Accepted: 9 December 2022
Published: 11 December 2022

Publisher's Note: MDPI stays neutral with regard to jurisdictional claims in published maps and institutional affiliations.

Abstract: Steroid receptor RNA activator gene (SRA1) emerges as a player in pathophysiological responses of adipose tissue (AT) in metabolic disorders such as obesity and type 2 diabetes (T2D). We previously showed association of the AT SRA1 expression with inflammatory cytokines/chemokines involved in metabolic derangement. However, the relationship between altered adipose expression of SRA1 and the innate immune Toll-like receptors (TLRs) as players in nutrient sensing and metabolic inflammation as well as their downstream signaling partners, including interferon regulatory factors (IRFs), remains elusive. Herein, we investigated the association of AT SRA1 expression with TLRs, IRFs, and other TLR-downstream signaling mediators in a cohort of 108 individuals, classified based on their body mass index (BMI) as persons with normal-weight (N = 12), overweight (N = 32), and obesity (N = 64), including 55 with and 53 without T2D. The gene expression of SRA1, TLRs-2,3,4,7,8,9,10 and their downstream signaling mediators including IRFs-3,4,5, myeloid differentiation factor 88 (MyD88), interleukin-1 receptor-associated kinase 1 (IRAK1), and nuclear factor-κB (NF-κB) were determined using qRT-PCR and SRA1 protein expression was determined by immunohistochemistry. AT SRA1 transcripts' expression was significantly correlated with TLRs-3,4,7, MyD88, NF-κB, and IRF5 expression in individuals with T2D, while it associated with TLR9 and TRAF6 expression in all individuals, with/without T2D. SRA1 expression associated with TLR2, IRAK1, and IRF3 expression only in individuals with obesity, regardless of diabetes status. Furthermore, TLR3/TLR7/IRAK1 and TLR3/TLR9 were identified as independent predictors of AT SRA1 expression in individuals with obesity and T2D, respectively. Overall, our data demonstrate a direct association between the AT SRA1 expression and the TLRs together with their downstream signaling partners and IRFs in individuals with obesity and/or T2D.

Keywords: steroid receptor RNA activator 1/SRA1; TLRs; IRFs; adipose tissue; obesity; type-2 diabetes; inflammation

---

1. Introduction

Obesity is known as a complex disease due to an excessive amount of body weight, and mainly the body fat associated with expansion and function of white adipose tissue. The expanded white adipose tissue in individuals with obesity produces a wide range of adipocytokines, such as proinflammatory cytokines, chemokines, hormones, and similar mediators [1–4]. Adipocytes, resident monocytes/macrophages, and other cell populations in the adipose tissue are actively involved in production and secretion of adipocytokines [5–8]. It is speculated that the mechanisms of obesity-induced insulin resistance are initiated and stimulated by these adipocytokines, mainly by proinflammatory cytokines in the white adipose tissue [9–13].

Toll-like receptors (TLRs) are the surface, innate immune receptors that identify the pathogen-associated molecular patterns (PAMPs) to activate and initiate inflammatory responses. In humans, 11 different TLRs have been so far identified [14,15]. Structurally, TLRs have an extracellular, leucine-rich repeat (LRR) domain which recognizes PAMPs, and a cytoplasmic Toll/IL-1 (TIR) domain that activates TLR-downstream signaling after ligand binding with its cognate TLR. LRR and TIR are involved in the recognition of PAMPs and activation of downstream adaptor proteins and signaling molecules, including myeloid differentiation factor 88 (MyD88), interleukin-1 (IL-1) receptor-associated kinases (IRAKs), and tumor necrosis factor receptor-associated factor (TRAF)-6, respectively [15]. After the activation of these adaptor proteins, stimulation of multiple pathways is initiated such as extracellular signal-regulated kinase (ERK), c-Jun N-terminal kinase (JNK), p38 mitogen-activated protein kinases (MAPK), NF-κB, and interferon regulatory factors (IRFs) pathways. The activation of NF-κB signaling results in the up-regulation of inflammatory markers including cytokines, chemokines, adhesion molecules, type-I interferons (IFN-α/β), tumor necrosis factor (TNF)-α, and IL-6) which offset the homeostasis between beneficial host innate immune responses and immunopathology [16–19].

TLRs such as TLR2, TLR3, TLR4, TLR7, TLR9, and TLR10 have been identified in multiple immune cell populations within the adipose tissue namely adipocytes, monocytes, and macrophages. Each of these TLRs has distinct ligands such as free fatty acids (FFAs), lipids, lipoproteins, nucleic acids, and proteins which indicate the potential role of innate immune TLRs as nutrient sensors involved in obesity [20,21]. The important roles of TLR signaling cascades in adipose tissue inflammation and impairment in obesity and type-2 diabetes (T2D) have been addressed and reported by us and others [22–24]. Expression changes in TLR1, TLR2, TLR4, TLR5, TLR6, TLR7, TLR8, TLR9, and TLR10 have been often reported in obesity, metabolic syndrome, inflammation and insulin resistance, T2D, and related complications such as cardiovascular disease, diabetic nephropathy, and atherosclerosis [23–34]. Specifically, TLR4/TLR2 have emerged as metabolic sensors of lipopolysaccharide (LPS) and saturated free fatty acids (sFFAs), both of which are abundantly found in individuals with obesity and T2D [35].

Steroid receptor RNA activator 1 (SRA1) was originally identified as an intergenic long non-coding RNA (lncRNA) which acts as an RNA coactivator of nuclear receptors to enhance steroid receptor-dependent gene expression [36]. It binds with DNA via interactions with other proteins that bind directly or indirectly with the DNA. Therefore, serving as a natural organizer by regulating physiological processes that dictate the epigenetic modifications including changes in chromatin and gene expression [37–39]. Increased SRA1 expression in the human liver, skeletal muscle, and in the white/brown adipose tissues as key organs regulating metabolic homeostasis, compared to other tissues, has been documented [36,40,41]. We recently showed that in individuals without T2D, adipose *SRA1* expression was significantly higher in obese people compared with normal weight people and the adipose tissue *SRA1* expression associated directly with metabolic markers including body mass index (BMI), percentage of body fat (PBF), serum insulin, homeostasis model assessment of insulin resistance (HOMA-IR), proinflammatory cytokines and chemokines or their receptors including C-X-C motif ligand-9 (CXCL9), CXCL10, CXCL11, TNFα, transforming growth factor-β (TGFβ), IL2RA, and IL18, but inversely with CCL19 and CCR2 expression. We further showed that TGFβ and IL18 were independent predictors of *SRA1* expression in individuals without T2D, while TNFα and IL2RA were the independent predictors in individuals with T2D. TNFα also predicted *SRA1* adipose expression in both normal weight and obese populations, regardless of diabetes status. Taken together, this study revealed specific association patterns of the adipose *SRA1* expression with diverse immune markers, most of them being inflammatory by nature [42].

In the current study, we examined whether the human adipose tissue SRA1 expression associated with TLRs expression, their adaptor proteins and TLR-downstream signaling molecules including NF-κB, IRF3, 4, and 5 in obesity and/or T2D.

2. Materials and Methods

2.1. Study Population and Anthropometry

This study included 108 participants who were classified based on BMI as persons with normal weight (NW) (BMI < 25 kg/m^2, n = 12), overweight (25 ≤ BMI < 30 kg/m^2, n = 32), or obese (BMI ≥ 30 kg/m^2, n = 64). Among these participants, 53 were with and 55 were without T2D. Diagnosis of T2D was based on fasting blood glucose and glycated hemoglobin (HbA1c) levels. HbA1c levels of <5.7, 5.7–6.4%, and >6.4% represented normal, prediabetes, and diabetes status, respectively. Height, weight, waist and hip circumferences, BMI, and PBF were measured and calculated as previously described [42]. For the assessment of insulin resistance (HOMA-IR) and insulin sensitivity, fasting blood glucose and insulin levels were used to determine HOMA index [43]. This study was approved by ethics committee of the Dasman Diabetes Institute, Kuwait in line with the ethical guidelines of the Declaration of Helsinki (Grant#: RA 2010-003). The written informed consent was obtained from each participant at the time of enrolment in the study. The individuals with chronic diseases of the heart, liver, kidney, lung, or those with type 1 diabetes, immune dysfunction, hematologic disorders, pregnancy, or malignancy were excluded previously stated [42,44]. Clinical and demographic features of the study cohort are presented in Supplementary Table S1.

2.2. Collection of Subcutaneous Adipose Tissue Samples

Biopsies of the human adipose tissue, about 0.5 g in size, were collected from the abdominal subcutaneous adipose tissue, next to the umbilicus by using sterile surgical technique as described [42,45]. The sample was further cut into smaller pieces, about 50–100 mg in size, and added to RNAlater (Sigma-Aldrich Chemie GmbH, Taufkirchen, Germany) and stored at −80 °C until use for RNA extraction.

2.3. Measurement of Metabolic Markers

Peripheral blood was collected from the individuals fasting overnight and the samples were analyzed for metabolic markers such as plasma glucose, insulin, and lipid profiles including triglycerides (TGL), low-density/high-density lipoproteins (LDL/HDL), and total cholesterol using Siemens Dimension RXL Analyzer (Diamond Diagnostics, Holliston, MA, USA). Glycated hemoglobin (HbA1c) was measured by Variant device (BioRad, Hercules, CA, USA).

2.4. Quantitative, Real-Time, Reverse-Transcription Polymerase Chain Reaction (RT-qPCR)

Fat samples were used for total RNA collection by using RNeasy kit (Qiagen, Valencia, CA, USA), following the protocol as recommended by the manufacturer. RNA template (0.5 μg) was used for cDNA synthesis using High-Capacity cDNA Reverse Transcription Kit (Applied Biosystems, Foster, CA, USA) as earlier stated [45]. Real-time qRT-PCR was carried out as per the protocol [42,45,46]. Briefly, 50 ng of cDNA was amplified using TaqMan Gene Expression Master Mix (Applied Biosystems, CA, USA) and gene-specific 20 X TaqMan gene expression assays including forward and reverse primers (Supplementary Table S2), target-specific TaqMan 5′FAM-labeled and 3′NFQ-labeled MGB probe, using 7500 Fast Real-Time PCR System (Applied Biosystems, CA, USA) as described elsewhere [42]. Target gene expression relative to control (NW fat samples) was determined using comparative Ct method [47] and data were normalized to GAPDH gene expression as described [48–52].

2.5. Immunohistochemistry (IHC)

Expression of SRA1, TLR4, IRAK1 or NF-kB was measured in the fat tissue by IHC as described elsewhere [42,53]. Briefly, subcutaneous fat tissue (paraffin-embedded) was cut in 4μm thick sections, deparaffinized by xylene, and rehydrated by serial immersions in 100%, 95%, and 75% ethanol in water. Antigen was retrieved by retrieval solution (pH 6.0; Dako, Glostrup, Denmark) with 8 min boiling and 15 min cooling steps. After 3 washes with PBS, internal peroxidase was blocked by 30 min treatment with 3% H_2O_2 and non-specific antibody binding was blocked by 1 h treatment, each, with 5% non-fat milk and 1% BSA. The primary antibody treatment was carried out overnight at room temperature, using anti-human SRA1 rabbit polyclonal antibody (1:800 dilution) (Thermo-Scientific PA5-62145, pH 6.0), rabbit monoclonal anti-IRAK antibody (ab302554, Abcam® Cambridge, UK), 1:400 dilution of rabbit monoclonal anti-IRF5 antibody (ab181553, Abcam® Cambridge, UK), 1:1000 dilution of rabbit polyclonal anti-NFkB antibody (ab16502, Abcam® Cambridge, UK), and 1:1000 dilution of mouse monoclonal anti-TLR4 antibody (ab13556, Abcam® Cambridge, UK). After washing 3 times using PBS (0.5% Tween), samples were incubated for 1 h with secondary antibody (HRP-conjugated goat anti-rabbit; EnVision™ Kit from Dako, Glostrup, Denmark) and the substrate (3,3′-DAB chromogen) was added to develop color. After washing 3 times in running water, samples were counterstained (Harris hematoxylin), dehydrated by immersion in 75%, 95%, and 100% ethanol in water, clarified by xylene, and mounted in DPX. Later, digital photomicrographs were taken (40X magnification) and regional heterogeneity was assessed in 4 different regions of tissue sample (PannoramicScan, 3DHistech, Budapest, Hungry). The data were expressed as the staining intensity measured in arbitrary units (AU) and analyzed using imageJ software (NIH, Bethesda, MD, USA).

In addition, we also performed H&E staining on lean and obese adipose tissue samples, 5 each, to represent sample quality for standard tissue staining (Supplementary Figure S1).

2.6. Statistical Analysis

The data obtained were presented as mean ± SEM values and analyzed using Graph-Pad Prism (GraphPad, San Diego, CA, USA) and SPSS (IBM SPSS Inc., Chicago, IL, USA) software. Means between two groups were compared using unpaired *t*-test, and between more than two groups were compared using One-way ANOVA, Kruskal-Wallis and Mann-Whitney tests. Spearman correlation and multivariate regression analyses were performed to determine associations between variables. All *p*-values ≤ 0.05 were considered significant. For multivariate linear regression, the Enter method was used, selecting the variables that significantly correlated with *SRA1* expression as predictor variables and were entered simultaneously to generate the model. *F*-test was used to test whether the independent variables collectively predicted the dependent variable. *R*-squared evaluated how much variance in the dependent variable was accounted for by the set of independent variables. The *p*-value assessed the significance and the β-value identified the magnitude of prediction for each independent variable.

3. Results

3.1. SRA1 Adipose Expression and Its Association with TLRs, Downstream Signaling Mediators and IRFs Expression in the Study Population

We sought to determine SRA1 protein expression in adipose tissue samples from NW individuals and those with obesity, 10 individuals each, and the data show that the expression was significantly higher in people with obesity compared to their NW counterparts, whether with or without T2D ($p < 0.0001$) (Figure 1).

We next sought to determine the adipose *SRA1* gene expression and assessed its relationship with typical inflammatory sensing and signaling components such as TLRs, their downstream signaling mediators, and IRFs. The expression of these markers was first compared among populations defined as those with NW, overweight, and obesity. To this end, adipose SRA1 mRNA expression differed significantly only between non-diabetic NW and those with obesity individuals ($p = 0.015$) whereas it differed non-significantly between all other BMI groups, with or without T2D (Table 1). Regarding expression of TLRs, their downstream signaling mediators and IRFs, adipose expression of TLR2

(p = 0.018), TLR3 (p = 0.010), TLR8 (p = 0.048), IRAK1 (p = 0.047), and IRF5 (p = 0.016) was significantly higher in participants with obesity compared to NW participants. IRF5 expression in both overweight (p = 0.031) and those with obesity (p = 0.016) differed significantly from that of NW participants. Expression of IRF3 (p = 0.047), IRF4 (p = 0.020), and IRF5 (p = 0.039) was significantly higher in individuals with T2D compared with those without T2D. Furthermore, adipose *SRA1* expression was associated with TLR2 (r = 0.218, p = 0.036), TLR3 (r = 0.218, p < 0.0001), TLR4 (r = 0.226, p = 0.027), TLR7 (r = 0.196, p = 0.045), NF-κB (r = 0.297, p = 0.002), and IRAK1 (r = 0.201, p = 0.044) expression in the total (N = 108) study population (Table 2; Figure 2A–F).

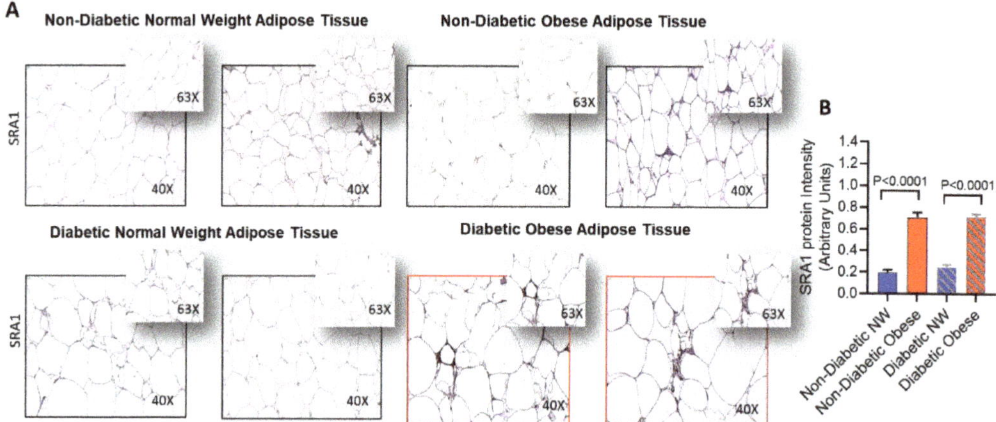

Figure 1. Adipose SRA1 protein expression in adipose tissue. SRA1 protein expression was determined in adipose tissue samples from non-diabetic individuals with NW and obesity, as well as from persons with type-2 diabetic (T2D), as NW and those with obesity, 10 each, using by immunohistochemistry (IHC) as described in Materials and Methods. IHC staining intensity was expressed as arbitrary units (AU) and the data (mean ± SEM) were compared between NW and obese populations, with or without T2D, using unpaired *t*-test and p < 0.05 was considered significant. (**A**) The representative IHC images are shown for non-diabetic NW/obese and T2D NW/obese individuals, one each. (**B**) IHC staining data (AU) show the elevated adipose SRA1 expression in persons with obesity compared to NW persons, whether with or without T2D (p < 0.0001).

Furthermore, to confirm the expression of protein levels in the adipose tissue, we performed immunohistochemistry analysis on TLR4, IRAK1 and NF-kB as representatives. Immunohistochemistry analysis showed that TLR4 (Figure 3A,B), IRAK1 (Figure 4A,B) and NF-kB (Figure 5A,B) were significantly upregulated in individuals with obesity. Our protein data show that SRA1 positively correlated with TLR4 protein (r^2 = 0.623; p = 0.0002; Figure 3C). IRAK1 protein (r^2 = 0.0649; p < 0.0001; Figure 4C) and NF-kB protein (r^2 = 0.379; p = 0.0085; Figure 5C).

Table 1. Adipose expression (mRNA) of TLRs, their downstream signaling mediators, IRFs, and SRA1 in participants with and without T2D.

Inflammatory Marker	Total Participants (N = 108)										
	Without T2D (N = 55)					With T2D (N = 53)					
	Normal Weight (N = 8)	Overweight (N = 19)	Obese (N = 28)	Normal Weight vs. Overweight (*p* Value) #	Normal Weight vs. Obese (*p* Value) #	Normal Weight (N = 4)	Overweight (N = 13)	Obese (N = 36)	Normal Weight vs. Overweight (*p* Value) #	Normal Weight vs. Obese (*p* Value) #	Without T2D (N = 55) vs. With T2D (N = 53) (*p* Value) \$
SRA1	1.7 ± 0.24	1.8 ± 0.61	2.2 ± 0.62	0.335	0.025 *	2.3 ± 0.95	1.7 ± 0.55	1.9 ± 1.17	0.175	0.436	0.373
TLR2	1.95 ± 1.04	2.88 ± 1.96	3.20 ± 1.31	0.197	0.018 *	2.18 ± 0.76	2.82 ± 1.17	3.27 ± 1.66	0.465	0.272	0.305
TLR3	1.49 ± 0.30	1.79 ± 1.17	2.57 ± 1.50	0.629	0.019 *	3.24 ± 2.20	1.74 ± 1.06	2.07 ± 0.59	0.168	0.616	0.618
TLR4	1.31 ± 0.23	1.42 ± 0.41	1.67 ± 0.65	0.661	0.267	2.04 ± 0.89	1.33 ± 0.31	1.57 ±0.45	0.144	0.441	0.609
TLR7	2.06 ± 0.83	3.15 ± 2.27	3.38 ± 2.17	0.502	0.141	2.89 ± 1.51	1.92 ± 1.36	3.78 ± 2.04	0.387	0.638	0.546
TLR8	1.68 ± 0.69	3.51 ± 2.90	3.31 ± 1.96	0.117	0.061 *	2.85 ± 0.75	2.71 ± 1.60	3.94 ± 2.78	0.776	0.655	0.202
TLR9	8.51 ± 3.91	7.85 ± 3.37	7.32 ± 3.87	0.803	0.292	7.02 ± 5.62	7.63 ± 2.54	7.91 ± 3.47	0.379	0.357	0.852
TLR10	21.88 ± 22.28	23.04 ± 22.15	29.88 ± 37.15	0.719	0.549	24.96 ± 15.52	23.97 ± 24.62	23.05 ± 20.07	0.625	0.65	0.976
NF-κB	1.18 ± 0.20	1.28 ± 0.33	1.38 ± 0.28	0.476	0.053	1.47 ± 0.29	1.28 ± 0.42	1.26 ± 0.37	0.342	0.282	0.559
MyD88	1.62 ± 0.47	1.86 ± 0.73	1.85 ± 0.57	0.421	0.346	2.71 ± 0.77	1.96 ± 0.57	1.92 ± 0.59	0.166	0.112	0.240
IRAK1	6.14 ± 2.83	8.19 ± 3.37	9.12 ± 3.28	0.119	0.042 *	8.01 ± 1.11	9.45 ± 3.46	9.69 ± 3.23	0.432	0.353	0.116
TRAF6	2.49 ± 0.35	2.26 ± 1.18	2.36 ± 0.98	0.113	0.255	2.12 ± 0.80	2.26 ± 0.75	2.33 ± 0.64	0.922	0.712	0.544
IRF3	1.81 ± 0.57	2.31 ± 0.68	2.54 ± 0.61	0.152	0.065	2.45 ± 0.80	2.73 ± 0.56	2.65 ± 0.59	0.4542	0.675	0.047 *
IRF4	2.21 ± 0.84	3.61 ± 2.18	3.40 ± 1.99	0.228	0.169	3.74 ± 3.41	4.32 ± 2.10	3.95 ±1.46	0.702	0.828	0.020 *
IRF5	1.53 ± 0.33	2.82 ± 1.44	2.73 ± 1.08	0.027 *	0.016 *	2.72 ± 0.93	2.90 ± 1.31	3.13 ±1.13	0.879	0.584	0.039 *

= One way Anova, Kruskal-Wallis test, \$ = Mann-Whitney test. * Significant.

Table 2. Spearman's correlation analysis of adipose gene expression of *SRA1* with TLRs, their signaling mediators, and IRFs in the total study population.

All participants (N = 108)		
Adipose Marker	*r*	*p* **Value**
TLR2	0.218	0.036 *
TLR3	0.406	<0.0001 ****
TLR4	0.226	0.027 *
TLR7	0.196	0.045 *
TLR8	0.071	0.481
TLR9	0.080	0.411
TLR10	0.074	0.459
NF-κB	0.297	0.002 **
MyD88	0.149	0.134
IRAK1	0.201	0.044 *
TRAF6	0.032	0.746
IRF3	0.139	0.187
IRF4	−0.009	0.931
IRF5	0.175	0.080

$p \leq 0.05$ *, $p < 0.01$ **, $p < 0.0001$ ****.

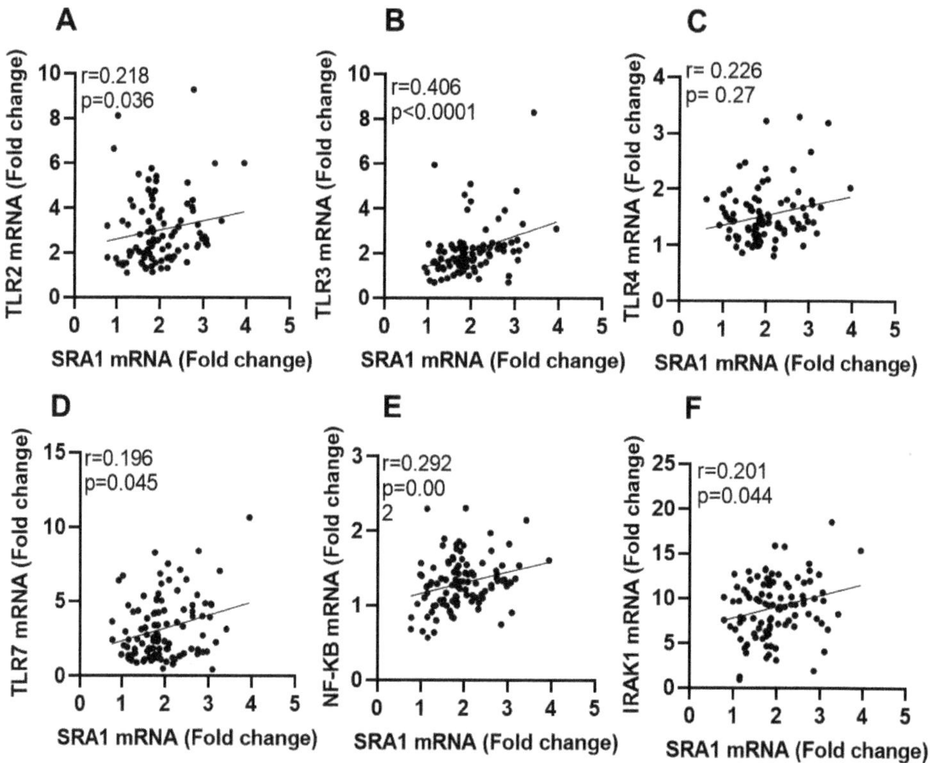

Figure 2. Adipose tissue SRA1 gene expression is correlated with TLRs and signaling molecules. Subcutaneous adipose tissues were obtained from lean, overweight, obese individuals. In this case, mRNA expression of SRA1, TLR2, TLR3, TLR4, TLR7, IRAK1 and NF-kB was detected by real-time RT-PCR and represented as fold change over controls. (**A–F**) In the studied population, SRA1 transcript levels, in adipose tissue, are positively correlated with TLRs and their signaling molecules.

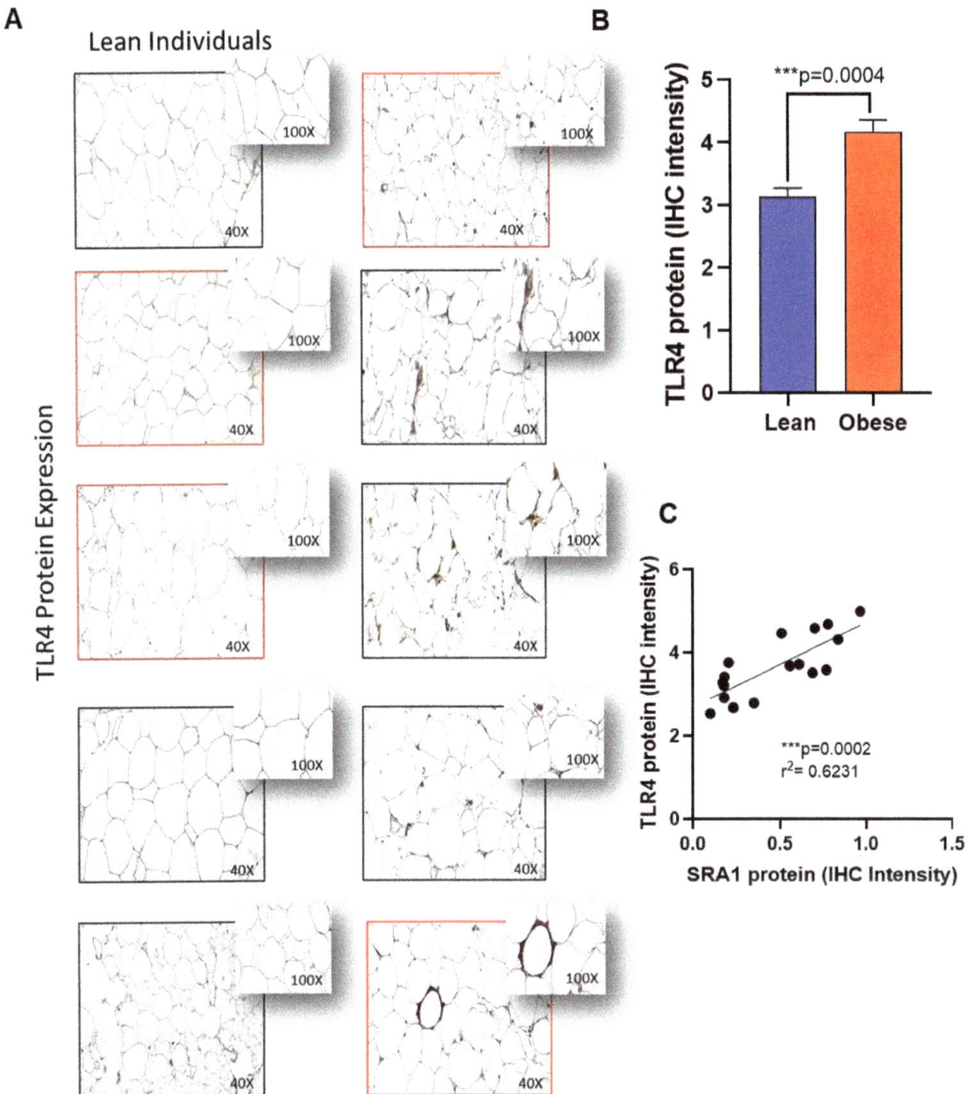

Figure 3. Adipose TLR4 protein expression in adipose tissue and its correlation with SRA1 protein expression. TLR4 protein expression was determined in adipose tissue samples from lean (normal weight; NW) and obese individuals (n = 8 for each group) using by immunohistochemistry (IHC) as described in Materials and Methods. IHC staining intensity was expressed as arbitrary units (AU) and the data (mean ± SEM) were compared between NW and obese populations using unpaired *t*-test and $p < 0.05$ was considered significant. (**A**) The representative IHC images are shown for NW and obese individuals. (**B**) IHC staining intensity data (AU). (**C**) Correlation between TLR4 protein expression with SRA1.

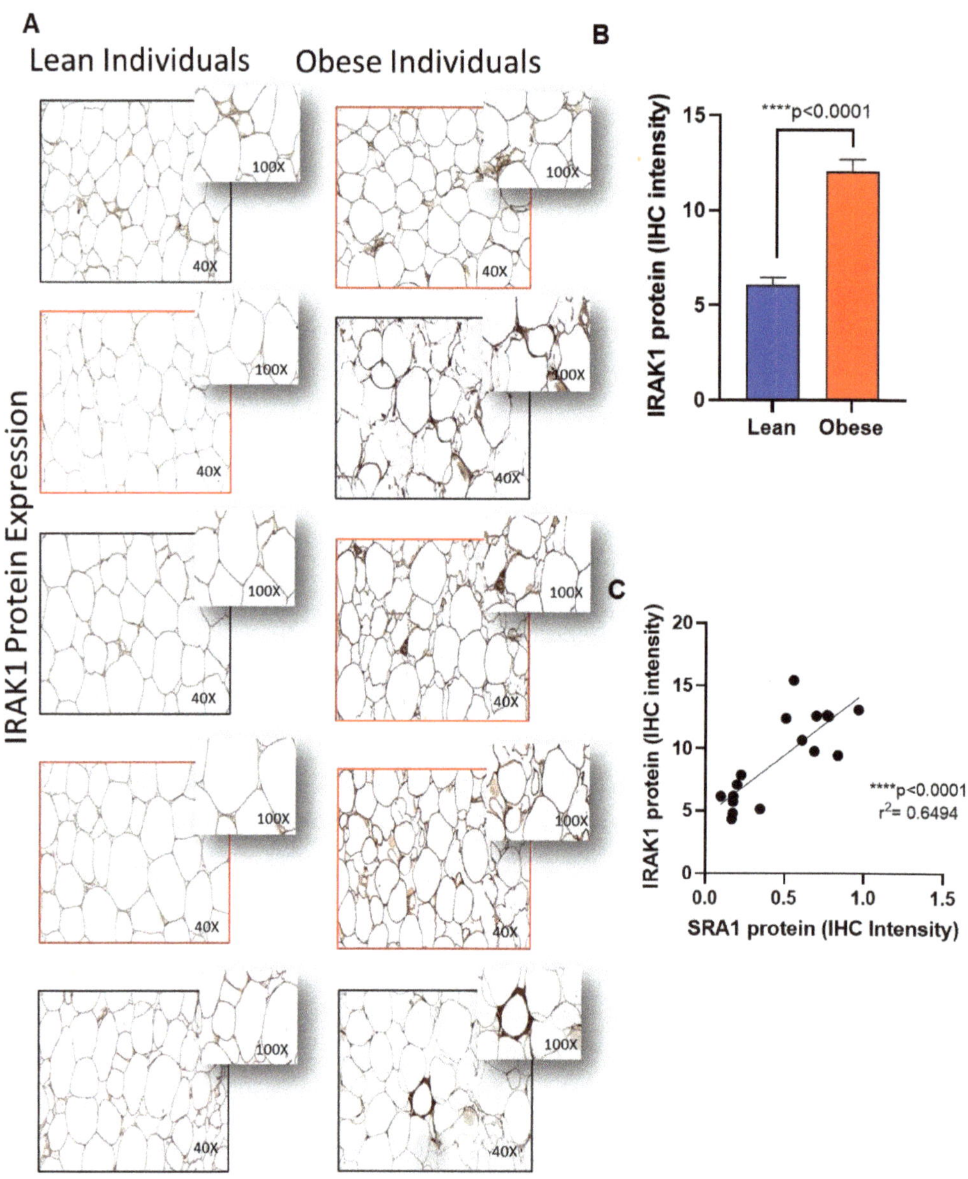

Figure 4. Adipose IRAK1 protein expression in adipose tissue and its correlation with SRA1 protein expression. IRAK1 protein expression was determined in adipose tissue samples from lean (normal weight; NW) and obese individuals (n = 8 for each group) using by immunohistochemistry (IHC) as described in Materials and Methods. IHC staining intensity was expressed as arbitrary units (AU) and the data (mean ± SEM) were compared between NW and obese populations using unpaired *t*-test and $p < 0.05$ was considered significant. (**A**) The representative IHC images are shown for NW and obese individuals. (**B**) IHC staining intensity data (AU). (**C**) Correlation between IRAK1 protein expression with SRA1.

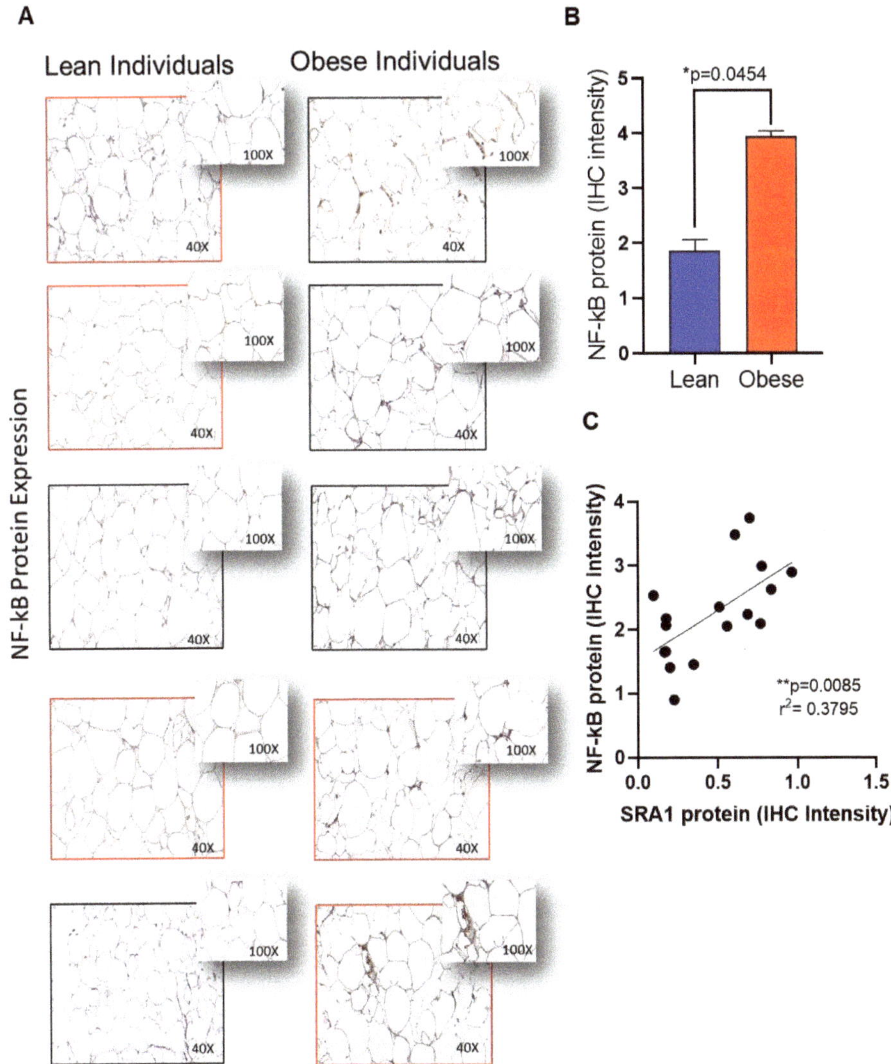

Figure 5. Adipose NF-kB protein expression in adipose tissue and its correlation with SRA1 protein expression. NF-kB protein expression was determined in adipose tissue samples from lean (normal weight; NW) and obese individuals (n= 8 for each group) using by immunohistochemistry (IHC) as described in Materials and Methods. IHC staining intensity was expressed as arbitrary units (AU) and the data (mean ± SEM) were compared between NW and obese populations using unpaired *t*-test and $p < 0.05$ was considered significant. (**A**) The representative IHC images are shown for NW and obese individuals. (**B**) IHC staining intensity data (AU). (**C**) Correlation between NF-kB protein expression with SRA1.

3.2. Association of Adipose SRA1 Expression (mRNA) with TLRs, Their Signaling Mediators, and IRFs in Individuals Classified as Those with NW, Overweight, and Obesity

Regarding the association between adipose *SRA1* gene expression and meta-inflammatory markers studied (Table 3), we found that *SRA1* correlated with TLR2 (r = 0.317, $p = 0.017$), TLR3 (r = 0.531, $p < 0.0001$), TLR4 (r = 0.311, $p = 0.022$), TLR7 (r = 0.305, $p = 0.015$),

TLR9 (r = 0.374, *p* = 0.002), MyD88 (r = 0.324, *p* = 0.010), IRAK1 (r = 0.255, *p* = 0.044), NF-κB (r = 0.454, *p* < 0.001), IRF3 (r = 0.290, *p* = 0.030) and IRF5 (r = 0.321, *p* = 0.010) expression in individuals with obesity. *SRA1* associated inversely with TLR9 expression (r = −0.489, *p* = 0.005) only in overweight group. In NW participants, *SRA1* was associated with MyD88 (r = 648, *p* = 0.043), IRF3 (r = 0.0857, *p* = 0.014), and IRF5 (r = 0.929, *p* = 0.003) expression. Heat map shown in Figure 6.

Table 3. Spearman's correlation analysis of adipose gene expression of *SRA1* with TLRs, their signaling mediators, and IRFs in individuals differing by obesity levels.

Obesity Level	Normal Weight		Overweight		Obese	
	(N = 12)		(N = 32)		(N = 64)	
Adipose Marker	*r*	*p* **Value**	*r*	*p* **Value**	*r*	*p* **Value**
TLR2	0.095	0.823	−0.026	0.895	0.317	0.017 *
TLR3	0.310	0.456	−0.016	0.934	0.531	<0.0001 ****
TLR4	0.405	0.320	−0.164	0.386	0.311	0.022 *
TLR7	0.224	0.533	−0.190	0.297	0.305	0.015 *
TLR8	0.267	0.488	−0.347	0.060	0.229	0.071
TLR9	0.098	0.762	−0.489	0.005 **	0.374	0.002 **
TLR10	0.105	0.746	−0.248	0.194	0.192	0.134
NF-κB	0.266	0.404	−0.071	0.699	0.454	<0.001 ***
MyD88	0.648	0.043 *	−0.347	0.060	0.324	0.010 *
IRAK1	0.429	0.289	0.044	0.818	0.255	0.044 *
TRAF6	−0.378	0.226	−0.327	0.073	0.233	0.064
IRF3	0.857	0.014 *	−0.322	0.088	0.290	0.030 *
IRF4	0.429	0.397	−0.330	0.075	0.195	0.136
IRF5	0.929	0.003 **	−0.105	0.575	0.321	0.010 *

$p \leq 0.05$ *, $p < 0.01$ **, $p < 0.001$ ***, $p < 0.0001$ ****.

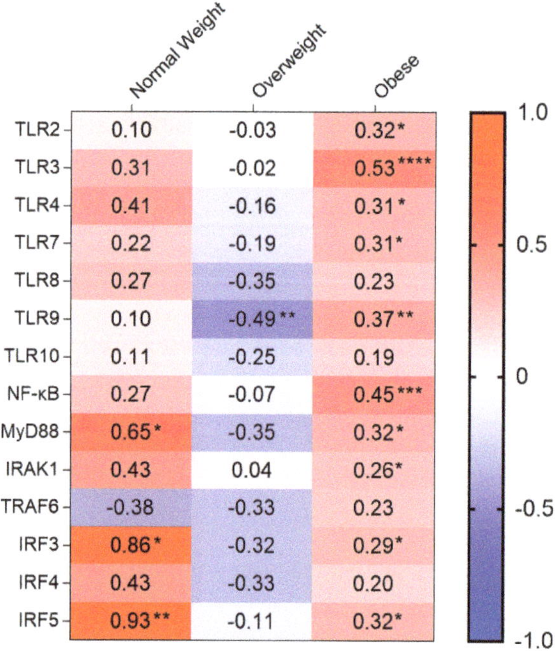

Figure 6. Heat map of the correlation of SRA1 expression with TLRs and their signaling molecules in adipose tissues obtained from NW, Overweight and Obese. $p \leq 0.05$ *, $p < 0.01$ **, $p < 0.001$ ***, $p < 0.0001$ ****.

3.3. Association of Adipose SRA1 Expression (mRNA) with TLRs, Their Signaling Mediators, and IRFs in Individuals with/without T2D

In the analysis whether diabetic status affected associations between adipose *SRA1* expression and other markers of inflammation, we found that in people with T2D, *SRA1* expression was associated with TLR3 (r = 0.555, p < 0.0001), TLR4 (r = 0.302, p = 0.044), TLR7 (r = 0.292, p = 0.040), TLR9 (r = 0.398, p = 0.003), NF-κB (r = 0.381, p = 0.005, MyD88 (r = 0.311, p = 0.030), TRAF6 (r = 0.286, p = 0.038), and IRF5 (r = 0.288, p = 0.043) expression (Table 4; Figure 7). On the other hand, in participants without T2D, *SRA1* was inversely associated with TLR9 (r = −0.290, p = 0.034) and TRAF6 (r = −0.318, p = 0.019) expression (Table 4).

Table 4. Spearman's correlation analysis of adipose gene expression of *SRA1* with TLRs, their signaling mediators, and IRFs in individuals with/without T2D.

Diabetes Status	Without T2D		With T2D	
	(N = 55)		(N = 53)	
Adipose Marker	r	p **Value**	r	p **Value**
TLR2	0.206	0.165	0.223	0.136
TLR3	0.247	0.080	0.555	<0.0001 ****
TLR4	0.096	0.522	0.302	0.044 *
TLR7	0.140	0.309	0.292	0.040 *
TLR8	0.028	0.843	0.180	0.210
TLR9	−0.290	0.034 *	0.398	0.003 **
TLR10	0.062	0.667	0.084	0.556
NF-κB	0.191	0.162	0.381	0.005 **
MyD88	0.014	0.918	0.311	0.030 *
IRAK1	0.220	0.117	0.239	0.099
TRAF6	−0.318	0.019 *	0.286	0.038 *
IRF3	0.051	0.739	0.220	0.137
IRF4	0.066	0.649	−0.017	0.909
IRF5	0.167	0.243	0.288	0.043 *

$p \leq 0.05$ *, $p < 0.01$ **, $p < 0.0001$ ****.

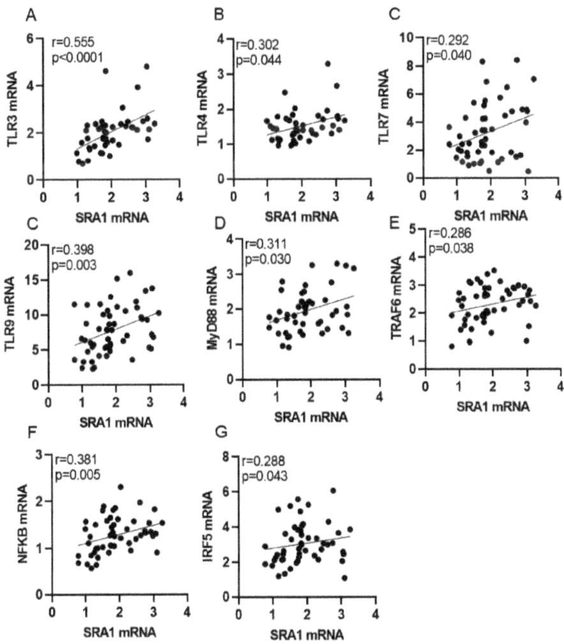

Figure 7. SRA1 gene expression is correlated with TLR3, TLR4, TLR7, TLR9, Myd88, TRAF6, NF-kB or IRF5 in the adipose tissues obtained from individuals with T2D (**A–G**).

3.4. Analysis of the Independent Associations between SRA1 and TLRs, Their Signaling Mediators, and IRFs

In order to determine independent associations, the markers showing significant associations with adipose *SRA1* expression were further assessed by multivariable stepwise linear regression analysis (Table 5). We found that in the total population (N = 108), TLR3 and IRAK1 independently predicted the adipose SRA1 expression. In participants with T2D (N = 53), TLR3 and TLR9 were identified as the independent predictors of adipose *SRA1* expression, but not in participants without T2D (N = 55). Regression analysis stratified by obesity status revealed the independent association of adipose SRA1 expression with MyD88 in NW (N = 12), and with TLR9 in overweight (N = 32) participants. In individuals with obesity (N = 64), TLR3, TLR7 and IRAK1 were detected as the independent predictors of adipose SRA1 expression.

Table 5. The multiple linear regression analysis.

All participants (N = 108)	ANOVA		$r^2 = 0.21$	p value < 0.0001
	Predictor Variable	TLR3	β value = 0.392	p value < 0.0001
		IRAK1	β value = 0.279	p value = 0004
With T2D (N = 53)	ANOVA		$r^2 = 0.35$	p value < 0.0001
	Predictor Variable	TLR3	β value = 0.499	p value < 0.001
		TLR9	β value = 0.329	p value = 0.010
Normal Weight (N = 12)	ANOVA		$r^2 = 0.64$	p value = 0.005
	Predictor Variable	MyD88	β value = 0.801	p value = 0.005
Overweight (N = 32)	ANOVA		$r^2 = 0.25$	p value = 0.004
	Predictor Variable	TLR9	β value = −0.499	p value = 0.004
Obese (N = 64)	ANOVA		$r^2 = 0.38$	p value < 0.0001
	Predictor Variable	TLR3	β value = 0.464	p value < 0.0001
		IRAK1	β value = 0.317	p value = 0.005
		TLR7	β value = 0.246	p value = 0.027

Multiple linear regression analysis is performed to identify TLRs and their signaling molecules associated with SRA1 as predictor variables.

4. Discussion

There is a relative lack of data highlighting the role and significance of long noncoding RNAs in metabolic inflammation and insulin resistance. In the present study, we show that the adipose tissue SRA1 protein expression is significantly increased in individuals with obesity, as compared to their normal weight counterparts, regardless of T2D status. While *SRA1* mRNA expression is significantly elevated, only in people with obesity people without T2D as compared to normal-weight counterparts. We further demonstrate that adipose *SRA1* gene expression is associated with the expression of TLRs, their signaling mediators and IRFs including TLR2, TLR3, TLR4, TLR7, TLR9, MyD88, IRAK1, NF-κB, IRF3, and IRF5. There is a general consensus that obesity is associated with low-grade chronic inflammation, marked by the abnormal production of pro- and anti-inflammatory cytokines and adipokines by the expanding white adipose tissue, which influences changes in adipose tissue expression of the innate immune receptors such as TLRs. The growing evidence now supports that TLRs play a role in inducing a systemic acute phase response characterized by chronic inflammation and oxidative stress [1–4]. Apart from PAMPs, FFAs also act as TLR agonists which consolidates emerging role of TLRs as metabolic sensors and as receptors of immune-metabolic significance [20]. Regarding nutrient sensing mechanism and ensuing inflammatory responses, TLR signaling is initiated in the TIR domain and the TLR-downstream signaling is propagated through the pathways dependent or independent of MyD88 adaptor protein [16,17]. MyD88-dependent signaling is induced after TLR-ligand engagement and TLR dimerization, leading to recruitment of MyD88 and IRAKs to the

TIR domain [54,55]. IRAKs are known as the death domain-containing serine/threonine kinases and adapter proteins which play key roles in signaling pathways of the IL-1 family receptors and TLRs. IRAK1 is activated after phosphorylation by IRAK4 and associates with TRAF6. The IRAK1/TRAF6 complex then engages with the TGFβ-activated kinase (TAK)-1 and TAB-1/2 adapter proteins to yield a macromolecular complex [56]. IRAK1, following its phosphorylation, disengages from the signaling complex and TAK1 activates the inhibitor of NF-κB kinase alpha/beta (IKKα/β), which leads to the phosphorylation, ubiquitination, and degradation of IκBα to allow nuclear translocation of the p65 NF-κB complexes [57]. In parallel, TAK1 phosphorylates MAPKs such as MKK4, MKK3, or MKK6 which, in turn, activate the ERK, JNK, p38 MAPK, NF-κB, and IRFs cascades. Activation of these signaling pathways leads to the increased expression of inflammatory markers including cytokines, chemokines, and adhesion molecules [16–19,58].

TLRs, especially, the TLR4 and TLR2 have emerged as key players in metabolic inflammation by nutrient sensing of LPS as well as sFFAs, both of which are abundantly found in individuals with obesity/T2D [35]. Increased expression of TLR2 and TLR4 has been found in patients with T2D, highlighting their associations with the pathogenesis of diabetes [24,28]. We previously showed that the elevated adipose gene expression of TLR4, TLR2, and MyD88 in people with obesity/T2D was associated with the IRAK1 gene expression [59]. Overall, several studies have reported the altered expression of TLRs (TLRs 1/2, TLRs 4-10) in people with obesity and/or T2D [25–27].

LncRNAs are emerging as key players in inflammatory and innate immune signaling cascades and are involved in pre- and post-transcriptional gene regulation [60,61]. Many lncRNAs are known to be differentially expressed in different tissues from individuals with obesity. SRA1 expression is more notable in the tissues or organs that have high energy demand such as heart, adipose tissue, skeletal muscle, and liver [36,62,63]. Furthermore, in these tissues/organs, SRA1 regulates several pathophysiological processes such as myocyte/adipocyte differentiation, hepatic steatosis, stem cell function, steroidogenesis, mammary gland development, and tumorigenesis [64]. Increased *SRA1* expression has been reported, in order, in adipocyte fractions from white adipose tissue, brown adipose tissue, and in preadipocytes [40,41]. *SRA1* is involved in regulation of adipocyte differentiation as well as in glucose homeostasis and insulin sensitivity of adipocytes [37,40,41].

Using a mouse model, Liu et al. demonstrated that *SRA1* knockout mice, compared to wild-type controls, had reduced body weight, an increased percentage of lean mass, and less fat percentage, less epididymal/subcutaneous white fat mass, and less liver mass [40]. Two different studies using this mouse model present congruent data showing better insulin sensitivity, less proclivity for obesity development under high-fat diet feeding, reduced hepatic steatosis, and improved glucose tolerance together with less inflammation and lower plasma TNFα levels as well as reduced expression of inflammatory genes (*Tnfα*, and *Ccl2*) in the white fat [40,41]. We also reported that the adipose *SRA1* expression was higher in non-diabetic persons with obesity compared with non-diabetic lean participants, and that the changes associated directly with BMI, PBF, fasting serum insulin, HOMA-IR, certain inflammatory markers but inversely with HbA1c level [42]. The results from previous studies collectively suggest that SRA1 may play a role in the adipose tissue function, pathobiology, as well as in inflammatory cascades implicated with insulin resistance, presenting it as a new target for therapeutic intervention of metabolic inflammation and insulin resistance.

Our data further show significant association of *SRA1* expression with that of TLR3, TLR4, TLR7, TLR9, NF-κB, IRF5, MyD88, and TRAF6 in the adipose tissue. Interestingly, TLR9 and TRAF6 associated differentially with the adipose *SRA1* expression depending on the diabetic status: TLR9/TRAF6 associated positively with *SRA1* expression in individuals with T2D but associated negatively with *SRA1* expression in individuals without T2D. We speculate that the negative association between these factors and adipose *SRA1* expression in the absence of diabetes factor imparts tenacity to the positive relationship of these factors with *SRA1* expression in the cohort with diabetes. These observations need to be further validated in larger cohorts with more diverse populations.

Our data further show that TLR3/TLR9 expression was associated independently with the expression of *SRA1* in individuals with T2D, while TLR3, TLR7 and IRAK1 were identified as the independent predictors of adipose *SRA1* expression in individuals with obesity. Taken together, TLR3 remains the independent predictor of adipose *SRA1* expression in settings of both obesity and T2D. Overall, these data support the plausibility of adipose *SRA1* expression to be considered as a novel, surrogate biomarker of adipose inflammation in obesity/T2D. However, caution will be needed for interpreting results of this primarily correlative study and further investigations will be required to determine: (i) which immunometabolic insults such as adipocytokines, chemokines, as well as gluco-lipotoxic and oxidative stresses may lead to the induction or upregulation of adipose *SRA1* expression, (ii) how does the *SRA1* expression differ between adipocytes and stromal vascular fraction, especially in monocytes/macrophages, and (iii) how will *SRA1* gene silencing and overexpression in adipocytes and monocytes/macrophages affect the insulin-stimulated glucose uptake as a measure of cellular function modulation.

5. Conclusions

In conclusion, our data show that the AT *SRA1* expression correlates with TLRs-3,4,7,9, MyD88, NF-κB, and IRF5 expression in individuals with T2D and with TLR2, IRAK1, and IRF3 expression in individuals with obesity. AT-*SRA1* expression was independently predicted by TLR3/TLR7 and IRAK1 in those with obesity and by TLR3/TLR9 in individuals with T2D. An association between the increased *SRA1* expression in the adipose tissue and the markers of immune signaling derangement in this compartment suggest that SRA1 molecules may have significance as new surrogate biomarkers of metabolic inflammation.

Supplementary Materials: The following supporting information can be downloaded at: https://www.mdpi.com/article/10.3390/cells11244007/s1, Figure S1: H&E staining of human adipose tissue samples; Table S1: Demographic and clinical characteristics of study population; Table S2: List of TaqMan gene expression assays.

Author Contributions: Conceptualization, S.K. and R.A.; data curation, S.K., R.T. and T.J.; formal analysis, H.A., S.S. (Sardar Sindhu), R.T., T.J., A.A.-S., S.S. (Steve Shenouda), H.A.K., F.A.-M., J.T. and R.A.; funding acquisition, F.A.-R. and R.A.; investigation, S.K., H.A., S.S. (Sardar Sindhu), H.A.K., J.T. and R.A.; methodology, S.K., T.J. and J.T.; resources, R.A.; writing—original draft, H.A. and S.S. (Sardar Sindhu); writing—review and editing, S.K., A.A.-S., S.S. (Steve Shenouda), F.A.-R., H.A.K., F.A.-M., J.T. and R.A. All authors have read and agreed to the published version of the manuscript.

Funding: This study was supported by Kuwait Foundation for Advancement of Sciences (KFAS) (Grant #: RA-2010-003; RA AM 2016-007).

Institutional Review Board Statement: The study was conducted according to the guidelines of the Declaration of Helsinki, reviewed and approved by the ethics committee of Dasman Diabetes Institute, Kuwait (No.: RA-2010-003).

Informed Consent Statement: Informed consent was obtained from all subjects involved in the study.

Data Availability Statement: The data presented in this study are available on request from the corresponding author.

Acknowledgments: The authors also thank Fahad Al-Ghamlas for help with patient recruitment.

Conflicts of Interest: The authors declare no conflict of interest.

References

1. Calder, P.C.; Ahluwalia, N.; Brouns, F.; Buetler, T.; Clement, K.; Cunningham, K.; Esposito, K.; Jönsson, L.S.; Kolb, H.; Lansink, M.; et al. Dietary factors and low-grade inflammation in relation to overweight and obesity. *Br. J. Nutr.* **2011**, *106* (Suppl. S3), S5–S78. [CrossRef] [PubMed]

2. Dandona, P.; Aljada, A.; Bandyopadhyay, A. Inflammation: The link between insulin resistance, obesity and diabetes. *Trends Immunol.* **2004**, *25*, 4–7. [CrossRef] [PubMed]

3. Khaodhiar, L.; Ling, P.R.; Blackburn, G.L.; Bistrian, B.R. Serum levels of interleukin-6 and c-reactive protein correlate with body mass index across the broad range of obesity. *JPEN J. Parenter. Enter. Nutr.* **2004**, *28*, 410–415. [CrossRef] [PubMed]

4. Moschen, A.R.; Kaser, A.; Enrich, B.; Mosheimer, B.; Theurl, M.; Niederegger, H.; Tilg, H. Visfatin, an adipocytokine with proinflammatory and immunomodulating properties. *J. Immunol.* **2007**, *178*, 1748–1758. [CrossRef] [PubMed]

5. Gustafson, B. Adipose tissue, inflammation and atherosclerosis. *J. Atheroscler. Thromb.* **2010**, *17*, 332–341. [CrossRef]

6. Lyon, C.J.; Law, R.E.; Hsueh, W.A. Minireview: Adiposity, inflammation, and atherogenesis. *Endocrinology* **2003**, *144*, 2195–2200. [CrossRef]

7. Greenberg, A.S.; Obin, M.S. Obesity and the role of adipose tissue in inflammation and metabolism. *Am. J. Clin. Nutr.* **2006**, *83*, 461s–465s. [CrossRef]

8. Shoelson, S.E.; Lee, J.; Goldfine, A.B. Inflammation and insulin resistance. *J. Clin. Investig.* **2006**, *116*, 1793–1801. [CrossRef]

9. Curat, C.A.; Miranville, A.; Sengenès, C.; Diehl, M.; Tonus, C.; Busse, R.; Bouloumié, A. From blood monocytes to adipose tissue-resident macrophages. *Induction Diapedesis Hum. Matur. Adipocytes* **2004**, *53*, 1285–1292. [CrossRef]

10. Bigornia, S.J.; Farb, M.G.; Mott, M.M.; Hess, D.T.; Carmine, B.; Fiscale, A.; Joseph, L.; Apovian, C.M.; Gokce, N. Relation of depot-specific adipose inflammation to insulin resistance in human obesity. *Nutr. Diabetes* **2012**, *2*, e30. [CrossRef]

11. Morris, D.L.; Singer, K.; Lumeng, C.N. Adipose tissue macrophages: Phenotypic plasticity and diversity in lean and obese states. *Curr. Opin. Clin. Nutr. Metab. Care* **2011**, *14*, 341–346. [CrossRef]

12. Sell, H.; Eckel, J. Adipose tissue inflammation: Novel insight into the role of macrophages and lymphocytes. *Curr. Opin. Clin. Nutr. Metab. Care* **2010**, *13*, 366–370. [CrossRef]

13. Weisberg, S.P.; McCann, D.; Desai, M.; Rosenbaum, M.; Leibel, R.L.; Ferrante, A.W., Jr. Obesity is associated with macrophage accumulation in adipose tissue. *J. Clin. Investig.* **2003**, *112*, 1796–1808. [CrossRef]

14. Trudler, D.; Farfara, D.; Frenkel, D. Toll-like receptors expression and signaling in glia cells in neuro-amyloidogenic diseases: Towards future therapeutic application. *Mediat. Inflamm.* **2010**, *2010*, 497987. [CrossRef]

15. Akira, S.; Takeda, K. Toll-like receptor signalling. *Nat. Rev. Immunol.* **2004**, *4*, 499–511. [CrossRef]

16. Burns, K.; Martinon, F.; Esslinger, C.; Pahl, H.; Schneider, P.; Bodmer, J.L.; Di Marco, F.; French, L.; Tschopp, J. Myd88, an adapter protein involved in interleukin-1 signaling. *J. Biol. Chem.* **1998**, *273*, 12203–12209. [CrossRef]

17. Kawai, T.; Akira, S. The role of pattern-recognition receptors in innate immunity: Update on toll-like receptors. *Nat. Immunol.* **2010**, *11*, 373–384. [CrossRef]

18. Ospelt, C.; Gay, S. Tlrs and chronic inflammation. *Int. J. Biochem. Cell Biol.* **2010**, *42*, 495–505. [CrossRef]

19. Lee, S.M.; Kok, K.H.; Jaume, M.; Cheung, T.K.; Yip, T.F.; Lai, J.C.; Guan, Y.; Webster, R.G.; Jin, D.Y.; Peiris, J.S. Toll-like receptor 10 is involved in induction of innate immune responses to influenza virus infection. *Proc. Natl. Acad. Sci. USA* **2014**, *111*, 3793–3798. [CrossRef]

20. O'Neill, L.A. How toll-like receptors signal: What we know and what we don't know. *Curr. Opin. Immunol.* **2006**, *18*, 3–9. [CrossRef]

21. Narayanankutty, A. Toll-like receptors as a novel therapeutic target for natural products against chronic diseases. *Curr. Drug Targets* **2019**, *20*, 1068–1080. [CrossRef] [PubMed]

22. Kim, S.J.; Choi, Y.; Choi, Y.H.; Park, T. Obesity activates toll-like receptor-mediated proinflammatory signaling cascades in the adipose tissue of mice. *J. Nutr. Biochem.* **2012**, *23*, 113–122. [CrossRef] [PubMed]

23. Ahmad, R.; Al-Mass, A.; Atizado, V.; Al-Hubail, A.; Al-Ghimlas, F.; Al-Arouj, M.; Bennakhi, A.; Dermime, S.; Behbehani, K. Elevated expression of the toll like receptors 2 and 4 in obese individuals: Its significance for obesity-induced inflammation. *J. Inflamm.* **2012**, *9*, 48. [CrossRef] [PubMed]

24. Shi, H.; Kokoeva, M.V.; Inouye, K.; Tzameli, I.; Yin, H.; Flier, J.S. Tlr4 links innate immunity and fatty acid-induced insulin resistance. *J. Clin. Investig.* **2006**, *116*, 3015–3025. [CrossRef] [PubMed]

25. Dasu, M.R.; Devaraj, S.; Park, S.; Jialal, I. Increased toll-like receptor (tlr) activation and tlr ligands in recently diagnosed type 2 diabetic subjects. *Diabetes Care* **2010**, *33*, 861–868. [CrossRef]

26. Yin, J.; Peng, Y.; Wu, J.; Wang, Y.; Yao, L. Toll-like receptor 2/4 links to free fatty acid-induced inflammation and β-cell dysfunction. *J. Leukoc. Biol.* **2014**, *95*, 47–52. [CrossRef]

27. Könner, A.C.; Brüning, J.C. Toll-like receptors: Linking inflammation to metabolism. *Trends Endocrinol. Metab. TEM* **2011**, *22*, 16–23. [CrossRef]

28. Wong, F.S.; Wen, L. Toll-like receptors and diabetes. *Ann. N. Y. Acad. Sci.* **2008**, *1150*, 123–132. [CrossRef]

29. Curtiss, L.K.; Tobias, P.S. Emerging role of toll-like receptors in atherosclerosis. *J. Lipid Res.* **2009**, *50*, S340–S345. [CrossRef]

30. Devaraj, S.; Dasu, M.R.; Rockwood, J.; Winter, W.; Griffen, S.C.; Jialal, I. Increased toll-like receptor (tlr) 2 and tlr4 expression in monocytes from patients with type 1 diabetes: Further evidence of a proinflammatory state. *J. Clin. Endocrinol. Metab.* **2008**, *93*, 578–583. [CrossRef]

31. Reyna, S.M.; Ghosh, S.; Tantiwong, P.; Meka, C.S.; Eagan, P.; Jenkinson, C.P.; Cersosimo, E.; Defronzo, R.A.; Coletta, D.K.; Sriwijitkamol, A.; et al. Elevated toll-like receptor 4 expression and signaling in muscle from insulin-resistant subjects. *Diabetes* **2008**, *57*, 2595–2602. [CrossRef]

32. Creely, S.J.; McTernan, P.G.; Kusminski, C.M.; Fisher f, M.; Da Silva, N.F.; Khanolkar, M.; Evans, M.; Harte, A.L.; Kumar, S. Lipopolysaccharide activates an innate immune system response in human adipose tissue in obesity and type 2 diabetes. *Am. J. Physiol. Endocrinol. Metab.* **2007**, *292*, E740–E747. [CrossRef]
33. Kaur, H.; Chien, A.; Jialal, I. Hyperglycemia induces toll like receptor 4 expression and activity in mouse mesangial cells: Relevance to diabetic nephropathy. *Am. J. Physiol. Ren. Physiol.* **2012**, *303*, F1145–F1150. [CrossRef]
34. Tencerová, M.; Kračmerová, J.; Krauzová, E.; Mališová, L.; Kováčová, Z.; Wedellová, Z.; Šiklová, M.; Štich, V.; Rossmeislová, L. Experimental hyperglycemia induces an increase of monocyte and t-lymphocyte content in adipose tissue of healthy obese women. *PLoS ONE* **2015**, *10*, e0122872. [CrossRef]
35. Kim, J.J.; Sears, D.D. Tlr4 and insulin resistance. *Gastroenterol. Res. Pract.* **2010**, *2010*, 212563. [CrossRef]
36. Lanz, R.B.; McKenna, N.J.; Onate, S.A.; Albrecht, U.; Wong, J.; Tsai, S.Y.; Tsai, M.-J.; O'Malley, B.W. A steroid receptor coactivator, sra, functions as an rna and is present in an src-1 complex. *Cell* **1999**, *97*, 17–27. [CrossRef]
37. Colley, S.M.; Leedman, P.J. Steroid receptor rna activator—A nuclear receptor coregulator with multiple partners: Insights and challenges. *Biochimie* **2011**, *93*, 1966–1972. [CrossRef]
38. Beato, M.; Vicent, G.P. A new role for an old player: Steroid receptor rna activator (sra) represses hormone inducible genes. *Transcription* **2013**, *4*, 167–171. [CrossRef]
39. Leygue, E. Steroid receptor rna activator (sra1): Unusual bifaceted gene products with suspected relevance to breast cancer. *Nucl. Recept. Signal.* **2007**, *5*, e006. [CrossRef]
40. Liu, S.; Sheng, L.; Miao, H.; Saunders, T.L.; MacDougald, O.A.; Koenig, R.J.; Xu, B. Sra gene knockout protects against diet-induced obesity and improves glucose tolerance*. *J. Biol. Chem.* **2014**, *289*, 13000–13009. [CrossRef]
41. Xu, B.; Gerin, I.; Miao, H.; Vu-Phan, D.; Johnson, C.N.; Xu, R.; Chen, X.W.; Cawthorn, W.P.; MacDougald, O.A.; Koenig, R.J. Multiple roles for the non-coding rna sra in regulation of adipogenesis and insulin sensitivity. *PLoS ONE* **2010**, *5*, e14199. [CrossRef] [PubMed]
42. Kochumon, S.; Arefanian, H.; Sindhu, S.; Shenouda, S.; Thomas, R.; Al-Mulla, F.; Tuomilehto, J.; Ahmad, R. Adipose tissue steroid receptor rna activator 1 (sra1) expression is associated with obesity, insulin resistance, and inflammation. *Cells* **2021**, *10*, 2602. [CrossRef] [PubMed]
43. Hill, N.R.; Levy, J.C.; Matthews, D.R. Expansion of the homeostasis model assessment of β-cell function and insulin resistance to enable clinical trial outcome modeling through the interactive adjustment of physiology and treatment effects: Ihoma2. *Diabetes Care* **2013**, *36*, 2324–2330. [CrossRef] [PubMed]
44. Sindhu, S.; Thomas, R.; Kochumon, S.; Wilson, A.; Abu-Farha, M.; Bennakhi, A.; Al-Mulla, F.; Ahmad, R. Increased adipose tissue expression of interferon regulatory factor (irf)-5 in obesity: Association with metabolic inflammation. *Cells* **2019**, *8*, 1418. [CrossRef] [PubMed]
45. Kochumon, S.; Al-Rashed, F.; Abu-Farha, M.; Devarajan, S.; Tuomilehto, J.; Ahmad, R. Adipose tissue expression of ccl19 chemokine is positively associated with insulin resistance. *Diabetes Metab. Res. Rev.* **2019**, *35*, e3087. [CrossRef]
46. Kochumon, S.; Jacob, T.; Koshy, M.; Al-Rashed, F.; Sindhu, S.; Al-Ozairi, E.; Al-Mulla, F.; Rosen, E.D.; Ahmad, R. Palmitate potentiates lipopolysaccharide-induced il-6 production via coordinated acetylation of h3k9/h3k18, p300, and rna polymerase ii. *J. Immunol.* **2022**, *209*, 731–741.
47. Khadir, A.; Tiss, A.; Abubaker, J.; Abu-Farha, M.; Al-Khairi, I.; Cherian, P.; John, J.; Kavalakatt, S.; Warsame, S.; Al-Madhoun, A.; et al. Map kinase phosphatase dusp1 is overexpressed in obese humans and modulated by physical exercise. *Am. J. Physiol. Endocrinol. Metab.* **2015**, *308*, E71–E83. [CrossRef]
48. Ahmad, R.; Al-Mass, A.; Al-Ghawas, D.; Shareif, N.; Zghoul, N.; Melhem, M.; Hasan, A.; Al-Ghimlas, F.; Dermime, S.; Behbehani, K. Interaction of osteopontin with il-18 in obese individuals: Implications for insulin resistance. *PLoS ONE* **2013**, *8*, e63944. [CrossRef]
49. Al Madhoun, A.; Kochumon, S.; Al-Rashed, F.; Sindhu, S.; Thomas, R.; Miranda, L.; Al-Mulla, F.; Ahmad, R. Dectin-1 as a potential inflammatory biomarker for metabolic inflammation in adipose tissue of individuals with obesity. *Cells* **2022**, *11*, 2879. [CrossRef]
50. Haider, M.J.A.; Albaqsumi, Z.; Al-Mulla, F.; Ahmad, R.; Al-Rashed, F. Socs3 regulates dectin-2-induced inflammation in pbmcs of diabetic patients. *Cells* **2022**, *11*, 2670. [CrossRef]
51. Akhter, N.; Kochumon, S.; Hasan, A.; Wilson, A.; Nizam, R.; Al Madhoun, A.; Al-Rashed, F.; Arefanian, H.; Alzaid, F.; Sindhu, S.; et al. Ifn-γ and lps induce synergistic expression of ccl2 in monocytic cells via h3k27 acetylation. *J. Inflamm. Res.* **2022**, *15*, 4291–4302. [CrossRef]
52. Ahmad, R.; Al-Roub, A.; Kochumon, S.; Akther, N.; Thomas, R.; Kumari, M.; Koshy, M.S.; Tiss, A.; Hannun, Y.A.; Tuomilehto, J.; et al. The synergy between palmitate and tnf-alpha for ccl2 production is dependent on the trif/irf3 pathway: Implications for metabolic inflammation. *J. Immunol.* **2018**, *200*, 3599–3611. [CrossRef]
53. Sindhu, S.; Kochumon, S.; Thomas, R.; Bennakhi, A.; Al-Mulla, F.; Ahmad, R. Enhanced adipose expression of interferon regulatory factor (irf)-5 associates with the signatures of metabolic inflammation in diabetic obese patients. *Cells* **2020**, *9*, 730. [CrossRef]
54. Burns, K.; Clatworthy, J.; Martin, L.; Martinon, F.; Plumpton, C.; Maschera, B.; Lewis, A.; Ray, K.; Tschopp, J.; Volpe, F. Tollip, a new component of the il-1ri pathway, links irak to the il-1 receptor. *Nat. Cell Biol.* **2000**, *2*, 346–351. [CrossRef]
55. Huang, J.; Gao, X.; Li, S.; Cao, Z. Recruitment of irak to the interleukin 1 receptor complex requires interleukin 1 receptor accessory protein. *Proc. Natl. Acad. Sci. USA* **1997**, *94*, 12829–12832. [CrossRef]

56. Qian, Y.; Commane, M.; Ninomiya-Tsuji, J.; Matsumoto, K.; Li, X. Irak-mediated translocation of traf6 and tab2 in the interleukin-1-induced activation of nfkappa b. *J. Biol. Chem.* **2001**, *276*, 41661–41667. [CrossRef]
57. Takaesu, G.; Kishida, S.; Hiyama, A.; Yamaguchi, K.; Shibuya, H.; Irie, K.; Ninomiya-Tsuji, J.; Matsumoto, K. Tab2, a novel adaptor protein, mediates activation of tak1 mapkkk by linking tak1 to traf6 in the il-1 signal transduction pathway. *Mol. Cell* **2000**, *5*, 649–658. [CrossRef]
58. Baccala, R.; Hoebe, K.; Kono, D.H.; Beutler, B.; Theofilopoulos, A.N. Tlr-dependent and tlr-independent pathways of type i interferon induction in systemic autoimmunity. *Nat. Med.* **2007**, *13*, 543–551. [CrossRef]
59. Ahmad, R.; Shihab, P.K.; Thomas, R.; Alghanim, M.; Hasan, A.; Sindhu, S.; Behbehani, K. Increased expression of the interleukin-1 receptor-associated kinase (irak)-1 is associated with adipose tissue inflammatory state in obesity. *Diabetol. Metab. Syndr.* **2015**, *7*, 71. [CrossRef]
60. Walther, K.; Schulte, L.N. The role of lncrnas in innate immunity and inflammation. *RNA Biol.* **2021**, *18*, 587–603. [CrossRef]
61. Yarani, R.; Mirza, A.H.; Kaur, S.; Pociot, F. The emerging role of lncrnas in inflammatory bowel disease. *Exp. Mol. Med.* **2018**, *50*, 1–14. [CrossRef] [PubMed]
62. Emberley, E.; Huang, G.-J.; Hamedani, M.K.; Czosnek, A.; Ali, D.; Grolla, A.; Lu, B.; Watson, P.H.; Murphy, L.C.; Leygue, E. Identification of new human coding steroid receptor rna activator isoforms. *Biochem. Biophys. Res. Commun.* **2003**, *301*, 509–515. [CrossRef] [PubMed]
63. Friedrichs, F.; Zugck, C.; Rauch, G.J.; Ivandic, B.; Weichenhan, D.; Müller-Bardorff, M.; Meder, B.; Mokhtari, N.E.E.; Regitz-Zagrosek, V.; Hetzer, R.; et al. Hbegf, sra1, and ik: Three cosegregating genes as determinants of cardiomyopathy. *Genome Res.* **2009**, *19*, 395–403. [CrossRef] [PubMed]
64. Sheng, L.; Ye, L.; Zhang, D.; Cawthorn, W.P.; Xu, B. New insights into the long non-coding rna sra: Physiological functions and mechanisms of action. *Front. Med.* **2018**, *5*, 244. [CrossRef]

Article

Physical Exercise Affects Adipose Tissue Profile and Prevents Arterial Thrombosis in *BDNF* Val66Met Mice

Leonardo Sandrini [1,2], Alessandro Ieraci [2], Patrizia Amadio [1], Marta Zarà [1], Nico Mitro [2], Francis S. Lee [3], Elena Tremoli [1] and Silvia Stella Barbieri [1,*]

[1] Centro Cardiologico Monzino IRCCS, 20138 Milano, Italy
[2] Dipartimento di Scienze Farmacologiche e Biomolecolari, Università degli Studi di Milano, 20133 Milano, Italy
[3] Department of Psychiatry, Weill Cornell Medical College of Cornell University, New York, NY 10065, USA
* Correspondence: silvia.barbieri@ccfm.it; Tel.: +39-02-58002021

Received: 15 July 2019; Accepted: 10 August 2019; Published: 11 August 2019

Abstract: Adipose tissue accumulation is an independent and modifiable risk factor for cardiovascular disease (CVD). The recent CVD European Guidelines strongly recommend regular physical exercise (PE) as a management strategy for prevention and treatment of CVD associated with metabolic disorders and obesity. Although mutations as well as common genetic variants, including the *brain-derived neurotrophic factor (BDNF)* Val66Met polymorphism, are associated with increased body weight, eating and neuropsychiatric disorders, and myocardial infarction, the effect of this polymorphism on adipose tissue accumulation and regulation as well as its relation to obesity/thrombosis remains to be elucidated. Here, we showed that white adipose tissue (WAT) of humanized knock-in BDNFVal66Met (BDNF$^{Met/Met}$) mice is characterized by an altered morphology and an enhanced inflammatory profile compared to wild-type BDNF$^{Val/Val}$. Four weeks of voluntary PE restored the adipocyte size distribution, counteracted the inflammatory profile of adipose tissue, and prevented the prothrombotic phenotype displayed, per se, by BDNF$^{Met/Met}$ mice. C3H10T1/2 cells treated with the Pro-BDNFMet peptide well recapitulated the gene alterations observed in BDNF$^{Met/Met}$ WAT mice. In conclusion, these data indicate the strong impact of lifestyle, in particular of the beneficial effect of PE, on the management of arterial thrombosis and inflammation associated with obesity in relation to the specific BDNF Val66Met mutation.

Keywords: BDNF; Val66Met polymorphism; adipose tissue; adipogenesis; arterial thrombosis; physical exercise

---

1. Introduction

Despite the huge growth in knowledge and advances in the prevention and treatment of cardiovascular disease (CVD), this pathology is still the leading cause of morbidity and mortality in the world and is predicted to reach 23.3 million by 2030 [1]. It is well known that an important modifiable risk factor for CVD mortality and morbidity is represented by excessive weight [2], and several follow-up studies demonstrated that a body mass index (BMI) >25 (>75th percentile based on percentile curves of BMI in the US reference population) is associated with a higher mortality rate [3,4]. Excessive body weight may influence CVD through its effect on risk factors such as hypertension, glucose intolerance, and dyslipidemia and may contribute through not already identified mechanisms [5]. In overweight and obese patients, adipose tissue accumulation is associated with a low-grade inflammatory profile and a higher secretion of cytokines and chemokines in the circulation compared to normal weight people [6]. The resulting subclinical inflammation is associated, among others, with hypercoagulability and increased thrombotic risk due to the enhanced platelet and leukocyte numbers and reactivities [7–9].

International guidelines, including 2016 European Guidelines on CVD prevention in clinical practice [10], strongly recommended regular physical exercise (PE) as management for the prevention and treatment of CVD, in particular when related to obesity and metabolic disorders. Regular PE reduces adipose-derived systemic inflammation, improves endothelial function, decreases platelet and leukocyte activation, and halts the progression of coronary stenosis in both obese and normal-weight individuals [8,11–14].

Starting from the discovery that several rare forms of obesity, called monogenic obesity, result from a mutation in single genes primarily located in the leptin–melanocortin pathway [15,16], recent evidence has identified additional selected genes associated with obesity, providing that the genetic background can play a pivotal role in causing or triggering susceptibility to the pathology when associated with environmental factors such as overeating and PE reduction [17,18]. Of note, *brain-derived neurotrophic factor* (*BDNF*) is included among these genes. Genome-wide association studies (GWAS) have shown a strong association between the *BDNF* locus and anorexia nervosa, bulimia nervosa [19], or obesity [20,21]. Indeed, it is well known that *BDNF* plays an important role in energy metabolism food intake and weight control [22,23].

In this context, the common human *BDNF* Val66Met variant through reduction of the activity-dependent secretion and signaling of mature BDNF, is associated not only to neuro-psychiatric disorders [24] and CVD [25] but also to eating disorders and obesity in humans [26–30]. Interestingly, a knock-in mouse carrying the human *BDNF* Val66Met polymorphism has a significantly higher body weight than wild-type littermates [31], associated with a proinflammatory and prothrombotic phenotype [25]. The frequency of the Met allele has a wide range of values: in Asians, Met allele frequency is nearly 50% heterozygous, while is about 20%–30% homozygous [32,33]. In the Caucasian population the Met allele is less frequent, with a frequency of 20%–30% heterozygous and only about 4% homozygous [33,34].

The aim of the present study was to investigate the relationship between the *BDNF* Val66Met polymorphism, obesity, and thrombosis, by analyzing the adipose tissue profile in BDNF$^{Met/Met}$ mice, and to evaluate the ability of PE to affect adipose tissue and reduce the prothrombotic phenotype in *BDNF* Val66Met knock-in mice. Finally, in vitro studies were performed to investigate the functional relevance of *BDNF* Val66Met polymorphism on adipogenesis.

2. Materials and Methods

2.1. Mice

All experiments were performed in adult (3–4 months old) wild-type BDNF$^{Val/Val}$ and BDNF$^{Met/Met}$ littermate mice, generated by Chen Z-Y et al. [31]. Only male mice were used to avoid the potential impact of hormones involved in the ovarian cycle in adipose tissue present in female mice. All experiments were approved by the National Ministry of Health-University of Milan Committee and of DGSA (12/2015 and 349/2015). Surgical procedures were performed in mice anesthetized with ketamine chlorhydrate (75 mg/kg; Intervet, Segrate, Milan, Italy) and medetomidine (1 mg/kg; Virbac, Milan, Italy). Mice were housed under standard conditions (20–22 °C, 12 h light/dark cycle, light on at 7 a.m.) with water and food ad libitum. All efforts were made to minimize animal distress and to reduce the numbers of animals used in this study.

2.2. Voluntary Physical Exercise (PE) Protocol

Mice underwent voluntary PE protocol as previously described [35,36]. Briefly, BDNF$^{Val/Val}$ and BDNF$^{Met/Met}$ mice were weighed and allocated randomly into running (RUN) and control (sedentary, SED) groups in cages equipped with or without running wheels, respectively, for 4 weeks with free access to food and water. Four sedentary control mice were housed in a standard polypropylene mice cage. Four runner mice were housed in standard polypropylene rat cages, with free access to two running wheels (12 cm diameter and 5.5 cm width). The greater dimensions of cages for runner mice were necessary for an adequate setup of running wheels. Running wheels were connected to an electronic counter, and the total distance ran was recorded daily. The average distance ran by a single mouse was calculated by dividing by 2 the total distance recorded per wheel (two running wheels × cage × four mice). The average distance ran by a single mouse, in our model, was comparable with the average distance reported by others [35–38].

2.3. Arterial Thrombosis Model

Experimental arterial thrombosis was induced as previously described [39]. Briefly, the left carotid artery of anesthetized mice was freely dissected, and a flow probe (model 0.7 VB, Transonic System, Ithaca, NY, USA) connected to a transonic flowmeter (TransonicT106) was used to measure blood flow. When blood flow was constant for 7 min (at least 0.8 mL/s), a strip of filter paper (Whatman N°1) soaked with $FeCl_3$ (7% solution; Sigma-Aldrich, Saint Louis, MO, USA) was applied for 3 min, and the flow was recorded for 30 min. An occlusion was considered to be total and stable when the flow was reduced by >90% from baseline until the 30 min observation time.

2.4. Whole Blood Counts

Blood was collected by cardiac venipuncture into 3.8% sodium citrate (1:10 vol:vol) from anesthetized mice, and white blood cells and platelets were counted optically.

2.5. Platelet–Leukocyte Aggregate Analysis

Platelet/leukocyte aggregates were analyzed as previously described [40]. Briefly, citrated blood was stimulated with 5 µM ADP (Sigma-Aldrich, Saint Louis, MO, USA) for 5 min, red blood cells were lysed by FACS Lysing solution, and samples were stained with anti-CD45 and anti-CD41 and analyzed by flow FACS "Novocyte 3000". A minimum of 5000 events were collected in the anti-CD45$^+$ gate.

2.6. Cell Culture, Treatment, and Differentiation

The C3H10T1/2 cell line has been used to evaluate the effect of different compounds on adipogenesis processes, as previously shown [41–43]. C3H10T1/2 cells (ThermoFisher Scientific, Paislay, Scotland, UK) were cultured in DMEM medium supplemented with 100 U/mL penicillin (Gibco, Rodano, Milan, Italy), 100 µg/mL streptomycin (Gibco, Rodano, Milan, Italy) and 10% FBS at 37 °C in 5% CO_2/95% air atmosphere. Cells were plated in 6-well plates at a concentration of 3.5×10^4 cells/mL, and when they reached 80% confluence (day –2), they were treated with 10 ng/mL of ProBDNFVal or ProBDNFMet synthetic peptide (Alomone Labs, Jerusalem, Israel) [44–46] to simulate the kinetics of *BDNF* expression occurring in physiological conditions during adipogenesis [47]. Forty-eight hours later (day 0), cells were treated with adipogenic commitment mix (5 µg/mL insulin, 2 µg/mL dexamethasone, 0.5 mM IBMX, and 5 µM rosiglitazone; all from Cayman Chemical, Arcore, Italy). Insulin (5 µg/mL) was added at days 3, 5, and 7, and complete differentiation of the cells was reached at day 9.

2.7. Adipogenesis Evaluation by Flow Cytometry and Oil-Red-O

After ProBDNFVal or ProBDNFMet treatment, C3H10T1/2 cells were analyzed during adipogenesis by flow cytometry, as previously described [48]. Briefly, at days 3, 5, and 9, cells were harvested in ice-cold PBS, analyzed by flow cytometry, and, according to granularity (SSC-H), were divided into four categories that correlated with the increased level of cell lipid accumulation after adipogenic commitment. In particular, noninduced cells were detected in the R1 gate, while cells with increasing granularity were identified in the regions from R2 to R4.

Oil-Red-O staining was performed as already described [49] on day 9. Lipid content was quantified as absorbance at a wavelength of 518 nm using a Tecan Infinite M1000 plate reader spectrophotometer (TECAN, Männedorf, Switzerland).

2.8. Quantitative Real-Time Polymerase Chain Reaction (RT-qPCR)

Total RNA was isolated from mouse adipose tissue or C3H10T1/2 cells with TRIzol Reagent (Sigma-Aldrich, Saint Louis, MO, USA) and a Direct-zol RNA extraction kit (Zymo Research, Irvine, CA, USA) according to the manufacturer's instructions. One microgram of RNA was reverse-transcribed using an iScript Advanced cDNA Synthesis Kit (Bio-Rad Laboratories, Segrate, Milan, Italy).

Samples of cDNA were incubated in 15 μL Luna® Universal qPCR mix containing the specific primers and fluorescent dye SYBR Green (New England Biolabs, Pero, Milan, Italy). RT-qPCR was carried out in triplicate for each sample on the CFX Connect real-time System (Bio-Rad Laboratories, Segrate, Milan, Italy) as previously described [39]. Gene expression was analyzed using parameters available in CFX Manager Software 3.1 (Bio-Rad Laboratories, Segrate, Milan, Italy). qPCR was then carried out using the primer sequences shown in Table S1. In particular, the expression of genes relevant in adipogenesis, inflammation, and the *BDNF* pathway were assessed (*Pparγ, C/ebp-α and C/ebp-β, Adipoq, Fabp4, Adra2a, Il-6, Mcp-1, Tnf-α, Tgf-β, Pai-1, Tf, CD163, CD80, Sorl1, Sirt1, Bdnf, TrkB* full and *TrkB-T1*).

2.9. Adipose Tissue Histology and Quantification of Adipocyte Size and Number

Immunocytochemistry and the analysis of adipocytes were performed in inguinal (ingWAT) and epididymal (epiWAT) white adipose tissue. Tissues were fixed overnight in 4% formalin, embedded in paraffin, cut at 5 μm, and mounted on polarized slides [50]. Five sections at three different levels along the whole length of epiWAT for each animal were analyzed. In particular, the mean values for each group were obtained from a total of 90 sections (5 sections × 3 points × 6 animals/group). The tissue contiguous to the epididymis were excluded from the analyses since its structure is different from that of general adipose tissue [51].

The number and size of adipocytes were evaluated in hematoxylin and eosin stained sections by counting five 5× microscopic fields for each tissue sample using the ImageJ-Macro Adipocytes Tool. Images were taken with a Zeiss Axioskop (Carl Zeiss, Milan, Italy) equipped with an intensified charge-coupled device (CCD) camera system (Photometrics, Tucson, AZ, USA).

2.10. Statistical Analysis

Statistical analyses were performed with GraphPad Prism 7.0 and SAS versus 9.4 software (SASA Institute, Cary, NC, USA). Data were analyzed by Student's t-test, two-way or three-way ANOVA with or without repeated measures for main effects of genotype and treatment or time and stimuli, as reported in every graph, followed by a Bonferroni post hoc analysis as appropriate. When data distribution was not normal, the variables were included in the analyses after logarithmic transformation. Values of $p < 0.05$ were considered statistically significant. Data are expressed as mean ± SEM.

3. Results

3.1. Characterization of the White Adipose Tissue Depots in BDNF^{Met/Met} Mice

As previously shown, BDNF^{Met/Met} mice have a significantly greater body weight compared to wild-type BDNF^{Val/Val} littermates (Figure 1A). In addition, we observed that the percentage of both inguinal white adipose tissue (ingWAT) and epididymal white adipose tissue (epiWAT) on total body weight were significantly enhanced in BDNF^{Met/Met} mice compared to BDNF^{Val/Val} (Figure 1B,C).

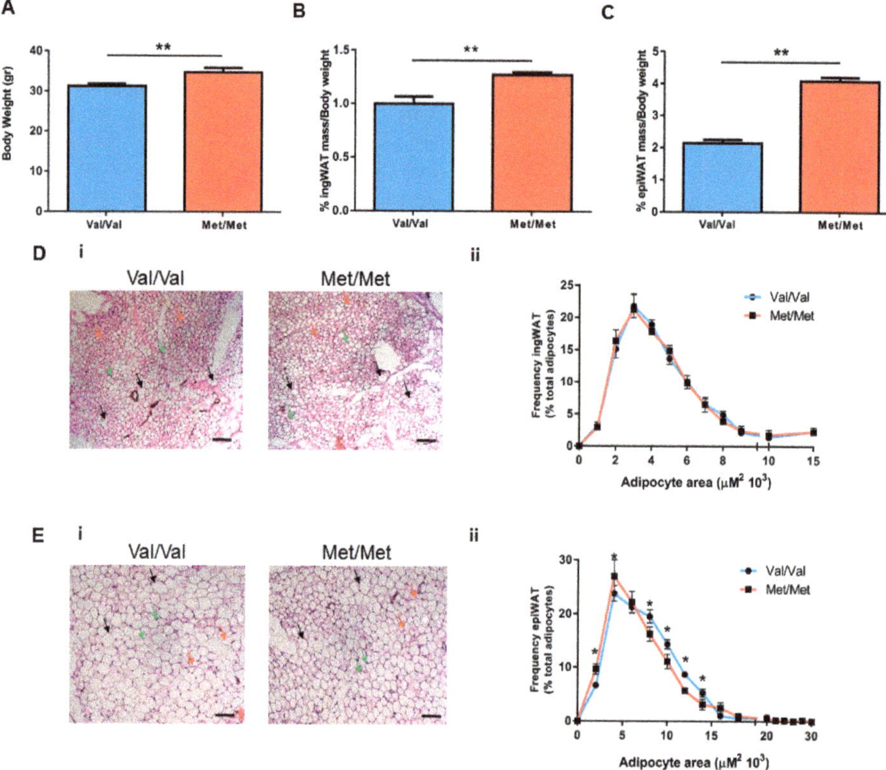

Figure 1. Characterization of white adipose tissue depots in BDNF$^{Val/Val}$ and BDNF$^{Met/Met}$ mice. (**A**) Body weight, percentage of (**B**) inguinal (ingWAT) and (**C**) epidydimal (epiWAT) white adipose tissue on total mouse body weight. (i) Representative hematoxylin and eosin (H&E) staining images and (ii) analysis of the frequency distribution of adipocyte sizes in (**D**) ingWAT and (**E**) epiWAT. Size bar: 100 μm. Black arrow: large adipocytes, green arrow: medium adipocytes, and red arrow: small adipocytes. Data are expressed as mean ± SEM. n = 6 mice/group. (**A–C**) Student's t-test and (**D,E**) two-way ANOVA followed by Bonferroni post hoc analysis. * $p < 0.05$, ** $p < 0.01$.

The histological examination of adipose depots revealed no difference in the frequency distribution of adipocyte sizes in ingWAT, while the BDNF$^{Met/Met}$ mice showed enrichment in small-size and a reduction in middle-size adipocytes in the epiWAT when compared to BDNF$^{Val/Val}$ (Figure 1D,E).

Then, the molecular signatures underlying the distinct morphological features of the epiWAT were investigated. Mutant mice had significantly lower levels of *Ppary*, *C/ebp-α* and *C/ebp-β* genes involved in the adipogenic program, as well as *Adipoq*, but a similar expression of *Fabp4* compared to BDNF$^{Val/Val}$ mice (Figure 2A). Interestingly, the *BDNF* Val66Met polymorphism affected also the expression of *Adra2a*, *Sirt1*, and *Sorl1*, genes involved in both adipose tissue energy balance and adipocyte morphology (Figure 2A–C).

In addition, a significant increase in the expression of Il-6, Tnf-α, Tgf-β, Mcp-1, and Pai-1 in BDNF$^{Met/Met}$ mice compared to BDNF$^{Val/Val}$ was found, whereas similar levels of TF between the two groups were found (Figure 2B). The enhanced inflammatory profile of BDNF$^{Met/Met}$ epiWAT was associated with a greater expression of CD80, an M1 inflammatory macrophage marker, and with a reduction of CD163, an alternatively activated M2 macrophage marker (Figure 2B).

Finally, BDNF$^{Met/Met}$ mice had a higher *BDNF* mRNA level in epiWAT, whereas no differences in the expression of both *TrkB*-full length and the truncated isoform *TrkB-T1* were found (Figure 2C).

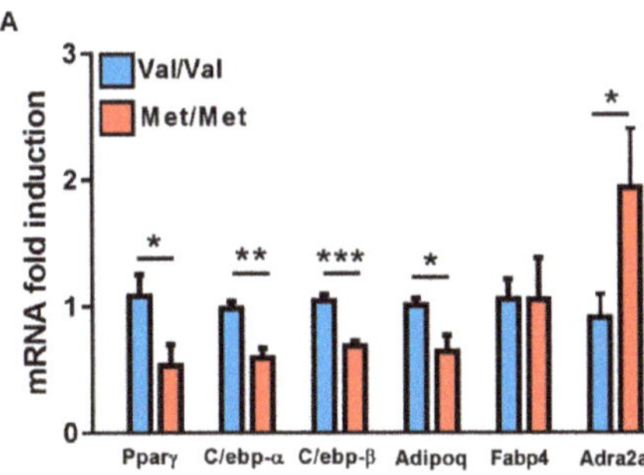

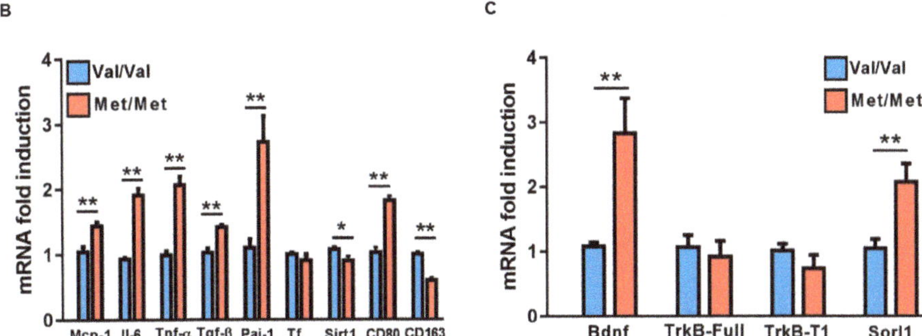

Figure 2. Gene expression profile of epidydimal white adipose tissue (epiWAT) in BDNF^Val/Val and BDNF^Met/Met mice. mRNA levels of genes related to (**A**) adipogenesis, (**B**) inflammation, and (**C**) BDNF/TrkB pathway in epidydimal white adipose tissue (epiWAT) of BDNF^Val/Val and BDNF^Met/Met mice. Data are expressed as mean ± SEM. n = 6 mice/group. Student's t-test. * $p < 0.05$, ** $p < 0.01$, and *** $p < 0.005$.

3.2. Evaluation of the Role of Mutant BDNF Val66Met Protein on Adipogenesis

Next, in vitro studies were performed to investigate the role of the BDNF Val66Met protein on adipogenesis. Pre-confluent C3H10Ts1/2 murine mesenchymal stem cells were exposed to ProBDNFVal or to ProBDNFMet synthetic peptides before inducing the adipocyte differentiation program. Synthetic peptide treatment did not affect cell number and morphology (Figure S1).

Notably, gene expression analysis at late (day 9) stages of differentiation showed that pretreatment with the peptide carrying the Met mutation determined a significant down-regulation of adipogenic genes, including *Ppar*γ, *C/ebp*α and *C/ebp*β mRNA levels (Figure 3A). In addition, ProBDNFMet treatment decreased the percentage of cells with low granularity (noninduced; R1) and increased those with high granularity (R4) both at 3 and 9 d post-induction (Figure 3B and Figure S2). However, at day 9, as provided by the oil-red-O staining, a similar accumulation of lipid droplets was detected in both samples (Figure 3C).

In this experimental condition, among the genes that were previously modulated in epiWAT of BDNF^Met/Met mice, only *Sorl1* was enhanced by ProBDNFMet treatment at late stages of differentiation (day 9) (Figure 3A and Figure S3).

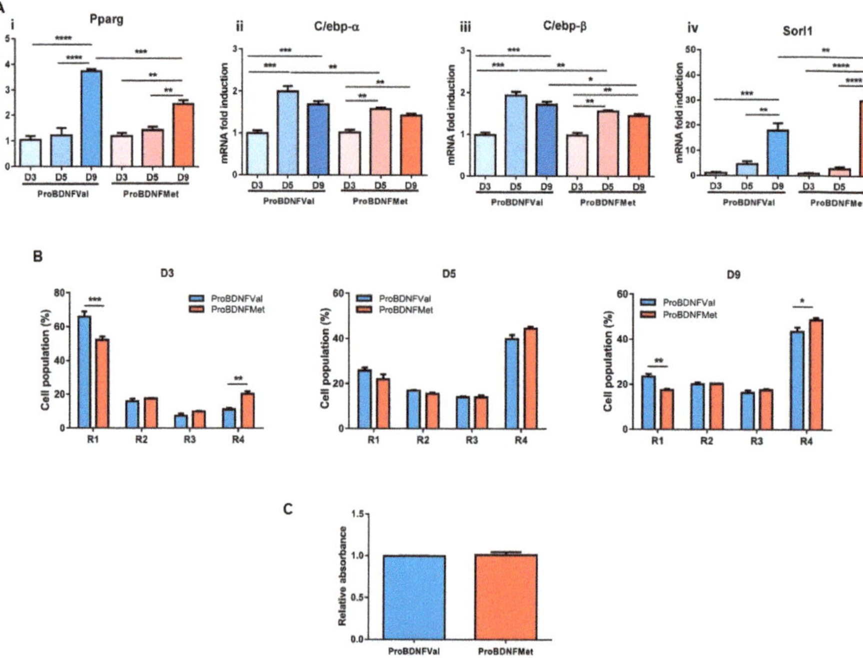

Figure 3. Effect of proBDNFMet on adipogenic differentiation of C3H10T1/2 cells. (**A**) mRNA levels of (i) *Pparγ*, (ii) *C/ebp-α*, (iii) *C/ebp-β*, and (iv) *Sorl1*. (**B**) Percentage of different cell populations based on their granularity profile analyzed by flow cytometry (R1: noninduced, R2-R3: growing granularity, and R4: high granularity) at day 3 (D3), day 5 (D5), and day 9 (D9) of differentiation, and (**C**) Oil-Red-O staining absorbance measurement in C3H10T1/2 cells. Data are expressed as mean ± SEM. *n* = 5 independent experiments/group. (**A**) Two-way ANOVA followed by Bonferroni post hoc analysis. (**B,C**) Student's t-test. * $p < 0.05$, ** $p < 0.01$, *** $p < 0.005$, and **** $p < 0.001$.

3.3. Effect of Physical Exercise (PE) on Adipose Tissue Phenotype of BDNF Val66Met Mice

According to international cardiovascular guidelines [10] that recommend regular PE as management for the prevention and treatment of CVD, we evaluated the potential beneficial effect of PE on adipose tissue and on prothrombotic phenotypes in *BDNF* Val66Met knock-in mice.

$BDNF^{Val/Val}$ and $BDNF^{Met/Met}$ mice underwent 4 weeks of voluntary PE in cages equipped with a running wheel. As previously reported [35], no difference in the daily running distance was found between $BDNF^{Val/Val}$ and $BDNF^{Met/Met}$ mice ($BDNF^{Val/Val}$: 6.676 ± 0.720 Km/d and $BDNF^{Met/Met}$ 6.657 ± 0.602 Km/d; $p = 0.9837$) in our experimental setting. In addition, we showed that PE did not affect the percentage of ingWAT and epiWAT on the total body weight in both $BDNF^{Val/Val}$ and $BDNF^{Met/Met}$ mice, compared to sedentary mice, whereas the morphology of adipose depots was modified as provided by histological analyses (Figure 4).

PE induced a change in the profile of the frequency distribution of adipocyte sizes in the ingWAT of both genotypes; however, this effect was more evident in $BDNF^{Val/Val}$ than in $BDNF^{Met/Met}$ mice (Figure 4A).

Interestingly, in the epiWAT, $BDNF^{Val/Val}$ running mice displayed a significant enrichment in small-size adipocytes and a reduction in medium-size ones compared to sedentary mice, whereas $BDNF^{Met/Met}$ mice showed an opposite trend, even if less marked (Figure 4B).

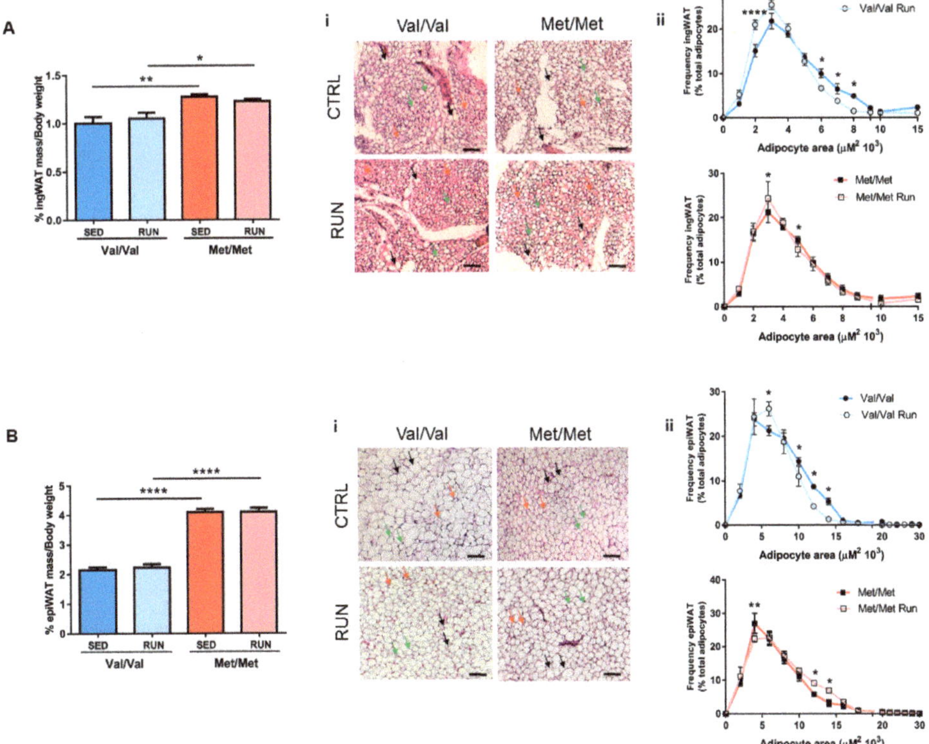

Figure 4. Impact of voluntary physical exercise (PE) on epiWAT morphology. (**A**) Inguinal (ingWAT) and (**B**) epidydimal (epiWAT) white adipose tissue on total mouse body weight. (i) Representative hematoxylin and eosin (H&E) staining images and (ii) analysis of the frequency distribution of adipocyte sizes in (**A**) ingWAT and (**B**) epiWAT. Size bar: 100 μm. Black arrow: large adipocytes, green arrow: medium adipocytes, and red arrow: small adipocytes. Data are expressed as mean ± SEM. $n = 6$ mice/group. Two-way ANOVA followed by Bonferroni post hoc analysis. * $p < 0.05$, ** $p < 0.01$, and **** $p < 0.001$.

Notably, PE strongly influenced the gene expression profile of epiWAT. In particular, in BDNF[Val/Val], 4 weeks of PE enhanced mRNA levels of *Adipoq*, whereas it did not modify the expression of genes involved in the adipogenic program (Figure 5A and Figure S4) and in inflammation compared to the sedentary mice. In BDNF[Met/Met] mice, PE was not sufficient to affect the expression of adipogenic genes, but it was sufficient to improve the inflammatory profile, decreasing the expression of Il-6, Tnf-α, Tgf-β, Mcp-1, and Pai-1, and to switch M1/M2 macrophage polarization, reducing the expression of CD80 and increasing the expression of CD163, (Figure 5B).

In addition, the expression of *Sorl1* was markedly reduced by PE in both BDNF[Val/Val] and BDNF[Met/Met] mice, whereas *Adra2a* and *Sirt1* were only slightly, but not significantly, decreased in BDNF[Met/Met] running mice (Figure 5C and Figure S4).

Conversely, PE modulated the *BDNF* expression in the two groups of mice. In particular, *BDNF* mRNA levels increased in BDNF[Val/Val] running mice and reduced in BDNF[Met/Met] running mice when compared to their respective sedentary controls (Figure 5C). Of note, the expression of both *TrkB* full length and the *TrkB-T1* isoform were slightly, but not significantly, increased in both groups of mice after PE (Figure S4).

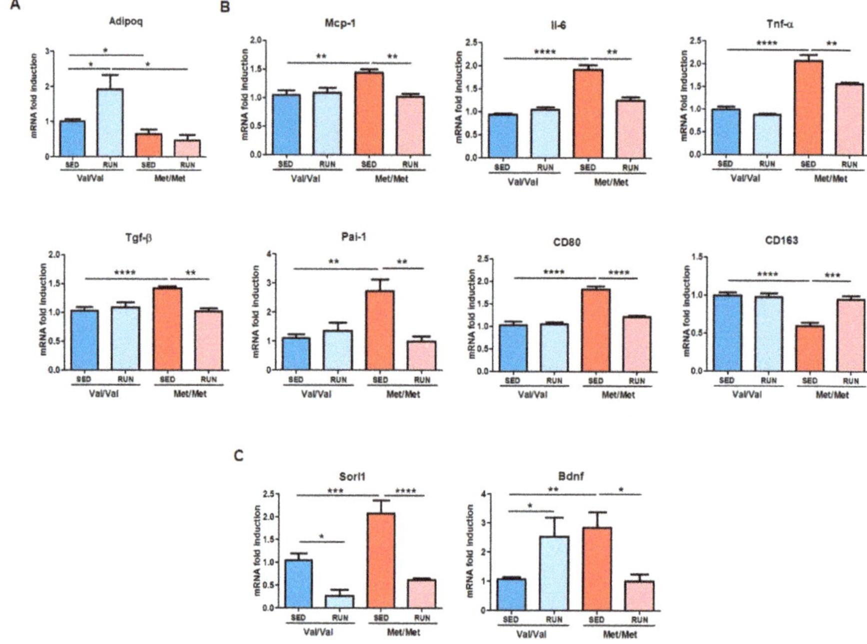

Figure 5. Impact of voluntary physical exercise (PE) on the gene expression profile of adipose tissue isolated from BDNF$^{Val/Val}$ and BDNF$^{Met/Met}$ mice. (**A**) Adipogenesis, (**B**) inflammation, and (**C**) BDNF/TrkB pathway related to mRNA levels in epiWAT of sedentary and running BDNF$^{Val/Val}$ and BDNF$^{Met/Met}$ mice. Data are expressed as mean ± SEM. n = 6 mice/group. Two-way ANOVA followed by Bonferroni post hoc analysis. * $p < 0.05$, ** $p < 0.01$, *** $p < 0.005$, and **** $p < 0.001$.

3.4. Effect of Physical Exercise (PE) on the Pro-Thrombotic Phenotype in BDNF$^{Met/Met}$ Mice

Finally, we investigated the ability of 4 weeks of PE to improve the prothrombotic phenotype already observed in BDNF$^{Met/Met}$ [25], in terms of platelet and leukocyte aggregates and FeCl$_3$-induced arterial thrombosis.

As previously shown, in the BDNF$^{Met/Met}$ mice there was a higher number of circulating blood cells, a higher platelet activation state, and enhanced arterial thrombosis [25]. PE restored the physiological number of platelets and leukocytes, and the natural percentage of platelet/leukocyte aggregates in response to ADP in BDNF$^{Met/Met}$ mice, without affecting significantly these parameters in BDNF$^{Val/Val}$ mice (Figure 6A–C).

Application of FeCl$_3$ to the carotid artery reduced the blood flow in all BDNF$^{Met/Met}$ sedentary mice, leading to a stable occlusion in 100% of mice, whereas only a slight reduction was observed in BNDF$^{Val/Val}$ mice. Of note, PE ameliorated arterial thrombosis, preventing completely the occlusion of the carotid artery in BDNF$^{Met/Met}$ mouse group (Figure 6D). In addition, no statistical differences were observed among sedentary BNDF$^{Val/Val}$ mice and running BNDF$^{Val/Val}$ and/or running BDNF$^{Met/Met}$ mice in terms of carotid artery occlusion (Figure 6D). In line with these data, total occlusion (flow reduction >90%) was reached only in sedentary BDNF$^{Met/Met}$ mice after an average time of 15 min (Figure 6E).

Overall, these data show that a paradigm of 4 weeks of voluntary PE is able to prevent the prothrombotic phenotype of BDNF$^{Met/Met}$ mice.

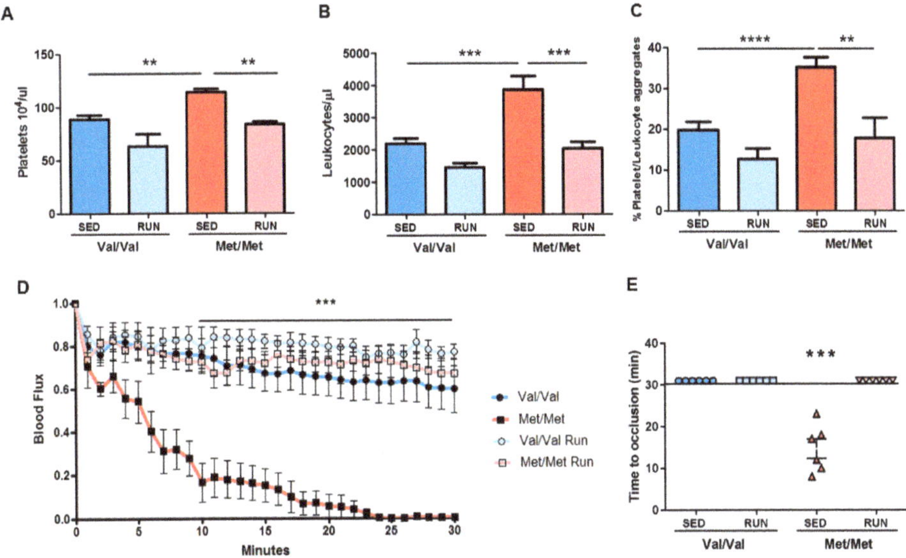

Figure 6. Effect of voluntary physical exercise (PE) on the prothrombotic phenotype of BDNF[Val/Val] and BDNF[Met/Met] mice. Numbers of circulating (**A**) platelets and (**B**) leukocytes. (**C**) Percentage of platelet/leukocytes in whole blood analyzed by flow cytometry. Arterial thrombosis induced by topical application of $FeCl_3$ to the carotid artery: (**D**) blood flow and (**E**) time to occlusion measured in sedentary and running BDNF[Val/Val] and BDNF[Met/Met] mice. $n = 6$ mice/group. (**A–C** and **E**) Two-way ANOVA followed by Bonferroni post hoc analysis. (**D**) Three-way ANOVA with repeated measures followed by Bonferroni post hoc analysis. ** $p < 0.01$, *** $p < 0.005$.

4. Discussion

Although mutations, as well as genetic variants, including *BDNF* Val66Met polymorphism, have been associated with increased body weight and eating disorders in both human and animal models [19–23,31,35,52–55], the factors and mechanisms involved in the development of obesity in presence of the *BDNF* Met homozygosity remain to be elucidated. It is only known that *BDNF-to-TrkB* signaling is an important downstream target of MC4R-mediated signaling involved in the regulation of energy balance and food intake [55–57].

Using a knock-in *BDNF* Val66Met mouse model, here we confirmed that BDNF[Met/Met] mice had a higher body weight when compared to BDNF[Val/Val] [31], and we showed that this increase was related to the enhanced percentage of epiWAT and ingWAT. In particular, adipocytes from epiWAT of mutant mice had a different size distribution, with an enrichment in the percentage of small-sized adipocytes. The presence of small adipocytes in epiWAT of BDNF[Met/Met] might trace back to hyperplasia or expansion of the small cell population, which are mechanisms of defense that the adipose tissue can undergo in obesity after a threshold of hypertrophy is reached [58–60]. This hypothesis is also supported by the higher expression of *Adra2a* and *Sorl1* found in epiWAT of BDNF[Met/Met]. Indeed, overexpression of *Adra2a* in animal models has been associated with adipose tissue hyperplasia [61]. In addition, it is well known that the activation of *Adra2a* has an antilipolytic effect, and the increased alpha/beta adrenoreceptor ratio as well as the gain of function mutations of *Adra2* have been associated with obesity in humans [62–64]. Similarly, upregulation of the expression of *Sorl1*, which encodes for the protein Sorla, has been related to reduced lipolytic activity in adipocytes [65], and GWAS analyses have associated *Sorl1* with obesity in humans and in mouse models [21,66], suggesting its key role in metabolic diseases.

The adipose tissue accumulation found in BDNF$^{Met/Met}$ mice was accompanied by a higher expression of the M1 proinflammatory marker CD80, of the monocyte chemoattractant protein-1 (Mcp-1) and of the mediators of inflammation such as Pai-1, Tnf-alpha, and Il-6, which is in line with the well-established paradigm that overweight and obesity are related to adipose tissue inflammation [67]. In addition, the higher levels of these inflammatory transcripts, concomitant with the lower expression of *Adipoq* measured in the epiWAT of BDNF$^{Met/Met}$ mice and the higher number of circulating leukocytes and platelets as well as their activation state, might well summarize the relationship between adipose tissue inflammation and thrombosis. Indeed, the inflammatory profile of adipose tissue in obese subjects as well as the increased presence of these proteins in the circulation have a direct role in the onset and progression of the pathology [68–70], enhancing platelet activation and ability of leukocytes to produce, in turn, inflammatory factors such as Il-6, Tnf, and Cox-2 [9,68,71–74]. All these findings thoroughly summarize data obtained in human adipose tissue samples. Indeed, a positive correlation between proinflammatory cytokines, including IL-6, TNF-α and MCP-1, and adipocyte size was found. Interestingly, the small adipocytes expressed anti-inflammatory factors such as IL-10 and IL-8 [75].

Of note, the reduced levels of *Pparγ* along with those of adiponectin found in *BDNF* mutant mice might also contribute to the observed adipose tissue inflammation. It is well known that *PPARγ*, alongside the role of master regulator of adipogenesis, is also involved in the regulation of adipose tissue inflammation. In particular, it was demonstrated that *PPARγ* downregulates inflammatory adipokines in WAT. Specifically, *PPARγ* activation downregulates the expression of inflammatory markers such as MCP-1 and TNFα and, thus, reduces inflammation in activated macrophages [56,76–78]. Moreover, *PPARγ* activation induces adiponectin expression, thus further contributing to the reduction of chronic inflammation [79].

Remarkably, *BDNF* expression was markedly greater in epiWAT of mutant mice, supporting our hypothesis that the *BDNF* Val66Met polymorphism contributes to adipose tissue pathophysiology.

Indeed, studies performed using *BDNF*-(si)RNA-mediated knockdown in the 3T3 cell line showed a reduced adipogenic differentiation ability, supporting the hypothesis that *BDNF* expression is of functional relevance for adipogenesis. In addition, it was reported that *BDNF* expression is dramatically downregulated during adipocyte differentiation, and mature adipocytes only marginally contribute to the production of BDNF in the adipose tissue [80].

Interestingly, we showed that the treatment of C3H10T1/2 cells with Pro-BDNFMet before cell commitment well recapitulated the expression profile of genes that were found altered in the epiWAT of mutant mice. Pro-BDNFMet reduced *Pparγ* and upregulated *Sorl1* expression, and it increased the percentage of mature adipocytes evaluated in the flow cytometry analysis, suggesting a direct role of the *BDNF* Val66Met polymorphism in the regulation of adipogenesis. However, Pro-BDNFMet was not able to affect *Adipoq* and *Adra2a* as well as *Pai-1* expression, leading us to hypothesize a more complex process that may involve the fraction stromal vascular cells. Indeed, it is suggested that mesenchymal progenitor/stem cells, preadipocytes, endothelial cells, pericytes, T cells, and macrophages, and not mature adipocytes, are the main source of adipokines and PAI-1 in adipose tissue. Of note, the stromal vascular fraction in adipose tissue increases with an increasing degree of obesity [81].

Adipose tissue accumulation represents an independent and modifiable risk factor for CVD [5], and regular PE was recently recognized and strongly recommended as a valuable management strategy for the prevention and treatment of CVD and metabolic disorders from the European Guidelines of cardiology [10,82].

In the present study, we provide evidence that, in mutant BDNF$^{Met/Met}$ mice, four weeks of PE was sufficient to change epiWAT morphology and the inflammatory profile with a concomitant reversion of the prothrombotic phenotype. In particular, the change in adipose tissue morphology observed in BDNF$^{Met/Met}$ running mice was accompanied with a reduction in *Sorl1* and *Adra2a* expression, thus suggesting that PE might improve the metabolic profile of mutant mice, ultimately affecting lipolysis [65,83,84].

The beneficial effect of PE has been provided in animal studies and human trials, showing an impact on both systemic [14,85] and specific reduction of visceral fat mass [86,87], protecting against chronic inflammation-associated disease [88]. Several mechanisms have been proposed to explain the beneficial anti-inflammatory effect of PE. By affecting AMPK and PGC-1α pathways, PE decreases mitochondrial dysfunction and reduces oxidative stress [89,90], with the consequent reduction of proinflammatory adipokines released from the visceral fat mass. Moreover, PE increased production of anti-inflammatory molecules from skeletal muscle and leukocytes [91]. PE decreases Toll-like receptors on monocytes and macrophages, thus preventing their infiltration into adipose tissue and inducing the M1 to M2 macrophage switching to limit macrophage M1 polarization [88].

In line with this evidence, we showed that PE in BDNF$^{Met/Met}$ mice reduced the levels of inflammation mediators, induced a switch in macrophage polarization, and decreased the number of circulating leukocytes and platelets, modifications that, in turn, occur to improve the prothrombotic phenotype observed in mutant mice. Interestingly, for the first time, we provide evidence that PE influenced differently the expression of *BDNF* in the two genotypes, increasing and decreasing its levels in BDNF$^{Val/Val}$ and BDNF$^{Met/Met}$, respectively. These results might be related to the intrinsic adipose tissue morphology of BDNF$^{Val/Val}$ and BDNF$^{Met/Met}$ mice, suggesting a strong relationship between adipocyte dimension and *BDNF* levels. In fact, the great number of small adipocytes was associated with high levels of *BDNF* (e.g., sedentary BDNF$^{Met/Met}$ and running BDNF$^{Val/Val}$), and conversely, low levels of transcript were measured in epiWAT when the mean adipocyte dimension was higher (e.g., sedentary BDNF$^{Val/Val}$ and running BDNF$^{Met/Met}$). The different involvement of the stromal vascular cell fraction in sustaining the adipocyte turnover, as well as the potential contribution of the peripheral nervous system, might explain the different mRNA levels of *BDNF* detected in our experimental setting [92–94]. In this regard, the inability of PE to enhance *BDNF* transcripts in the central nervous system of mutant mice [35] might have important consequences on the levels of BDNF in the peripheral nervous system, thus affecting their levels in epiWAT. Interestingly, it is worth mentioning that, contrary to data presented here related to CVD, the *BDNF* Val66Met polymorphism impairs the beneficial neurobiological changes induced by physical exercise in mice [35].

5. Conclusions

Cardiovascular disease still represents the first cause of mortality worldwide, and obesity is a well-known modifiable risk factor for this pathology. Of note, PE is highly recommended to manage the prevention and treatment of CVD and obesity, showing beneficial cardiometabolic effects.

In human subjects, the *BDNF* Val66Met polymorphism is known to be related to adipose tissue accumulation and cardiovascular risk.

Interestingly, our in vitro data well support the role of Pro-BDNFMet in adipogenesis, in line with data obtained in the BDNF$^{Met/Met}$ WAT mice.

Taking advantage of a mouse model carrying the human *BDNF* Val66Met polymorphism, we showed that 4 weeks of voluntary physical exercise was sufficient to induce positive morphological changes and reduce the inflammatory profile of the adipose tissue.

These beneficial effects might be the bases of the observed reduction in the prothrombotic phenotype detected in this animal model. Future studies are required to assess this relationship.

These data indicate the strong impact of lifestyle, in particular the beneficial effect of PE, on the management of arterial thrombosis and obesity-associated inflammation in relation to genetic mutations that predisposes one, per se, to these pathologies. Nevertheless, human studies need to support these results.

Supplementary Materials: The following are available online at http://www.mdpi.com/2073-4409/8/8/875/s1, Figure S1: Cell number and morphology are not altered by ProBDNFVal and ProBDNFMet treatment. Figure S2: Representative flow cytometry graphs showing gate selected for cell granularity analyses at day 3, 5 and 9. Figure S3: Evaluation of the functional relevance of BDNF Val66Met protein on C3H10T1/2 cells adipogenic differentiation. Figure S4: Impact of voluntary physical exercise on gene expression profile of adipose tissue.

Table S1: Primers sequences of the analyzed genes. Table S2: F and P values referred to each graph analyzed by Two-way ANOVA or Three-way ANOVA.

Author Contributions: Conceptualization, S.S.B.; formal analysis and data curation, L.S., A.I., P.A., S.S.B.; investigation, L.S., A.I., P.A.; original draft preparation, L.S. and S.S.B.; contributed to the discussion on the results from a biological point of view, review and editing, A.I., P.A., M.Z., N.M., F.S.L. and E.T.; supervision and funding acquisition, S.S.B. All authors read and approved the final manuscript.

Funding: This research was funded by the Italian Ministry of Health (Ricerca Corrente) grants number BIO35-2015: 2622789; BIO37-2016: 2613074; BIO37-2017: 2631213; MPP1.2A-2018: 2634597. Co-funding provided by the contribution of the Italian "5 × 1000" tax (2014, 2015 and 2016).

Acknowledgments: L.S. is supported by the 32nd cycle Ph.D. program in "Scienze farmacologiche sperimentali e cliniche", Università degli Studi di Milano. The authors would like to thank Alessandra Giannopulo Nicolini for the technical support.

Conflicts of Interest: The authors declare no conflict of interest.

References

1. Mathers, C.D.; Loncar, D. Projections of global mortality and burden of disease from 2002 to 2030. *PLoS Med.* **2006**, *3*, e442. [CrossRef] [PubMed]

2. Poirier, P.; Giles, T.D.; Bray, G.A.; Hong, Y.; Stern, J.S.; Pi-Sunyer, F.X.; Eckel, R.H.; American Heart Association; Obesity Committee of the Council on Nutrition, Physical Activity, and Metabolism. Obesity and cardiovascular disease: Pathophysiology, evaluation, and effect of weight loss: An update of the 1997 american heart association scientific statement on obesity and heart disease from the obesity committee of the council on nutrition, physical activity, and metabolism. *Circulation* **2006**, *113*, 898–918. [PubMed]

3. Engeland, A.; Bjørge, T.; Søgaard, A.J.; Tverdal, A. Body mass index in adolescence in relation to total mortality: 32-year follow-up of 227,000 norwegian boys and girls. *Am. J. Epidemiol.* **2003**, *157*, 517–523. [CrossRef] [PubMed]

4. Must, A.; Jacques, P.F.; Dallal, G.E.; Bajema, C.J.; Dietz, W.H. Long-term morbidity and mortality of overweight adolescents. A follow-up of the harvard growth study of 1922 to 1935. *N. Engl. J. Med.* **1992**, *327*, 1350–1355. [CrossRef]

5. Ortega, F.B.; Lavie, C.J.; Blair, S.N. Obesity and cardiovascular disease. *Circ. Res.* **2016**, *118*, 1752–1770. [CrossRef]

6. Fain, J.N.; Madan, A.K.; Hiler, M.L.; Cheema, P.; Bahouth, S.W. Comparison of the release of adipokines by adipose tissue, adipose tissue matrix, and adipocytes from visceral and subcutaneous abdominal adipose tissues of obese humans. *Endocrinology* **2004**, *145*, 2273–2282. [CrossRef]

7. Carvalheira, J.B.; Qiu, Y.; Chawla, A. Blood spotlight on leukocytes and obesity. *Blood* **2013**, *122*, 3263–3267. [CrossRef]

8. Ertek, S.; Cicero, A. Impact of physical activity on inflammation: Effects on cardiovascular disease risk and other inflammatory conditions. *Arch. Med. Sci* **2012**, *8*, 794–804. [CrossRef]

9. Santilli, F.; Vazzana, N.; Liani, R.; Guagnano, M.T.; Davì, G. Platelet activation in obesity and metabolic syndrome. *Obes Rev.* **2012**, *13*, 27–42. [CrossRef]

10. Piepoli, M.F.; Hoes, A.W.; Agewall, S.; Albus, C.; Brotons, C.; Catapano, A.L.; Cooney, M.T.; Corrà, U.; Cosyns, B.; Deaton, C.; et al. 2016 european guidelines on cardiovascular disease prevention in clinical practice: The sixth joint task force of the european society of cardiology and other societies on cardiovascular disease prevention in clinical practice (constituted by representatives of 10 societies and by invited experts)developed with the special contribution of the european association for cardiovascular prevention & rehabilitation (eacpr). *Eur. Heart J.* **2016**, *37*, 2315–2381.

11. Bruunsgaard, H. Physical activity and modulation of systemic low-level inflammation. *J. Leukoc Biol.* **2005**, *78*, 819–835. [CrossRef]

12. Schuler, G.; Adams, V.; Goto, Y. Role of exercise in the prevention of cardiovascular disease: Results, mechanisms, and new perspectives. *Eur. Heart J.* **2013**, *34*, 1790–1799. [CrossRef]

13. Winzer, E.B.; Woitek, F.; Linke, A. Physical activity in the prevention and treatment of coronary artery disease. *J. Am. Heart Assoc.* **2018**, *7*. [CrossRef]

14. You, T.; Arsenis, N.C.; Disanzo, B.L.; Lamonte, M.J. Effects of exercise training on chronic inflammation in obesity: Current evidence and potential mechanisms. *Sports Med.* **2013**, *43*, 243–256. [CrossRef]

15. Montague, C.T.; Farooqi, I.S.; Whitehead, J.P.; Soos, M.A.; Rau, H.; Wareham, N.J.; Sewter, C.P.; Digby, J.E.; Mohammed, S.N.; Hurst, J.A.; et al. Congenital leptin deficiency is associated with severe early-onset obesity in humans. *Nature* **1997**, *387*, 903–908. [CrossRef]

16. Thaker, V.V. Genetic and epigenetic causes of obesity. *Adolesc. Med. State Art Rev.* **2017**, *28*, 379–405.

17. Fairbrother, U.; Kidd, E.; Malagamuwa, T.; Walley, A. Genetics of severe obesity. *Curr. Diab. Rep.* **2018**, *18*, 85. [CrossRef]

18. Rankinen, T.; Zuberi, A.; Chagnon, Y.C.; Weisnagel, S.J.; Argyropoulos, G.; Walts, B.; Pérusse, L.; Bouchard, C. The human obesity gene map: The 2005 update. *Obesity (Silver Spring)* **2006**, *14*, 529–644. [CrossRef]

19. Monteleone, P.; Maj, M. Dysfunctions of leptin, ghrelin, bdnf and endocannabinoids in eating disorders: Beyond the homeostatic control of food intake. *Psychoneuroendocrinology* **2013**, *38*, 312–330. [CrossRef]

20. Speliotes, E.K.; Willer, C.J.; Berndt, S.I.; Monda, K.L.; Thorleifsson, G.; Jackson, A.U.; Lango Allen, H.; Lindgren, C.M.; Luan, J.; Mägi, R.; et al. Association analyses of 249,796 individuals reveal 18 new loci associated with body mass index. *Nat. Genet.* **2010**, *42*, 937–948. [CrossRef]

21. Thorleifsson, G.; Walters, G.B.; Gudbjartsson, D.F.; Steinthorsdottir, V.; Sulem, P.; Helgadottir, A.; Styrkarsdottir, U.; Gretarsdottir, S.; Thorlacius, S.; Jonsdottir, I.; et al. Genome-wide association yields new sequence variants at seven loci that associate with measures of obesity. *Nat. Genet.* **2009**, *41*, 18–24. [CrossRef]

22. Nakazato, M.; Hashimoto, K.; Shimizu, E.; Niitsu, T.; Iyo, M. Possible involvement of brain-derived neurotrophic factor in eating disorders. *Iubmb. Life* **2012**, *64*, 355–361. [CrossRef]

23. Noble, E.E.; Billington, C.J.; Kotz, C.M.; Wang, C. The lighter side of bdnf. *Am. J. Physiol. Regul. Integr. Comp. Physiol.* **2011**, *300*, R1053–R1069. [CrossRef]

24. Tsai, S.J. Critical issues in *BDNF* Val66Met Genetic Studies of Neuropsychiatric Disorders. *Front. Mol. Neurosci.* **2018**, *11*, 156. [CrossRef]

25. Amadio, P.; Colombo, G.I.; Tarantino, E.; Gianellini, S.; Ieraci, A.; Brioschi, M.; Banfi, C.; Werba, J.P.; Parolari, A.; Lee, F.S.; et al. Bdnfval66met polymorphism: A potential bridge between depression and thrombosis. *Eur. Heart J.* **2017**, *38*, 1426–1435.

26. Beckers, S.; Peeters, A.; Zegers, D.; Mertens, I.; Van Gaal, L.; Van Hul, W. Association of the bdnf val66met variation with obesity in women. *Mol. Genet. Metab.* **2008**, *95*, 110–112. [CrossRef]

27. Wu, L.; Xi, B.; Zhang, M.; Shen, Y.; Zhao, X.; Cheng, H.; Hou, D.; Sun, D.; Ott, J.; Wang, X.; et al. Associations of six single nucleotide polymorphisms in obesity-related genes with bmi and risk of obesity in chinese children. *Diabetes* **2010**, *59*, 3085–3089. [CrossRef]

28. Xi, B.; Cheng, H.; Shen, Y.; Chandak, G.R.; Zhao, X.; Hou, D.; Wu, L.; Wang, X.; Mi, J. Study of 11 bmi-associated loci identified in gwas for associations with central obesity in the chinese children. *PLoS ONE* **2013**, *8*, e56472. [CrossRef]

29. Zhang, M.; Zhao, X.; Xi, B.; Shen, Y.; Wu, L.; Cheng, H.; Hou, D.; Mi, J. [impact of obesity-related gene polymorphism on risk of obesity and metabolic disorder in childhood]. *Zhonghua Yu Fang Yi Xue Za Zhi* **2014**, *48*, 776–783.

30. Zhao, X.; Xi, B.; Shen, Y.; Wu, L.; Hou, D.; Cheng, H.; Mi, J. An obesity genetic risk score is associated with metabolic syndrome in chinese children. *Gene* **2014**, *535*, 299–302. [CrossRef]

31. Chen, Z.Y.; Jing, D.; Bath, K.G.; Ieraci, A.; Khan, T.; Siao, C.J.; Herrera, D.G.; Toth, M.; Yang, C.; McEwen, B.S.; et al. Genetic variant bdnf (val66met) polymorphism alters anxiety-related behavior. *Science* **2006**, *314*, 140–143. [CrossRef]

32. Naoe, Y.; Shinkai, T.; Hori, H.; Fukunaka, Y.; Utsunomiya, K.; Sakata, S.; Matsumoto, C.; Shimizu, K.; Hwang, R.; Ohmori, O.; et al. No association between the brain-derived neurotrophic factor (bdnf) val66met polymorphism and schizophrenia in asian populations: Evidence from a case-control study and meta-analysis. *Neurosci. Lett.* **2007**, *415*, 108–112. [CrossRef]

33. Shimizu, E.; Hashimoto, K.; Iyo, M. Ethnic difference of the bdnf 196g/a (val66met) polymorphism frequencies: The possibility to explain ethnic mental traits. *Am. J. Med. Genet. B Neuropsychiatr Genet.* **2004**, *126B*, 122–123. [CrossRef]

34. Egan, M.F.; Kojima, M.; Callicott, J.H.; Goldberg, T.E.; Kolachana, B.S.; Bertolino, A.; Zaitsev, E.; Gold, B.; Goldman, D.; Dean, M.; et al. The bdnf val66met polymorphism affects activity-dependent secretion of bdnf and human memory and hippocampal function. *Cell* **2003**, *112*, 257–269. [CrossRef]

35. Ieraci, A.; Madaio, A.I.; Mallei, A.; Lee, F.S.; Popoli, M. Brain-derived neurotrophic factor val66met human polymorphism impairs the beneficial exercise-induced neurobiological changes in mice. *Neuropsychopharmacology* **2016**, *41*, 3070–3079. [CrossRef]

36. Ieraci, A.; Mallei, A.; Musazzi, L.; Popoli, M. Physical exercise and acute restraint stress differentially modulate hippocampal brain-derived neurotrophic factor transcripts and epigenetic mechanisms in mice. *Hippocampus* **2015**, *25*, 1380–1392. [CrossRef]

37. Duman, C.H.; Schlesinger, L.; Russell, D.S.; Duman, R.S. Voluntary exercise produces antidepressant and anxiolytic behavioral effects in mice. *Brain Res.* **2008**, *1199*, 148–158. [CrossRef]

38. Sartori, C.R.; Vieira, A.S.; Ferrari, E.M.; Langone, F.; Tongiorgi, E.; Parada, C.A. The antidepressive effect of the physical exercise correlates with increased levels of mature bdnf, and probdnf proteolytic cleavage-related genes, p11 and tpa. *Neuroscience* **2011**, *180*, 9–18. [CrossRef]

39. Sandrini, L.; Ieraci, A.; Amadio, P.; Veglia, F.; Popoli, M.; Lee, F.S.; Tremoli, E.; Barbieri, S.S. Sub-chronic stress exacerbates the pro-thrombotic phenotype in bdnf. *Int. J. Mol. Sci.* **2018**, *19*, 3235. [CrossRef]

40. Sandrini, L.; Ieraci, A.; Amadio, P.; Popoli, M.; Tremoli, E.; Barbieri, S.S. Apocynin prevents abnormal megakaryopoiesis and platelet activation induced by chronic stress. *Oxid Med. Cell Longev.* **2017**, *2017*, 9258937. [CrossRef]

41. Beg, M.; Chauhan, P.; Varshney, S.; Shankar, K.; Rajan, S.; Saini, D.; Srivastava, M.N.; Yadav, P.P.; Gaikwad, A.N. A withanolide coagulin-l inhibits adipogenesis modulating wnt/beta-catenin pathway and cell cycle in mitotic clonal expansion. *Phytomedicine* **2014**, *21*, 406–414. [CrossRef]

42. Biemann, R.; Fischer, B.; Bluher, M.; Navarrete Santos, A. Tributyltin affects adipogenic cell fate commitment in mesenchymal stem cells by a ppargamma independent mechanism. *Chem. Biol. Interact.* **2014**, *214*, 1–9. [CrossRef]

43. Ruiz-Ojeda, F.J.; Ruperez, A.I.; Gomez-Llorente, C.; Gil, A.; Aguilera, C.M. Cell models and their application for studying adipogenic differentiation in relation to obesity: A review. *Int. J. Mol. Sci.* **2016**, *17*, 1040. [CrossRef]

44. Baho, E.; Chattopadhyaya, B.; Lavertu-Jolin, M.; Mazziotti, R.; Awad, P.N.; Chehrazi, P.; Groleau, M.; Jahannault-Talignani, C.; Vaucher, E.; Ango, F.; et al. P75 neurotrophin receptor activation regulates the timing of the maturation of cortical parvalbumin interneuron connectivity and promotes juvenile-like plasticity in adult visual cortex. *J. Neurosci. Off. J. Soc. Neurosci.* **2019**, *39*, 4489–4510. [CrossRef]

45. Guo, J.; Ji, Y.; Ding, Y.; Jiang, W.; Sun, Y.; Lu, B.; Nagappan, G. Bdnf pro-peptide regulates dendritic spines via caspase-3. *Cell Death Dis.* **2016**, *7*, e2264. [CrossRef]

46. Mast, T.G.; Fadool, D.A. Mature and precursor brain-derived neurotrophic factor have individual roles in the mouse olfactory bulb. *PLoS ONE* **2012**, *7*, e31978. [CrossRef]

47. Cheung, K.J.; Tzameli, I.; Pissios, P.; Rovira, I.; Gavrilova, O.; Ohtsubo, T.; Chen, Z.; Finkel, T.; Flier, J.S.; Friedman, J.M. Xanthine oxidoreductase is a regulator of adipogenesis and ppargamma activity. *Cell Metab.* **2007**, *5*, 115–128. [CrossRef]

48. Lee, Y.H.; Chen, S.Y.; Wiesner, R.J.; Huang, Y.F. Simple flow cytometric method used to assess lipid accumulation in fat cells. *J. Lipid Res.* **2004**, *45*, 1162–1167. [CrossRef]

49. Kraus, N.A.; Ehebauer, F.; Zapp, B.; Rudolphi, B.; Kraus, B.J.; Kraus, D. Quantitative assessment of adipocyte differentiation in cell culture. *Adipocyte* **2016**, *5*, 351–358. [CrossRef]

50. Tarantino, E.; Amadio, P.; Squellerio, I.; Porro, B.; Sandrini, L.; Turnu, L.; Cavalca, V.; Tremoli, E.; Barbieri, S.S. Role of thromboxane-dependent platelet activation in venous thrombosis: Aspirin effects in mouse model. *Pharm. Res.* **2016**, *107*, 415–425. [CrossRef]

51. Parlee, S.D.; Wang, Y.; Poirier, P.; Lapointe, M.; Martin, J.; Bastien, M.; Cianflone, K.; Goralski, K.B. Biliopancreatic diversion with duodenal switch modifies plasma chemerin in early and late post-operative periods. *Obesity (Silver Spring)* **2015**, *23*, 1201–1208. [CrossRef]

52. Kernie, S.G.; Liebl, D.J.; Parada, L.F. Bdnf regulates eating behavior and locomotor activity in mice. *EMBO J.* **2000**, *19*, 1290–1300. [CrossRef]

53. Lebrun, B.; Bariohay, B.; Moyse, E.; Jean, A. Brain-derived neurotrophic factor (bdnf) and food intake regulation: A minireview. *Auton. Neurosci.* **2006**, *126–127*. [CrossRef]

54. Lyons, W.E.; Mamounas, L.A.; Ricaurte, G.A.; Coppola, V.; Reid, S.W.; Bora, S.H.; Wihler, C.; Koliatsos, V.E.; Tessarollo, L. Brain-derived neurotrophic factor-deficient mice develop aggressiveness and hyperphagia in conjunction with brain serotonergic abnormalities. *Proc. Natl. Acad. Sci. USA* **1999**, *96*, 15239–15244. [CrossRef]

55. Rosas-Vargas, H.; Martínez-Ezquerro, J.D.; Bienvenu, T. Brain-derived neurotrophic factor, food intake regulation, and obesity. *Arch. Med. Res.* **2011**, *42*, 482–494. [CrossRef]

56. Xu, H.; Barnes, G.T.; Yang, Q.; Tan, G.; Yang, D.; Chou, C.J.; Sole, J.; Nichols, A.; Ross, J.S.; Tartaglia, L.A.; et al. Chronic inflammation in fat plays a crucial role in the development of obesity-related insulin resistance. *J. Clin. Investig.* **2003**, *112*, 1821–1830. [CrossRef]

57. Yeo, G.S.; Connie Hung, C.C.; Rochford, J.; Keogh, J.; Gray, J.; Sivaramakrishnan, S.; O'Rahilly, S.; Farooqi, I.S. A de novo mutation affecting human trkb associated with severe obesity and developmental delay. *Nat. Neurosci.* **2004**, *7*, 1187–1189. [CrossRef]

58. Choe, S.S.; Huh, J.Y.; Hwang, I.J.; Kim, J.I.; Kim, J.B. Adipose tissue remodeling: Its role in energy metabolism and metabolic disorders. *Front. Endocrinol. (Lausanne)* **2016**, *7*, 30. [CrossRef]

59. Heilbronn, L.; Smith, S.R.; Ravussin, E. Failure of fat cell proliferation, mitochondrial function and fat oxidation results in ectopic fat storage, insulin resistance and type ii diabetes mellitus. *Int. J. Obes. Relat. Metab. Disord.* **2004**, *28*, S12–S21. [CrossRef]

60. Jernås, M.; Palming, J.; Sjöholm, K.; Jennische, E.; Svensson, P.A.; Gabrielsson, B.G.; Levin, M.; Sjögren, A.; Rudemo, M.; Lystig, T.C.; et al. Separation of human adipocytes by size: Hypertrophic fat cells display distinct gene expression. *FASEB J.* **2006**, *20*, 1540–1542.

61. Valet, P.; Grujic, D.; Wade, J.; Ito, M.; Zingaretti, M.C.; Soloveva, V.; Ross, S.R.; Graves, R.A.; Cinti, S.; Lafontan, M.; et al. Expression of human alpha 2-adrenergic receptors in adipose tissue of beta 3-adrenergic receptor-deficient mice promotes diet-induced obesity. *J. Biol. Chem.* **2000**, *275*, 34797–34802. [CrossRef]

62. Lafontan, M.; Berlan, M.; Stich, V.; Crampes, F.; Rivière, D.; De Glisezinski, I.; Sengenes, C.; Galitzky, J. Recent data on the regulation of lipolysis by catecholamines and natriuretic peptides. *Ann. Endocrinol. (Paris)* **2002**, *63*, 86–90.

63. Långberg, E.C.; Seed Ahmed, M.; Efendic, S.; Gu, H.F.; Östenson, C.G. Genetic association of adrenergic receptor alpha 2a with obesity and type 2 diabetes. *Obesity (Silver Spring)* **2013**, *21*, 1720–1725.

64. Moro, C.; Polak, J.; Richterova, B.; Sengenès, C.; Pelikanova, T.; Galitzky, J.; Stich, V.; Lafontan, M.; Berlan, M. Differential regulation of atrial natriuretic peptide- and adrenergic receptor-dependent lipolytic pathways in human adipose tissue. *Metabolism* **2005**, *54*, 122–131. [CrossRef]

65. Schmidt, V.; Schulz, N.; Yan, X.; Schürmann, A.; Kempa, S.; Kern, M.; Blüher, M.; Poy, M.N.; Olivecrona, G.; Willnow, T.E. Sorla facilitates insulin receptor signaling in adipocytes and exacerbates obesity. *J. Clin. Investig.* **2016**, *126*, 2706–2720. [CrossRef]

66. Smith, E.N.; Chen, W.; Kähönen, M.; Kettunen, J.; Lehtimäki, T.; Peltonen, L.; Raitakari, O.T.; Salem, R.M.; Schork, N.J.; Shaw, M.; et al. Longitudinal genome-wide association of cardiovascular disease risk factors in the bogalusa heart study. *PLoS Genet.* **2010**, *6*, e1001094. [CrossRef]

67. Ellulu, M.S.; Patimah, I.; Khaza'ai, H.; Rahmat, A.; Abed, Y. Obesity and inflammation: The linking mechanism and the complications. *Arch. Med. Sci.* **2017**, *13*, 851–863. [CrossRef]

68. Berg, A.H.; Scherer, P.E. Adipose tissue, inflammation, and cardiovascular disease. *Circ. Res.* **2005**, *96*, 939–949. [CrossRef]

69. Bodary, P.F. Links between adipose tissue and thrombosis in the mouse. *Arter. Thromb. Vasc. Biol.* **2007**, *27*, 2284–2291. [CrossRef]

70. Odrowaz-Sypniewska, G. Markers of pro-inflammatory and pro-thrombotic state in the diagnosis of metabolic syndrome. *Adv. Med. Sci.* **2007**, *52*, 246–250.

71. Davì, G.; Guagnano, M.T.; Ciabattoni, G.; Basili, S.; Falco, A.; Marinopiccoli, M.; Nutini, M.; Sensi, S.; Patrono, C. Platelet activation in obese women: Role of inflammation and oxidant stress. *JAMA* **2002**, *288*, 2008–2014. [CrossRef]

72. Freedman, J.E.; Larson, M.G.; Tanriverdi, K.; O'Donnell, C.J.; Morin, K.; Hakanson, A.S.; Vasan, R.S.; Johnson, A.D.; Iafrati, M.D.; Benjamin, E.J. Relation of platelet and leukocyte inflammatory transcripts to body mass index in the framingham heart study. *Circulation* **2010**, *122*, 119–129. [CrossRef]

73. Furuncuoğlu, Y.; Tulgar, S.; Dogan, A.N.; Cakar, S.; Tulgar, Y.K.; Cakiroglu, B. How obesity affects the neutrophil/lymphocyte and platelet/lymphocyte ratio, systemic immune-inflammatory index and platelet indices: A retrospective study. *Eur. Rev. Med. Pharm. Sci.* **2016**, *20*, 1300–1306.

74. Vilahur, G.; Ben-Aicha, S.; Badimon, L. New insights into the role of adipose tissue in thrombosis. *Cardiovasc. Res.* **2017**, *113*, 1046–1054. [CrossRef]

75. Skurk, T.; Alberti-Huber, C.; Herder, C.; Hauner, H. Relationship between adipocyte size and adipokine expression and secretion. *J. Clin. Endocrinol. Metab.* **2007**, *92*, 1023–1033. [CrossRef]

76. Li, A.C.; Binder, C.J.; Gutierrez, A.; Brown, K.K.; Plotkin, C.R.; Pattison, J.W.; Valledor, A.F.; Davis, R.A.; Willson, T.M.; Witztum, J.L.; et al. Differential inhibition of macrophage foam-cell formation and atherosclerosis in mice by pparalpha, beta/delta, and gamma. *J. Clin. Investig.* **2004**, *114*, 1564–1576. [CrossRef]

77. Okuno, A.; Tamemoto, H.; Tobe, K.; Ueki, K.; Mori, Y.; Iwamoto, K.; Umesono, K.; Akanuma, Y.; Fujiwara, T.; Horikoshi, H.; et al. Troglitazone increases the number of small adipocytes without the change of white adipose tissue mass in obese zucker rats. *J. Clin. Investig.* **1998**, *101*, 1354–1361. [CrossRef]

78. Ricote, M.; Li, A.C.; Willson, T.M.; Kelly, C.J.; Glass, C.K. The peroxisome proliferator-activated receptor-gamma is a negative regulator of macrophage activation. *Nature* **1998**, *391*, 79–82. [CrossRef]

79. Yamauchi, T.; Kamon, J.; Waki, H.; Murakami, K.; Motojima, K.; Komeda, K.; Ide, T.; Kubota, N.; Terauchi, Y.; Tobe, K.; et al. The mechanisms by which both heterozygous peroxisome proliferator-activated receptor gamma (ppargamma) deficiency and ppargamma agonist improve insulin resistance. *J. Biol. Chem.* **2001**, *276*, 41245–41254. [CrossRef]

80. Bernhard, F.; Landgraf, K.; Klöting, N.; Berthold, A.; Büttner, P.; Friebe, D.; Kiess, W.; Kovacs, P.; Blüher, M.; Körner, A. Functional relevance of genes implicated by obesity genome-wide association study signals for human adipocyte biology. *Diabetologia* **2013**, *56*, 311–322. [CrossRef]

81. Cancello, R.; Clément, K. Is obesity an inflammatory illness? Role of low-grade inflammation and macrophage infiltration in human white adipose tissue. *BJOG* **2006**, *113*, 1141–1147. [CrossRef]

82. Lee, C.D.; Jackson, A.S.; Blair, S.N. Us weight guidelines: Is it also important to consider cardiorespiratory fitness? *Int. J. Obes. Relat. Metab. Disord.* **1998**, *22*, S2–S7.

83. Polak, J.; Moro, C.; Klimcakova, E.; Hejnova, J.; Majercik, M.; Viguerie, N.; Langin, D.; Lafontan, M.; Stich, V.; Berlan, M. Dynamic strength training improves insulin sensitivity and functional balance between adrenergic alpha 2a and beta pathways in subcutaneous adipose tissue of obese subjects. *Diabetologia* **2005**, *48*, 2631–2640. [CrossRef]

84. Stanford, K.I.; Goodyear, L.J. Exercise regulation of adipose tissue. *Adipocyte* **2016**, *5*, 153–162. [CrossRef]

85. Woods, J.A.; Vieira, V.J.; Keylock, K.T. Exercise, inflammation, and innate immunity. *Immunol. Allergy Clin. North. Am.* **2009**, *29*, 381–393. [CrossRef]

86. Bruun, J.M.; Helge, J.W.; Richelsen, B.; Stallknecht, B. Diet and exercise reduce low-grade inflammation and macrophage infiltration in adipose tissue but not in skeletal muscle in severely obese subjects. *Am. J. Physiol. Endocrinol. Metab.* **2006**, *290*, E961–E967. [CrossRef]

87. Vieira, V.J.; Valentine, R.J.; Wilund, K.R.; Antao, N.; Baynard, T.; Woods, J.A. Effects of exercise and low-fat diet on adipose tissue inflammation and metabolic complications in obese mice. *Am. J. Physiol. Endocrinol. Metab.* **2009**, *296*, E1164–E1171. [CrossRef]

88. Goh, J.; Goh, K.P.; Abbasi, A. Exercise and adipose tissue macrophages: New frontiers in obesity research? *Front. Endocrinol. (Lausanne)* **2016**, *7*, 65. [CrossRef]

89. Kjøbsted, R.; Hingst, J.R.; Fentz, J.; Foretz, M.; Sanz, M.N.; Pehmøller, C.; Shum, M.; Marette, A.; Mounier, R.; Treebak, J.T.; et al. Ampk in skeletal muscle function and metabolism. *FASEB J.* **2018**, *32*, 1741–1777. [CrossRef]

90. Lira, V.A.; Benton, C.R.; Yan, Z.; Bonen, A. Pgc-1alpha regulation by exercise training and its influences on muscle function and insulin sensitivity. *Am. J. Physiol. Endocrinol. Metab.* **2010**, *299*, E145–E161. [CrossRef]

91. Leal, L.G.; Lopes, M.A.; Batista, M.L. Physical exercise-induced myokines and muscle-adipose tissue crosstalk: A review of current knowledge and the implications for health and metabolic diseases. *Front. Physiol.* **2018**, *9*, 1307. [CrossRef]

92. Kim, S.M.; Lun, M.; Wang, M.; Senyo, S.E.; Guillermier, C.; Patwari, P.; Steinhauser, M.L. Loss of white adipose hyperplastic potential is associated with enhanced susceptibility to insulin resistance. *Cell Metab.* **2014**, *20*, 1049–1058. [CrossRef]

93. Rigamonti, A.; Brennand, K.; Lau, F.; Cowan, C.A. Rapid cellular turnover in adipose tissue. *PLoS ONE* **2011**, *6*, e17637. [CrossRef]

94. White, U.; Ravussin, E. Dynamics of adipose tissue turnover in human metabolic health and disease. *Diabetologia* **2019**, *62*, 17–23. [CrossRef]

Article

Melatonin Supplementation Attenuates the Pro-Inflammatory Adipokines Expression in Visceral Fat from Obese Mice Induced by A High-Fat Diet

Talita da Silva Mendes de Farias, Regislane Ino da Paixao, Maysa Mariana Cruz, Roberta Dourado Cavalcante da Cunha de Sa, Jussara de Jesus Simão, Vitor Jaco Antraco and Maria Isabel Cardoso Alonso-Vale *

Department of Biological Sciences, Institute of Environmental Sciences, Chemical and Pharmaceutical, Federal University of Sao Paulo, Diadema 09913-130, Brazil
* Correspondence: alonsovale@gmail.com

Received: 13 July 2019; Accepted: 5 September 2019; Published: 6 September 2019

Abstract: Obesity is defined as a condition of abnormal or excessive fat accumulation in white adipose tissue that results from the exacerbated consumption of calories associated with low energy expenditure. Fat accumulation in both adipose tissue and other organs contributes to a systemic inflammation leading to the development of metabolic disorders such as type 2 diabetes, hypertension, and dyslipidemia. Melatonin is a potent antioxidant and improves inflammatory processes and energy metabolism. Using male mice fed a high-fat diet (HFD—59% fat from lard and soybean oil; 9:1) as an obesity model, we investigated the effects of melatonin supplementation on the prevention of obesity-associated complications through an analysis of plasma biochemical profile, body and fat depots mass, adipocytes size and inflammatory cytokines expression in epididymal (EPI) adipose depot. Melatonin prevented a gain of body weight and fat depot mass as well as adipocyte hypertrophy. Melatonin also reversed the increase of total cholesterol, triglycerides and LDL-cholesterol. In addition, this neurohormone was effective in completely decreasing the inflammatory cytokines leptin and resistin in plasma. In the EPI depot, melatonin reversed the increase of leptin, Il-6, *Mcp-1* and *Tnf-α* triggered by obesity. These data allow us to infer that melatonin presents an anti-obesity effect since it acts to prevent the progression of pro-inflammatory markers in the epididymal adipose tissue together with a reduction in adiposity.

Keywords: adipose tissue; leptin; adiponectin; Il-6; Mcp-1; Tnf-α

---

1. Introduction

Obesity results from an imbalance between energy consumption and energy expenditure, promoting an abnormal or excessive accumulation of fat in various regions of the body. Associated with the fat increase, a chronic low-grade inflammation that contributes to systemic metabolic disorders such as dislipidemia, hypertension, nonalcoholic fatty liver diseases, steatohepatitis, cardiovascular diseases, type-2 diabetes and even some cancers has been observed.

Despite the fact that the molecular mechanisms that associate obesity with higher incidences of these diseases are not yet fully defined, evidence indicates that white adipose tissue(s) is one of the first tissues to develop inflammatory responses in obesity conditions, as evidenced by the activation of classical proinflammatory pathways, exacerbated infiltration of macrophages, neutrophils and lymphocytes and a variety of pro-inflammatory mediators secretion [1–3]. It is now well established that white adipose tissue (WAT) is not only involved in energy storage but also functions as an endocrine organ that secretes various bioactive substances or adipo (cyto)kines such as leptin, adiponectin, tumor necrosis factor-alfa (TNF-α), interleukin-6 (IL-6), resistin and macrophage chemoattractant protein 1 (MCP-1), which are known to be involved in a wide range of physiological processes [4]. These adipokines play an important role in the pathophysiological link between increased adiposity and cardiometabolic alterations [4,5]. Indeed, it is also now established that an imbalance of pro- and anti-inflammatory adipokines secretion by WAT due to the expansion of fat mass in obesity exerts potential effects on obesity-linked metabolic disorders [6,7]. Since obesity-associated increases in these adipokines present a great contribution to the development of a dysfunctional WAT, characterized by the infiltration of pro-inflamatory immune cells in this tissue and unresolved inflammation, in addition to inappropriate extracellular matrix remodeling and impaired angiogenesis [8], all of which lead to the development of chronic low grade inflammation [9].

High levels of leptin resulting from leptin resistance and reduced adiponectin levels have been associated with obesity pathophysiology disorders [10]. On the other hand, insulin resistance has been associated with leptin resistance and a reduction of plasma adiponectin, which is reverted by the simultaneous administration of leptin and adiponectin [11,12]. Indeed, a dysregulated expression of these (and others) adipokines by WAT that occurs in obesity is one of the most important components that predisposes one to insulin resistance and that is predictive of the development of metabolic syndrome [13,14]. This occurs because a lot of adipokines interact with the insulin pathway and interfere in glucose and lipid metabolism.

Evidence in the literature indicates that melatonin has a modulatory effect on energy metabolism [15–17], insulin secretion and insulin action, as well as the glucose end lipid metabolism of adipose tissue from rats [18–21]. Therefore, we hypothesized that the melatonin used as a therapeutic strategy could be used as a way to improve the low-grade inflammation observed in obesity conditions. According to some studies, the use of melatonin, a neurohormone produced by the pineal gland only in the night phase and which is responsible for the synchronization of innumerable physiological effects, is related to beneficial effects on the control of obesity and its complications [22,23]. Additionally, chronobiological melatonin aspects and their interrelationship with cytokines produced by adipocytes such as leptin and adiponectin have been evaluated [10,24] and promising results in the prevention and control of complications caused by obesity have been suggested.

Though some studies have evaluated the effects of melatonin and/or pinealectomy on the repercussions of adiponectin and leptin gene expression [19,24], there are a lack of studies related with the role of melatonin on the expression of these and other adipokines involved in the inflammatory response and insulin signaling in WAT by fat depots. Thus, we herein investigate whether melatonin supplementation could prevent the characteristic increase of pro-inflammatory adipokines produced by epididymal WAT during the development of obesity in mice.

2. Materials and Methods

2.1. Animals and Melatonin Supplementation

The study was performed according to protocols approved by the Ethics Committee of the Federal University of São Paulo (CEUA 5998280515). Eight-week-old male C57BL/6 mice obtained from the Center for Development of Experimental Models (CEDEME), Federal University of São Paulo, were housed at 3 mice per cage in a room with light–dark cycle (12-h light, 12-h dark cycle, lights on at 0600) and temperature of 24 ± 1 °C. Mice were divided into three groups: (a) Control (low fat) diet (CO), (b) high-fat diet (Obese), and (c) high-fat diet supplemented with melatonin 1 mg/kg (Obese + Mel). The CO diet contained 76% carbohydrate, 15% protein and 9% fat, and the high-fat diet (HFD) contained 26% carbohydrate, 15% protein, and 59% fat, in % kcal. Lard and Soybean oil (9:1) was used as fat source The detailed composition of the diet and energy distribution was provided in our previous study [25].

During obesity induction, the animals were supplemented with melatonin (1 mg/kg) [26] in drinking water during the dark phase, daily, for 10 weeks. Body weight and food intake were measured weekly. After 10 weeks of the experimental protocol, 12-hour fasted mice were killed by cervical dislocation, which occurred between 9am and 11am, after isoflurane anesthesia. Blood samples were centrifuged at 1500 rpm for 20 min at 4 °C, and serum were stored at −80 °C. Adipose depots were collected and weighted, and epididymal adipose fat (EPI) was processed as described below.

2.2. Glucose and Insulin Tolerance Tests

An oral glucose tolerance test (oGTT) and an insulin tolerance test (ITT) were evaluated after a 6-hour fast. For oGTT analysis, we administrated by gavage a 20% glucose 20% solution (1 g/kg b.w.). The blood glucose measurements were performed at 0, 15, 30, 45, 60 or 90 min. For ITT, animals were injejected intraperitoneally. with insulin (Humulin R, Lilly, 0.75 UI/kg b.w.), and glucose measurements were performed at 0, 10, 20, 30, 40, 50 or 60 min after injection. In both tests, blood samples were collected from the tail vein. This method was not stressful, as indicated by the low basal levels of the stress hormone corticosterone. oGTT and ITT were determined by using a glucometer (One Touch Ultra, Johnson and Johnson, New Brunswick, NJ, USA). The assays were always performed in all groups concomitantly in order to avoid any interference in the obtained results.

2.3. Adipocyte Isolation

Adipocyte isolation was performed as previously described [27] with slight modifications [28]. Briefly, Epi fat pads were diced in small fragments in a flask containing 4 mL of DMEM supplemented with HEPES (20 mM), glucose (5 mM), bovine serum albumin (BSA, 1%), and collagenase type II (1 mg/mL) at pH 7.4 and incubated for approximately 40 min at 37 °C in an orbital shaker. Isolated adipocytes were filtered through a plastic mesh (150 μm) and washed three times in a fresh buffer without collagenase. After washing and brief spinning, the medium was thoroughly aspirated, and adipocytes were harvested. Aliquots of isolated adipocytes suspensions were placed in a microscope slide, and 6 fields were photographed under an optical microscope (×100 magnification) coupled to a microscope camera (AxioCam ERc5s; Zeiss, Oberkochen, Germany), and mean adipocyte volume ($4/3 \times \pi \times r^3$) was determined by measuring 100 cells using AxioVision LE64 software.

2.4. Blood Measurements

Triacylglycerol (TG) [29], fasting glucose, total cholesterol (TC), LDL-cholesterol [30], and HDL-cholesterol levels [31] were determined by colorimetric assays (Labtest Diagnostics, Lagoa Santa, MG, Brazil).

2.5. RNA Extraction and Quantitative Real-Time Polymerase Chain Reaction (qPCR)

Total RNA was extracted from an EPI depot, reverse transcribed, and destined for quantitative PCR analysis as previously described [25]. An analysis of real-time PCR data was performed using the $2^{-\Delta\Delta C_T}$ method [32]. Data are expressed as the ratio between the expression of the target gene and housekeeping gene (18S gene). Primers used are presented: *Adipoq* (5′-3′sense: GCAGAGATGGCACTCCTGGA; 5′-3′antisense: CCCTTCAGCTCCTGTCATTCC), *Tnf-α* (5′-3′sense: CCCTCACACTCAGATCATCTTCT; 5′-3′antisense: GCTACGACGTGGGCTACAG), *Il-6* (5′-3′sense: TTCTCTGGGAAATCGTGGAAA; 5′-3′antisense: TCAGAATTGCCATTGCACAAC), *Lep* (5′-3′sense: CATCTGCTGGCCTTCTCCAA; 5′-3′antisense: ATCCAGGCTCTCTGGCTTCTG), *Mcp-1* (5′-3′sense: GCCCCACTCACCTGCTGCTACT; 5′-3′antisense: CCTGCTGCTGGTGATCCTCTTGT) and *18S* (5′-3′sense: GGCCGTTCTTAGTTGGTGGAGCG; 5′-3′antisense: CTGAACGCCACTTGTCCCTC).

2.6. Adipokine Measurements

Lysates from the EPI depot and peripheral blood were used to perform the ELISA test. The concentrations of the adipokines IL-6, resistin, adiponectin and leptin were determined using specific commercially available DuoSet ELISA kits according to instructions supplied by the manufacturer (R&D

Systems, Minneapolis, MN, USA; Catalog numbers DY406, DY1069, DY1119, DY 498, respectively). The concentrations of the cytokines were expressed in ng per 100 mg of tissue or in ng/ml, as indicated.

2.7. Statistical Analysis

Data are presented as mean ± SEM. A one-way ANOVA and a Tukey post-test were used for the comparison between groups. GraphPad Prism 5.0 software (GraphPad Software, Inc., San Diego, CA, USA) was used for analysis. The level of significance was set at $p < 0.05$.

3. Results

3.1. Melatonin Supplementation Decreased Body Mass (BM), Adipose Depots Mass, Adipocytes Hypertrophy and Blood Biochemical Parameters Triggered by HFD-Induced Obesity

After 10 weeks of diet-induced obesity (DIO), it was found that the HFD was efficient in increasing the body mass of the animals (between approximately two- and five-fold). Concerning food, calories, and fat intake, as compared to CO diet, mice fed with the HFD presented a reduction (by 44%, $p < 0.05$) in food intake but an increase (by three-fold, $p < 0.05$) in fat intake, whereas a slight reduction was observed in calorie (by 20%) and water (by 22%) intake. However, mice that received the HFD associated with melatonin supplementation presented a lower body mass gain (between approximately three- and eight-fold compared to the control group). Though the mice supplemented with melatonin gained less body mass, when the food intake was mesured, both the Obese and Obese + Mel groups presented the same pattern of food, calories and fat consumption (Figure 1A and Table 1).

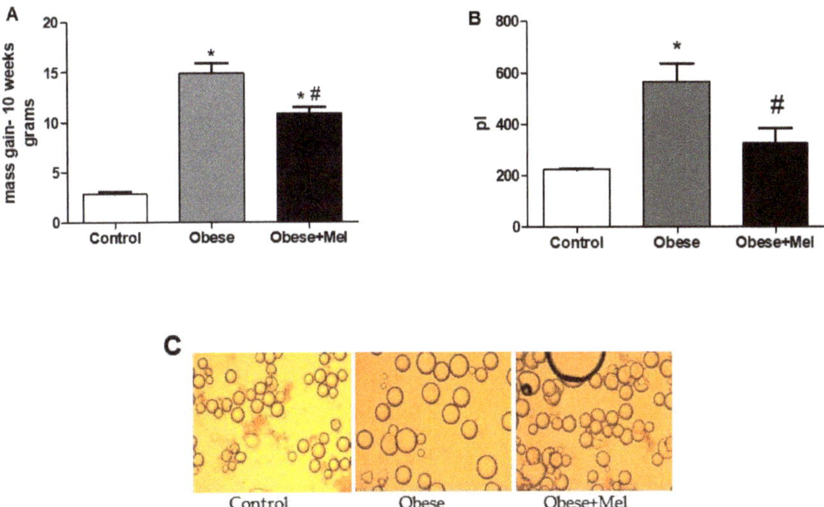

Figure 1. Effects of a high-fat diet (HFD) and melatonin supplementation (Mel, 1 mg/kg b.w., diluted in drinking water, daily, for 10 weeks) on body weight and adipocytes size. (**A**) Mice body mass gain at the end of experimental protocol. (**B**) Volume (in picoliters) of epididymal (EPI) isolated adipocytes. (**C**) Isolated EPI adipocytes photographed under optic microscope (×100 magnification). Adipocyte volume ($4/3 \times \pi \times r^3$) was determined by measuring 100 cells per animal (six fields for each slide). Results were analyzed by a one-way ANOVA and a Tukey post-test. Values are mean ± SEM (Control $n = 21$; Obese $n = 17$; Obese + Mel $n = 20$). * $p < 0.05$ vs. control; # $p < 0.05$ vs. obese.

Table 1. Body mass (BM), food intake, organ weights and blood biochemical parameters after 10 weeks of a high-fat diet feeding and melatonin suplementation in mice.

	Control	Obese	Obese + Mel
Initial BM (g)	22.70 ± 0.25	23.47 ± 0.29	22.56 ± 0.31
Final BM (g)	25.56 ± 0.30	38.22 ± 1.15*	33.44 ± 0.66*#
Water intake (ml/day/mice)	3.97 ± 0.20	3.08 ± 0.11*	3.20 ± 0.08*
Food Intake (g/day/mice)	4.76 ± 0.30	2.45 ± 0.20*	2.34 ± 0.22*
Calories intake	18.09 ± 0.32	13.08 ± 0.30*	12.51 ± 0.34*
Fat intake (g/day/mice)	0.43 ± 0.03	1.44 ± 0.12*	1.38 ± 0.13*
Relative ING weight (g/100g BM)	1.52 ± 0.07	4.67 ± 0.21*	3.86 ± 0.16*#
Relative EPI weight (g/100g BM)	2.04 ± 0.13	6.51 ± 0.20*	5.54 ± 0.17*#
Relative RP weight (g/100g BM)	0.45 ± 0.04	1.84 ± 0.05*	1.70 ± 0.04*
Relative BAT weight (g/100g BM)	0.25 ± 0.01	0.34 ± 0.02*	0.31 ± 0.01*
Fasting blood glucose (mg/dl)	175.64 ± 12.72	243.53 ± 8.20*	202.08 ± 13.61#
Total Cholesterol (mg/dl)	173.57 ± 13.19	265.50 ± 18.65*	209.30 ± 18.38
Triglycerides (mg/dl)	83.04 ±5.34	129.12 ±14.55*	94.87 ± 9.08
HDL (mg/dl)	69.56 ± 6.40	82.28 ±15.24	84.13 ± 11.62
LDL (mg/dl)	92.29 ± 9.74	148.34 ± 16.90*	114.22 ± 17.92

Results were analyzed by a one-way ANOVA and a Tukey post-test. Values are mean ± SEM. Control n = 20; Obese n = 17; Obese + Mel n = 21 to body mass, food, and fat intake and depots weight; Control n = 10; Obese n = 9; Obese + Mel n = 9 to blood biochemical parameters). *p < 0.05 vs. Control; #p < 0.05 vs. Obese.

Corroborating the lower body mass gain, we observed that the adipocyte size from the visceral (epididymal—EPI) region of the Obese + Mel group was 42% smaller than the obese group, thus preventing the hypertrophy triggered by the HFD (Figure 1B,C). In the same way, a significant reduction was observed in the EPI and inguinal (ING) depots mass from the animals supplemented with melatonin. The retroperitoneal (RP) and the brown fat (BAT) depots' mass did not present a significant reduction (Table 1).

The analysis of glucose and lipids serum concentrations indicated that the HFD significantly increased fasting glucose (21%), triglycerides (55%), total-cholesterol (52%) and LDL-cholesterol (60%) in the Obese group when compared to the Control group. This increase was prevented in animals that were supplemented with melatonin (Obese + Mel group), since the fasting glucose levels remained similar to the control group, and the serum levels of triglycerides, total cholesterol and LDL-cholesterol were reduced (a reduction of 26%, 21%, and 23%, respectively, in relation to the obese group). There were no significant differences in serum HDL-cholesterol between the groups (Table 1).

3.2. Melatonin Supplementation Did Not Alter Glycemic Curve after Glucose and Insulin Tolerance Test (GTT and ITT Test)

After glucose load, both groups receiving the HFD (Obese and Obese + Mel) presented higher blood glucose levels (Figure 2A). The HFD groups showed lower responsiveness to insulin compared to the control group (Figure 2B). Melatonin supplementation did not alter both the oGTT and ITT tests—that is, it did not prevent the development of insulin intolerance or the response to oral glucose loading.

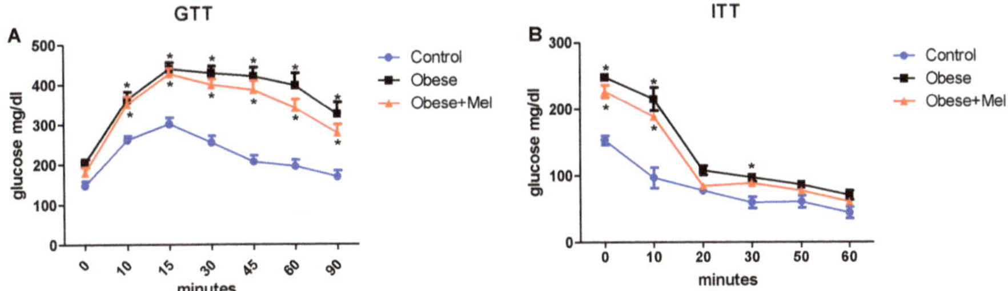

Figure 2. Effects of the high-fat diet (HFD) and melatonin supplementation (Mel, 1 mg/kg b.w., diluted in drinking water, daily, for 10 weeks) on glucose and insulin tolerance tests (GTT and ITT, respectively) in mice. (**A**) GTT or glucose concentration versus time after administration of glucose (2 g/kg b.w.); (**B**) ITT or glucose decay curve versus time after insulin administration (0.75 mU/g b.w.). Values are mean ± SEM (Control *n* = 11, Obese *n* = 7; Obese + Mel *n* = 11). * *p* < 0.05 versus Control.

3.3. Melatonin Supplementation Decreased the Gene Expression of Inflammatory Cytokines on EPI Depot

Whereas obesity is accompanied by a low-grade systemic inflammation, we evaluated the gene expression of the main adipokines produced in the visceral adipose depot from mice under an obesity condition. It was found that the HFD significantly increased the gene expression of *Lep, Il-6, Mcp-1* and *Tnf-α* ((Figure 3A–D) compared to the Control group. Melatonin was able to prevent some of these effects, since the Obese + Mel group presented a significant reduction in the expression of *Lep* (50%) and *Mcp-1* (55.8%) in relation to the Obese group. Moreover, the expressions of *Il-6* and *Tnf-α* in the EPI depot of mice supplemented with melatonin were partially prevented (44.6% and 44.8%, respectively), when compared to the Obese group. No change was observed in the *Adipoq* gene expression (Figure 3E).

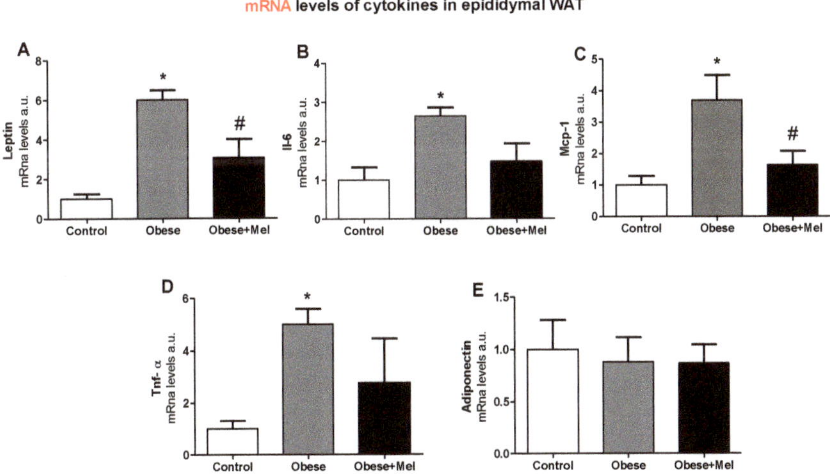

Figure 3. Effects of the high-fat diet (HFD) and melatonin supplementation (Mel, 1 mg/kg b.w., diluted in drinking water, daily, for 10 weeks) on mRNA levels of genes related to inflammation in epididymal (EPI) WAT from mice. (**A**) mRNA levels of *Lep*; (**B**) mRNA levels of *Il-6*; (**C**) mRNA levels of *Mcp-1*; (**D**) mRNA levels of *Tnf-α*; (**E**) mRNA levels of *Adipoq*. Results were analyzed by a one-way ANOVA and a Tukey post-test. Values are mean ± SEM (Control *n* = 9; Obese *n* = 8; Obese + Mel *n* = 10). * *p* < 0.05 vs. Control; # *p* < 0.05 vs. Obese.

3.4. Melatonin Supplementation Reduced the Protein Expression of Inflammatory Cytokines on EPI Depot and Peripheral Bood

The protein expression analysis by ELISA corroborated the data presented by the gene expression analysis. *Lep* expression in the EPI depot was significantly increased in the Obese group compared to the control group (82%, $p < 0.05$). In contrast, the Obese + Mel group showed a reduction by 30% ($p < 0.05$) compared to the Obese group (Figure 4A). Thus, melatonin supplementation partially prevented this increase, indicating its important action in reducing these adipokine levels. In plasma, we observed an even more pronounced effect, where the Obese group showed an increase in plasma leptin of approximately three-fold compared to the Control group, and melatonin supplementation reduced these levels by 28% (Obese vs Obese + Mel group, $p < 0.05$, Figure 4E).

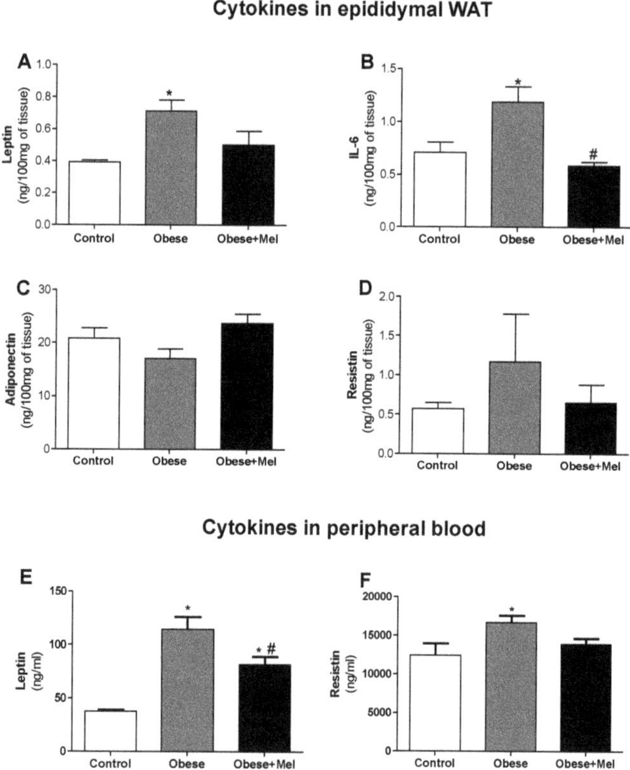

Figure 4. Effects of the high-fat diet (HFD) and melatonin supplementation (Mel, 1 mg/kg b.w., diluted in drinking water, daily, for 10 weeks) on ELISA analysis of cytokine levels in lysates from epididymal (EPI) WAT and peripheral blood from mice. (**A**) Leptin expression on the EPI depot, (**B**) Il-6 expression on the EPI depot, (**C**) adiponectin expression on the EPI depot, (**D**) resistin expression on the EPI depot, (**E**) leptin expression on peripheral blood, and (**F**) resistin expression on peripheral blood. Results were analyzed by a one-way ANOVA and a Tukey post-test. Values are mean ± SEM (Control $n = 10$; Obese $n = 9$; Obese + Mel n = 11). * $p < 0.05$ vs. Control; # $p < 0.05$ vs. Obese.

Melatonin supplementation also prevented the increase of IL-6 protein in the EPI depot triggered by the HFD since the Obese + Mel group showed a reduction of 51% ($p < 0.05$) when compared to the Obese group (Figure 4B). No statistical differences in adiponectin and resistin protein levels were observed in this adipose depot (Figure 4C,D). However, adiponectin expression showed a tendency ($p = 0.0629$) to increase in the Obese + Mel group (39% increase compared to the Obese group).

Finally, the HFD increased the resistin expression in plasma by 34%. Melatonin supplementation partially prevented this effect, since the Obese + Mel group presented an increase of only 11% compared to the Control group (Figure 4E).

4. Discussion

In this study, we evaluated the effects of melatonin on WAT inflammatory aspects in obese mice induced by an HFD. Initially, we verified that the experimental model adopted for DIO was efficient, since we observed that the animals fed an HFD showed a greater gain of body mass (with greater adiposity) and an increase in fasting glucose, triglycerides, total cholesterol and LDL-cholesterol in plasma, glucose intolerance and insulin resistance, corroborating the results obtained previously by our group [25]. Melatonin supplementation was effective in preventing most of the alterations triggered by the HFD, significantly hampering the gain of body mass and preventing the dyslipidemia progression.

There are, in the literature, some works showing the effect of melatonin decreasing body weight, and this hormone has therefore been considered as a possible therapeutic agent against obesity [33,34]. Herein, using a daily melatonin dose of 1 mg/kg (close to the endogenous physiological level), it was observed that the Obese + Mel group had an attenuation of their body mass gain in relation to the non-supplemented group. Taken together with a reduction in both the EPI and ING depots, melatonin suplemmentation performed a protective effect on the development of obesity and adiposity in animals fed an HFD.

Similarly to the findings reported here, using HFD-induced obese Wistar rats and supplementing them with melatonin (25 ug/ml) for 11 weeks, Rios Lugo et al. [34], observed a decrease in the body mass gain of these animals. Favero et al. [33], using a leptin-deficient (ob/ob) mouse supplemented with melatonin in drinking water (100 mg/kg) for eight weeks, also observed a decrease of approximately 5% in total body mass and a decrease in the weight of visceral and subcutaneous fat depots (53% and 41%, respectively) in animals supplemented with melatonin.

In contrast, melatonin supplementation in humans (3 mg daily for three months) did not promote changes in the body mass gain or loss of these individuals [35]. In the same way, Nduhirabandi et al. [36], using obese rats treated with melatonin (4 mg/kg), did not observe any decrease in the body mass despite having presented a cardioprotective effect. These controversial data may be due to the different doses and methods of melatonin administration.

It is believed that melatonin's effect on reducing body mass observed in rodents is probably mediated by central and peripheral target tissues, which leads to the synchronization of circadian rhythms and improved glucose uptake acting directly in adipocytes [24,34]. Here, we showed that dietary-induced obese animals supplemented with melatonin presented a significant reduction in fasting glycemia, indicating again the beneficial action of melatonin on glucose homeostasis, Corroborating these data, other studies employing pinealectomized animals have presented a significant reduction in the expression of the glucose transporter (GLUT 4), as well as glucose intolerance and insulin resistance, which were reverted by melatonin treatment [16,37]. The direct action of melatonin on the adipocytes and myocytes metabolism has been reported. In isolated adipocytes from the inguinal fat of rats, melatonin inhibited isoproterenol-stimulated lipolysis through the inhibition of the cAMP-PKA pathway [38]. In C2C12 skeletal muscle cells, melatonin activates the IRS-1 insulin receptor, thus stimulating GLUT4 expression [39], and these data corroborate a reduced glucose uptake in the skeletal muscle from melatonin receptor-1 knockout mice [40].

It is known that dyslipidemia triggered by obesity plays an important role in the development and worsening of the inflammatory state [41]. Here, we have shown that melatonin supplementation was effective in partially preventing the increase in total cholesterol, LDL-cholesterol and triglycerides characteristic of DIO. Previous studies have already shown this action of melatonin on lipid homeostasis [42–44]. Wistar rats induced to diabetes by streptozotocin and treated with melatonin

(10 mg/kg and 20 mg/kg) i.p. for two weeks also showed significant reductions in TG, TC and LDL-cholesterol levels [45]. However, in another study [46], C57Bl/6 mice induced to obesity by an HFD and treated with melatonin (10 mg/kg) for 12 weeks presented a reduction in body mass and LDL levels but not in TG levels. Considering melatonin supplementation in humans, it was observed that after three months of treatment, the daily use of melatonin (3 mg) was effective in significantly lowering TC and TG levels [35]. It is important to note that dyslipidemia is frequently observed in obese and/or diabetic individuals and that high plasma concentrations of total cholesterol and LDL-cholesterol are associated with an increased risk of cardiovascular disease [47,48]. On the other hand, the effects of melatonin on the cardiovascular system are well known. The removal of circulating melatonin causes hypertension in rats, and melatonin replacement prevents or reduces this effect [49,50]. Thus, the attenuation of serum triglyceride levels, total cholesterol and LDL-cholesterol reported here in animals treated with melatonin suggests a role for melatonin in the atherosclerosis prevention, one of the main complications of obesity.

There is a positive correlation between the increase in visceral WAT and hypertension, dyslipidemia, fasting glucose, non-alcoholic steatosis, age, and gender [51,52]. Thus, visceral obesity leads to increased risk of insulin resistance and cardiovascular disease. This fat depot displays a higher production of inflammatory cytokines in obese individuals [53,54].

We observed that the melatonin supplementation reduced the mass of the EPI depot by preventing cell hypertrophy because the adipocyte volume was reduced by 42%. Favero et al. [33], through histological and morphometric observations in WAT of animals supplemented with melatonin, observed that adipose depots from non-obese animals are composed of smaller and regular adipocytes, whereas in obese animals the adipocytes are larger and have a wide lipid droplet with presence of inflammatory infiltrate, macrophages and monocytes with degranulation signs which characterize the inflammatory tissue state (wherein there is a greater adipokines pro-inflammatory expression). Taken together, we suggest that melatonin contributes to the prevention of the inflammatory process in the visceral WAT (here represented by the EPI depot) because prevented the adipocytes hypertrophy.

As expected, we observed a significant increase in the expression of genes encoding proinflammatory cytokines, such as *IL-6, Lep, Mcp-1*, and *Tnf-α* in the EPI depot of animals with DIO. Melatonin fully reversed the increase of *Lep* and *Mcp-1* and partially reversed *Il-6* and *Tnf-α* gene expression. Melatonin supplementation also prevented an increase of leptin and IL-6 protein expression in the EPI depot triggered by the HFD. Moreover, melatonin was effective in decreasing the levels of leptin and resistin in plasma of animals induced to obesity by an HFD.

In the subcutaneous adipose tissue of ob/ob mice, immunofluorescence analyses also revealed melatonin's effects in reducing the expression of TNF-α, resistin, and visfatin, as well as in increasing the expression of adiponectin and its receptors [33]. In the liver of obese mice, Sun et al. [46] found that melatonin treatment also reduced the expression of proinflammatory markers *Tnf-α, Il-1β*, and *Il-6*. The same downregulation of these markers was observed in the liver of senescence accelerated prone male (SAMP8) mice [55]. In Wistar rats induced to obesity by an HFD, the plasma analyses demonstrated that melatonin attenuated the increase of the leptin observed in obese animals [34].

It is important to emphasize that the increase of these cytokines, such as leptin and IL-6, establishes a pathophysiological link between increased adiposity and obesity-associated cardiovascular disease, insulin resistance, type 2 diabetes, hypertension, and dyslipidemia [4,5,9]. On the other hand, obesity due to the ingestion of an HFD may be a consequence of a desynchronization in the biological rhythms of important metabolic processes [56,57] Consequently, the obese phenotype may be originated from this circadian desynchronization. Corroborating this, the clock genes and adipocytokines show circadian rhythmicity. The dysfunction of these genes is involved in the alteration of these adipokines during the development of obesity. Desynchronization between the central and peripheral clocks by an altered diet composition can lead to the uncoupling of peripheral clocks from the central pacemaker and to the development of metabolic disorders, leading to obesity. While CLOCK expression levels are increased with HDF-induced obesity, peroxisome proliferator-activated receptor (PPAR) alpha increases the transcriptional level of brain and muscle ARNT-like 1 (BMAL1) in obese subjects. Consequently, the disruption of clock genes results in dyslipidemia, insulin resistance and obesity [58]. Since melatonin is recognized by its important synchronization of diurnal and circadian rhythms [59], the supplementation with melatonin, respecting its physiological pattern of secretion (exclusively at night as performed in this work), is an important synchronizer to break changes triggered by diet [59] and could prevent changes observed in the obese phenotype.

Other assays need to be performed so that we can conclude in more detail how melatonin acts to reduce proinflammatory cytokines in adipose tissue. However, based on the facts that chronic inflammation in visceral WAT is one of the first steps in triggering obesity-associated diseases and that supplementation with melatonin for 10 weeks revealed significant effects in reducing body mass gain, adiposity and visceral adipocytes hypertrophy, plasma lipids and fasting glucose, as well as the expression of proinflammatory markers in visceral adipose tissue, we can infer here that melatonin must be considered to be a reliable therapeutic agent for the treatment of obesity.

Author Contributions: T.d.S.M.d.F.: Concept/design, design of experiments, acquisition of data, data analysis/interpretation, drafting and revision of the manuscript. R.I.d.P.: Concept/design, acquisition of data, revision of the manuscript. M.M.C.: Acquisition of data, revision of the manuscript. R.D.C.d.C.d.S.: Acquisition of data, revision of the manuscript. J.d.J.S.: Acquisition of data, revision of the manuscript. V.J.A.: Acquisition of data, revision of the manuscript. M.I.C.A.V.: Concept/design, data analysis/interpretation, revision of the manuscript, approval of the article.

Funding: This work was supported by a grant from FAPESP (2018/05485-6), TSMF is the recipient of Post-doc fellowship from FAPESP (2015/03554-2 and 2016/07638-9).

Conflicts of Interest: The authors declare no conflict of interest.

References

1. Hotamisligil, G.S. Inflammation and metabolic disorders. *Nature* **2006**, *444*, 860–867. [CrossRef]
2. Xu, H.; Barnes, G.T.; Yang, Q.; Tan, G.; Yang, D.; Chou, C.J.; Sole, J.; Nichols, A.; Ross, J.S.; Tartaglia, L.A.; et al. Chronic inflammation in fat plays a crucial role in the development of obesity-related insulin resistance. *J. Clin. Investig.* **2003**, *112*, 1821–1830. [CrossRef]
3. Weisberg, S.P.; McCann, D.; Desai, M.; Rosenbaum, M.; Leibel, R.L.; Ferrante, A.W. Obesity is associated with macrophage accumulation in adipose tissue. *J. Clin. Investig.* **2003**, *112*, 1796–1808. [CrossRef]
4. Lancha, A.; Frühbeck, G.; Gómez-Ambrosi, J. Peripheral signalling involved in energy homeostasis control. *Nutr. Res. Rev.* **2012**, *25*, 223–248. [CrossRef]
5. Gómez-Ambrosi, J.; Catalán, V.; Rodríguez, A.; Andrada, P.; Ramírez, B.; Ibanez, P.; Vilà, N.; Romero, S.; Margall, M.A.; Gil, M.J.; et al. Increased Cardiometabolic Risk Factors and Inflammation in Adipose Tissue in Obese Subjects Classified as Metabolically Healthy. *Diabetes Care* **2014**, *37*, 2813–2821. [CrossRef]
6. Fasshauer, M.; Blüher, M. Adipokines in health and disease. *Trends Pharmacol Sci.* **2015**, *36*, 461–470. [CrossRef]
7. Ouchi, N.; Parker, J.L.; Lugus, J.J.; Walsh, K. Adipokines in inflammation and metabolic disease. *Nat. Rev. Immunol.* **2011**, *11*, 85–97. [CrossRef]
8. Crewe, C.; An, Y.A.; Scherer, P.E. The ominous triad of adipose tissue dysfunction: inflammation, fibrosis, and impaired angiogenesis. *J. Clin. Investig.* **2017**, *127*, 74–82. [CrossRef]
9. McNelis, J.C.; Olefsky, J.M. Macrophages, Immunity, and Metabolic Disease. *Immunity* **2014**, *41*, 36–48. [CrossRef]
10. Szewczyk-Golec, K.; Wozniak, A.; Reiter, R.J. Inter-relationships of the chronobiotic, melatonin, with leptin and adiponectin: implications for obesity. *J. Pineal Res.* **2015**, *59*, 277–291. [CrossRef]
11. Yadav, A.; Kataria, M.A.; Saini, V.; Yadav, A. Role of leptin and adiponectin in insulin resistance. *Clin. Chim. Acta* **2013**, *417*, 80–84. [CrossRef]
12. López-Jaramillo, P.; Gomez-Arbelaez, D.; López-López, J.; Lopez-Lopez, C.; Martínez-Ortega, J.; Gómez-Rodríguez, A.; Triana-Cubillos, S. The role of leptin/adiponectin ratio in metabolic syndrome and diabetes. *Horm. Mol. Boil. Clin. Investig.* **2014**, *18*. [CrossRef]
13. Bastard, J.-P.; Maachi, M.; Lagathu, C.; Kim, M.J.; Caron, M.; Vidal, H.; Capeau, J.; Feve, B. Recent advances in the relationship between obesity, inflammation, and insulin resistance. *Eur. Cytokine Netw.* **2006**, *17*.

14. Greenberg, A.S.; Obin, M.S. Obesity and the role of adipose tissue in inflammation and metabolism. *Am. J. Clin. Nutr.* **2006**, *83*, 461S–465S. [CrossRef]

15. Lima, F.B.; Matsushita, D.H.; Hell, N.S.; Dolnikoff, M.S.; Okamoto, M.M.; Neto, J.C. The regulation of insulin action in isolated adipocytes. Role of the periodicity of food intake, time of day and melatonin. *Braz. J. Med Boil. Res.* **1994**, *27*.

16. Lima, F.B.; Machado, U.F.; Bartol, I.; Seraphim, P.M.; Sumida, D.H.; Moraes, S.M.F.; Hell, N.S.; Okamoto, M.M.; Saad, M.J.A.; Carvalho, C.R.O.; et al. Pinealectomy causes glucose intolerance and decreases adipose cell responsiveness to insulin in rats. *Am. J. Physiol. Metab.* **1998**, *275*. [CrossRef]

17. Zanquetta, M.M.; Corrêa-Giannella, M.L.; Monteiro, M.B.; Villares, S.M. Body weight, metabolism and clock genes. *Diabetol. Metab. Syndr.* **2010**, *2*, 53. [CrossRef]

18. Picinato, M.C.; Haber, E.P.; Carpinelli, A.R.; Cipolla-Neto, J.; Cipolla-Neto, J. Daily rhythm of glucose-induced insulin secretion by isolated islets from intact and pinealectomized rat. *J. Pineal Res.* **2002**, *33*, 172–177. [CrossRef]

19. Alonso-Vale, M.I.C.; Andreotti, S.; Peres, S.B.; Anhê, G.F.; Borges-Silva, C.D.N.; Neto, J.C.; Lima, F.B.; Alonso-Vale, M.I.C. Melatonin enhances leptin expression by rat adipocytes in the presence of insulin. *Am. J. Physiol. Metab.* **2005**, *288*, E805–E812. [CrossRef]

20. Alonso-Vale, M.I.; Borges-Silva, C.N.; Anhe, G.F.; Andreotti, S.; Machado, M.A.; Cipolla-Neto, J.; Lima, F.B. Light/Dark Cycle-dependent Metabolic Changes in Adipose Tissue of Pinealectomized Rats. *Horm. Metab. Res.* **2004**, *36*, 474–479. [CrossRef]

21. Takada, J.; Peres, S.B.; Andreotti, S.; Lima, F.B.; Borges-Silva, C.D.N.; Alonso-Vale, M.I.C.; Fonseca-Alaniz, M.H.; Cipolla-Neto, J.; Pithon-Curi, T.C.; Borges-Silva, C.D.N.; et al. Pinealectomy reduces hepatic and muscular glycogen content and attenuates aerobic power adaptability in trained rats. *J. Pineal Res.* **2007**, *43*, 96–103.

22. She, M.; Deng, X.; Guo, Z.; Laudon, M.; Hu, Z.; Liao, D.; Hu, X.; Luo, Y.; Shen, Q.; Su, Z. NEU-P11, a novel melatonin agonist, inhibits weight gain and improves insulin sensitivity in high-fat/high-sucrose-fed rats. *Pharmacol. Res.* **2009**, *59*, 248–253. [CrossRef]

23. Alamdari, N.M.; Mahdavi, R.; Roshanravan, N.; Yaghin, N.L.; Ostadrahimi, A.R.; Faramarzi, E. A double-blind, placebo-controlled trial related to the effects of melatonin on oxidative stress and inflammatory parameters of obese women. *Horm Metab Res.* **2015**, *47*, 504–508. [CrossRef]

24. Farias, T.D.S.M.D.; Chimin, P.; De Oliveira, A.C.; De Proença, A.R.G.; Leal, F.L.T.; Andreotti, S.; Lopes, A.B.; De Sousa, A.H.; Amaral, F.G.; Sertié, R.A.L.; et al. Pinealectomy interferes with the circadian clock genes expression in white adipose tissue. *J. Pineal Res.* **2015**, *58*, 251–261. [CrossRef]

25. de Sá, R.D.; Crisma, A.R.; Cruz, M.M.; Martins, A.R.; Masi, L.N.; do Amaral, C.L.; Curi, R.; Alonso-Vale, M.I. Fish oil prevents changes induced by a high-fat diet on metabolism and adipokine secretion in mice subcutaneous and visceral adipocytes. *J. Physiol.* **2016**, *594*, 6301–6317. [CrossRef]

26. De Oliveira, A.C.; Andreotti, S.; Farias, T.D.S.; Torres-Leal, F.L.; De Proença, A.R.; Campaña, A.B.; De Souza, A.H.; Sertié, R.A.; Carpinelli, A.R.; Cipolla-Neto, J.; et al. Metabolic Disorders and Adipose Tissue Insulin Responsiveness in Neonatally STZ-Induced Diabetic Rats Are Improved by Long-Term Melatonin Treatment. *Endocrinology* **2012**, *153*, 2178–2188. [CrossRef]

27. Rodbell, M. Metabolism of Isolated Fat Cells. I. Effects of Hormones on Glucose Metabolism and Lipolysis. *J. Biol. Chem.* **1964**, *239*, 375–380.

28. Bolsoni-Lopes, A.; Festuccia, W.T.; Farias, T.S.M.; Chimin, P.; Torres-Leal, F.L.; Derogis, P.B.M.; De Andrade, P.B.; Miyamoto, S.; Lima, F.B.; Curi, R.; et al. Palmitoleic acid (n-7) increases white adipocyte lipolysis and lipase content in a PPARα-dependent manner. *Am. J. Physiol. Metab.* **2013**, *305*, E1093–E1102. [CrossRef]

29. Bucolo, G.; David, H. Quantitative determination of serum triglycerides by the use of enzymes. *Clin. Chem.* **1973**, *19*, 476–482.

30. Postiglione, L.; Oriente, P.; Nastasi, L.; Cimmino, F.; Di Fraia, G.; Spanò, A. Determination of serum and lipoprotein phospholipids by a new enzymatic colorimetric method: characterization of a "reference range" in a sample of clinically healthy people. *Boll. della Soc. Ital. di Boil. Sper.* **1982**, *58*, 612–617.

31. Grillo, F.; Izzo, C.; Mazzotti, G.; Murador, E. Improved method for determination of high-density-lipoprotein cholesterol II. Enzymic determination of cholesterol in high-density lipoprotein fractions with a sensitive reagent. *Clin. Chem.* **1981**, *27*, 371–374.

32. Livak, K.J.; Schmittgen, T.D. Analysis of relative gene expression data using real-time quantitative PCR and the 2(-Delta Delta C(T)). *Method. Methods.* **2001**, *25*, 402–408.

33. Favero, G.; Stacchiotti, A.; Castrezzati, S.; Bonomini, F.; Albanese, M.; Rezzani, R.; Rodella, L.F. Melatonin reduces obesity and restores adipokine patterns and metabolism in obese (ob/ob) mice. *Nutr. Res.* **2015**, *35*, 891–900. [CrossRef]

34. Ríos-Lugo, M.J.; Cano, P.; Jiménez-Ortega, V.; Fernández-Mateos, M.P.; Scacchi, P.A.; Cardinali, D.P.; Esquifino, A.I.; Rios-Lugo, M.J.; Jimenez-Ortega, V.; Fernández-Mateos, M.P.; et al. Melatonin effect on plasma adiponectin, leptin, insulin, glucose, triglycerides and cholesterol in normal and high fat-fed rats. *J. Pineal Res.* **2010**, *49*, 342–348. [CrossRef]

35. Halpern, B.; Mancini, M.C.; Bueno, C.; Barcelos, I.P.; De Melo, M.E.; Lima, M.S.; Carneiro, C.G.; Sapienza, M.T.; Buchpiguel, C.A.; Amaral, F.G.D.; et al. Melatonin Increases Brown Adipose Tissue Volume and Activity in Patients With Melatonin Deficiency: A Proof-of-Concept Study. *Diabetes* **2019**, *68*, 947–952. [CrossRef]

36. Nduhirabandi, F.; Huisamen, B.; Strijdom, H.; Blackhurst, D.; Lochner, A. Short-term melatonin consumption protects the heart of obese rats independent of body weight change and visceral adiposity. *J. Pineal Res.* **2014**, *57*, 317–332. [CrossRef]

37. Nogueira, T.C.; Lellis-Santos, C.; Jesus, D.S.; Taneda, M.; Rodrigues, S.C.; Amaral, F.G.; Lopes, A.M.S.; Cipolla-Neto, J.; Bordin, S.; Anhê, G.F. Absence of Melatonin Induces Night-Time Hepatic Insulin Resistance and Increased Gluconeogenesis Due to Stimulation of Nocturnal Unfolded Protein Response. *Endocrinology* **2011**, *152*, 1253–1263. [CrossRef]

38. Zalatan, F.; Krause, J.A.; Blask, D.E. Inhibition of Isoproterenol-Induced Lipolysis in Rat Inguinal Adipocytes in Vitro by Physiological Melatonin via a Receptor-Mediated Mechanism. *Endocrinology* **2001**, *142*, 3783–3790. [CrossRef]

39. Ha, E.; Yim, S.-V.; Chung, J.-H.; Yoon, K.-S.; Kang, I.; Cho, Y.H.; Baik, H.H. Melatonin stimulates glucose transport via insulin receptor substrate-1/phosphatidylinositol 3-kinase pathway in C2C12 murine skeletal muscle cells. *J. Pineal Res.* **2006**, *41*, 67–72. [CrossRef]

40. Owino, S.; Sánchez-Bretaño, A.; Tchio, C.; Cecon, E.; Karamitri, A.; Dam, J.; Jockers, R.; Piccione, G.; Noh, H.L.; Kim, T.; et al. Nocturnal activation of melatonin receptor type 1 signaling modulates diurnal insulin sensitivity via regulation of PI3K activity. *J. Pineal Res.* **2018**, *64*, e12462. [CrossRef]

41. Loloei, S.; Sepidarkish, M.; Heydarian, A.; Tahvilian, N.; Khazdouz, M.; Heshmati, J.; Pouraram, H. The effect of melatonin supplementation on lipid profile and anthropometric indices: A systematic review and meta-analysis of clinical trials. *Diabetes Metab. Syndr. Clin. Res. Rev.* **2019**, *13*, 1901–1910. [CrossRef]

42. Hoyos, M.; Guerrero, J.M.; Perez-Cano, R.; Olivan, J.; Fabiani, F.; Garcia-Pergañeda, A.; Osuna, C. Serum cholesterol and lipid peroxidation are decreased by melatonin in diet-induced hypercholesterolemic rats. *J. Pineal Res.* **2000**, *28*, 150–155. [CrossRef]

43. Hussein, M.R.; Ahmed, O.G.; Hassan, A.F.; Ahmed, M.A. Intake of melatonin is associated with amelioration of physiological changes, both metabolic and morphological pathologies associated with obesity: An animal model. *Int. J. Exp. Pathol.* **2007**, *88*, 19–29. [CrossRef]

44. Pan, M.; Song, Y.-L.; Xu, J.-M.; Gan, H.-Z.; Song, Y.; Xu, J.; Gan, H. Melatonin ameliorates nonalcoholic fatty liver induced by high-fat diet in rats. *J. Pineal Res.* **2006**, *41*, 79–84. [CrossRef]

45. Hadjzadeh, M.A.R.; Alikhani, V.; Hosseinian, S.; Zarei, B.; Keshavarzi, Z. The Effect Of Melatonin Against Gastric Oxidative Stress And Dyslipidemia In Streptozotocin-Induced Diabetic Rats. *Acta Endocrinol. (Buchar.)* **2018**, *14*, 453–458. [CrossRef]

46. Sun, H.; Wang, X.; Chen, J.; Song, K.; Gusdon, A.M.; Li, L.; Bu, L.; Qu, S. Melatonin improves non-alcoholic fatty liver disease via MAPK-JNK/P38 signaling in high-fat-diet-induced obese mice. *Lipids Heal. Dis.* **2016**, *15*, 202. [CrossRef]

47. Franssen, R.; Monajemi, H.; Stroes, E.S.; Kastelein, J.J. Obesity and dyslipidemia. *Med. Clin. North Am.* **2011**, *95*, 893–902. [CrossRef]

48. Graham, I.; Cooney, M.-T.; Bradley, D.; Dudina, A.; Reiner, Z. Dyslipidemias in the Prevention of Cardiovascular Disease: Risks and Causality. *Curr. Cardiol. Rep.* **2012**, *14*, 709–720. [CrossRef]

49. Holmes, S.W.; Sugden, D. Proceedings: The effect of melatonin on pinealectomy-induced hypertension in the rat. *Br. J. Pharmacol.* **1976**, *56*, 360P–361P.

50. Zanoboni, A.; Forni, A.; Zanoboni-Muciaccia, W.; Zanussi, C. Effect of pinealectomy on arterial blood pressure and food and water intake in the rat. *J. Endocrinol. Investig.* **1978**, *1*, 125–130. [CrossRef]

51. Ayonrinde, O.T.; Olynyk, J.; Beilin, L.J.; Mori, T.A.; Pennell, C.E.; De Klerk, N.; Oddy, W.H.; Shipman, P.; Adams, L.A. Gender-specific differences in adipose distribution and adipocytokines influence adolescent nonalcoholic fatty liver disease. *Hepatol.* **2011**, *53*, 800–809. [CrossRef]

52. Fox, C.S.; Massaro, J.M.; Hoffmann, U.; Pou, K.; Maurovich-Horvat, P.; Liu, C. Abdominal visceral and subcutaneous adipose tissue compartments: Association with metabolic risk factors in the Framingham Heart Study. *J. Allergy Clin. Immunol.* **2007**, *116*, 39–48. [CrossRef]

53. Fontana, L.; Eagon, J.C.; Trujillo, M.E.; Scherer, P.E.; Klein, S. Visceral Fat Adipokine Secretion Is Associated With Systemic Inflammation in Obese Humans. *Diabetes* **2007**, *56*, 1010–1013. [CrossRef]

54. Cancello, R.; Tordjman, J.; Poitou, C.; Guilhem, G.; Bouillot, J.L.; Hugol, D.; Coussieu, C.; Basdevant, A.; Hen, A.B.; Bedossa, P.; et al. Increased Infiltration of Macrophages in Omental Adipose Tissue Is Associated With Marked Hepatic Lesions in Morbid Human Obesity. *Diabetes* **2006**, *55*, 1554–1561. [CrossRef]

55. Cuesta, S.; Kireev, R.; Forman, K.; García, C.; Escames, G.; Ariznavarreta, C.; Vara, E.; Tresguerres, J.A. Melatonin improves inflammation processes in liver of senescence-accelerated prone male mice (SAMP8). *Exp. Gerontol.* **2010**, *45*, 950–956. [CrossRef]

56. Kohsaka, A.; Laposky, A.D.; Ramsey, K.M.; Estrada, C.; Joshu, C.; Kobayashi, Y.; Turek, F.W.; Bass, J. High-Fat Diet Disrupts Behavioral and Molecular Circadian Rhythms in Mice. *Cell Metab.* **2007**, *6*, 414–421. [CrossRef]

57. Kaneko, K.; Yamada, T.; Tsukita, S.; Takahashi, K.; Ishigaki, Y.; Oka, Y.; Katagiri, H. Obesity alters circadian expressions of molecular clock genes in the brainstem. *Brain Res.* **2009**, *1263*, 58–68. [CrossRef]

58. Engin, A. Circadian Rhythms in Diet-Induced Obesity. *Results Probl. Cell Differ.* **2017**, *960*, 19–52.

59. Pfeffer, M.; Korf, H.-W.; Wicht, H. Synchronizing effects of melatonin on diurnal and circadian rhythms. *Gen. Comp. Endocrinol.* **2018**, *258*, 215–221. [CrossRef]

Article

Obesity as an Inflammatory Agent Can Cause Cellular Changes in Human Milk due to the Actions of the Adipokines Leptin and Adiponectin

Tassiane C. Morais [1,2], Luiz C. de Abreu [1,2,3,*], Ocilma B. de Quental [2], Rafael S. Pessoa [2,4], Mahmi Fujimori [2,4], Blanca E. G. Daboin [2], Eduardo L. França [2,4] and Adenilda C. Honorio-França [2,4,*]

1 Postgraduate Program in Public Health, School of Public Health, University of São Paulo (USP), São Paulo, SP 01246-904, Brazil; tassi.morais@usp.br
2 Laboratory of Scientific Writing, Department of Morphology and Physiology, Centro Universitário Saúde ABC (FMABC), Santo André, SP 09060-870, Brazil; ocilmaquental2011@hotmail.com (O.B.d.Q.); faelpessoa@gmail.com (R.S.P.); mahmi_fujimori@yahoo.com.br (M.F.); bgdaboin@yahoo.com (B.E.G.D.); dr.eduardo.franca@gmail.com (E.L.F.)
3 Postgraduate Program in Public Policies and Local Development, School of Sciences of Santa Casa de Misericordia de Vitoria (EMESCAM), Vitória, ES 29045-402, Brazil
4 Institute of Biological and Health Science, Federal University of Mato Grosso (UFMT), Barra do Garças, Mato Grosso, MT 78600-000, Brazil
* Correspondence: luizcarlos@usp.br (L.C.d.A.); adenildachf@gmail.com (A.C.H.-F.);
   Tel.: +55-(11)4993-7256 (L.C.d.A.); +55-663-405-5317 (A.C.H.-F.)

Received: 21 March 2019; Accepted: 7 May 2019; Published: 29 May 2019

Abstract: Adiponectin and leptin play roles in the hunger response, and they can induce the inflammatory process as the initial mechanism of the innate immune response. It is possible for alterations in the levels of these adipokines to compromise the functional activity of human colostrum phagocytes. Therefore, the objective of this study is to analyze the effects of adiponectin and leptin on colostrum mononuclear (MN) cells. Colostrum was collected from 80 healthy donors, who were divided into two groups: the control group and the high body mass index (BMI) group. MN cells were used to analyze phagocytosis by flow cytometry, and reactive oxygen species (ROS), intracellular calcium, and apoptosis were assessed by fluorimetry using a microplate reader. Adipokines restored the levels of phagocytosis to the high BMI group ($p < 0.05$), with a mechanism that is action-dependent on the release of ROS and intracellular calcium. However, adiponectin and leptin simultaneously contributed to better microbicidal activity, thus reflecting an increase in the apoptosis level ($p < 0.05$) in the high BMI group. Probably, the maintenance of the balance between adiponectin and leptin levels enhances the protection and decreases the indices of neonatal infection in the breastfeeding infants of women with high BMI values. Therefore, policies that support pre-gestational weight control should be encouraged.

Keywords: adiponectin; body mass index; colostrum; leptin; phagocytes; obesity; overweight; oxidative stress

---

1. Introduction

Obesity is considered a complex, recurrent, and progressive chronic disease. It can be characterized as an inflammatory condition involving elevated oxidative stress, and it is associated with other diseases [1,2] due to the risk of developing comorbidities such as asthma, musculoskeletal and sleep disorders, diabetes mellitus type 2, liver and kidney dysfunction, cardiovascular diseases, infertility, and cancer [1,3,4].

It is important to highlight that women are more affected by obesity than men, which has the potential to cause an impact on the health of future generations [5,6], whether through metabolic changes transmitted to the fetus during pregnancy [6] or through the changes in an infant's nutritional programming. Having adequate nutrition in the early stages of life is essential for the development of a child's metabolic programming, with human milk being the most recommended food for newborns, as it provides inclusive protection against metabolic changes associated with various conditions such as obesity and diabetes. In this way, breastfeeding represents an important pathway that impacts on the reduction of excess weight in both the mother and the child [7–12], because infants who are breastfed and mothers who breastfeed have lower rates of obesity [7,10].

For the mother, breastfeeding facilitates postpartum weight loss; it is positively associated with the lean mass index and inversely associated with visceral fat thickness [9,12]. Thus, this is a way to improve a woman's health after pregnancy, as it may help her to return to a normal metabolic profile and to lose the weight she gained during pregnancy. In infants, breastfeeding may protect against obesity through the components of human milk and behaviors related to infant feeding [8]. In any case, the programing of satiety control is one of the main breastfeeding mechanisms that helps to control obesity [12]. However, in the literature, the number of studies in this area is still scarce, so the mechanisms involved in this process have not yet been totally described.

Studies have demonstrated that metabolic changes due to maternal obesity can lead to changes in the constituents of colostrum and human milk, thereby modifying the different concentrations of hormones regulating appetite and metabolism, such as adiponectin and leptin, through breastfeeding [13–17]. In previously published results from this research group, it was found that the colostrum adiponectin concentration was 8.61 and 13.82 ng/mL, and the leptin concentration was 0.19 and 0.32 mg/dL in normal weight and obese groups, respectively. Furthermore, it was shown that changes in maternal serum constituents caused by maternal obesity are not necessarily reflected in colostrum constituents, especially the adiponectin levels, which were negatively correlated in colostrum and serum [15].

Adiponectin and leptin are mainly secreted by the adipose tissue and inhibit feeding by different mechanisms, acting on their respective receptors (AdipoRs and LepRs) located in neurons in the hypothalamus [18]. These adipokines levels are important indicators for the development of obesity and metabolic syndrome, since there is a reduction in the endogenous concentrations of adiponectin and an increase in leptin levels in overweight and obese individuals [19,20].

It is interesting to note that the actions of adiponectin and leptin are not restricted only to the hunger response. They also have binding sites in the cells of the immune system and activate monocytes/macrophages, triggering the inflammatory process in order to eliminate invading microorganisms. This process can also undergo alterations due to obesity and metabolic syndrome. It is emphasized that these adipokines usually act differently during the inflammatory process; adiponectin acts in a more anti-inflammatory manner, whereas leptin is more proinflammatory [3,21–25].

Several studies in the literature have reported that having a high maternal body mass index (BMI) is correlated with alterations in the constituents of colostrum and human milk [15,26–29]. However, the impact of maternal BMI on the functional activity of colostrum phagocytes is not entirely clear.

It is emphasized that the mother, through breastfeeding, transmits colostrum phagocytes to the infant, which are activated in the presence of microorganisms. So, the colostrum phagocytes encompass the invading particles, and during the process of phagocytosis, microbicidal activity develops via an oxidative burst, and consequently, there is an increase in the release of reactive oxygen species (ROS). Thus, colostrum phagocytes represent an additional mechanism of protection for the baby in case of neonatal infections until the infant's immune system is developed [30–32].

In this study, we investigate the actions of adiponectin and leptin on human colostrum phagocytes. It is possible that changes in adiponectin and leptin in human colostrum due to maternal overweightness can lead to alterations in the responses of mononuclear phagocytes in human colostrum and compromise this fundamental mechanism of infant protection, thus increasing the risk of neonatal infection.

Therefore, the aim of this study is to analyze the effects of exogenous adiponectin and leptin and their repercussions on human colostrum mononuclear phagocytes as a function of maternal body mass index (BMI).

2. Materials and Methods

2.1. Design and Sample

A cross-sectional study was carried out with the participation of 80 clinically healthy colostrum donors enrolled at the Hospital of the University of São Paulo, SP, Brazil, in 2017.The women were divided into two groups according to their pre-maternal BMI: normal BMI (18.5–24.9 kg/m^2) and high BMI ($\geq$30.0 kg/m^2).

The inclusion criteria of the study were as follows: aged from 18 to 35 years; pre-gestational weight known or measured until the end of the 13th gestational week; gestational age at delivery between 37 and 41$^{6/7}$ weeks; negative serological reactions for hepatitis, HIV, and syphilis; prenatal and non-food restrictions; and informed consent form signed. The exclusion criteria were as follows: gestational diabetes; twin pregnancy; fetal malformations; and delivery before the 36th week of gestation.

The study was approved by the Institutional Committee for Ethics in Research of the Hospital of the University of São Paulo (HU/USP) (CAAE 46643515.0.3001.0076), and all the subjects gave informed written consent before entering the experimental protocol.

2.2. Obtaining Colostrum and Cell Separation

About 5 mL of colostrum was collected manually from both breasts of each woman into sterile plastic tubes between 48–72 h postpartum. Colostrum was collected between feeding intervals, in the daytime period (between 10:00 and 12:00). The samples were stored at –80 °C until analysis. The experimental assays were carried out in 2017 and 2018.

The samples were thawed and then centrifuged for 10 min (160× g, 4 °C), which separated colostrum into three different phases: cell pellet, an intermediate aqueous phase, and an upper fat layer. The upper fat layer and the aqueous supernatant were discarded, and the cell pellet was separated using the Ficoll–Paque concentration gradient (Pharmacia, Uppsala, Sweden). The cells were resuspended in Medium 199 (Gibco, Grand Island, NE, USA) at a concentration of 1×10^6 cells/mL [30–32] and immediately used in the assays.

2.3. Treatment of Mononuclear Cells with Adipokines and Zymosan

The activation of mononuclear (MN) cells was performed by incubation with Zymosan in the presence and absence of the exogenous adipokines human adiponectin (Sigma, St Louis, MO, USA) and human leptin (Thermo Fisher, Carlsbad, CA, USA), each at a concentration of 100 ng/mL. The concentrations were in accordance with data from the scientific literature [25], and preliminary pilot tests were conducted to standardize the concentrations used.

The MN cells were incubated with Zymosan (for 2 h at 37 °C under gentle shaking) and treated with 199 medium (negative control), adiponectin, leptin, and adiponectin+leptin.

The phagocytosis assays were performed with Zymosan pHrodo Green™ (Thermo Fisher, Carlsbad, CA, USA), because it emits green fluorescence in the presence of an acidic pH during the phagocytosis process. Free radical release, apoptosis, and intracellular calcium assays were performed with Zymosan (Sigma, St Louis, MO, USA) without conjugated fluorochrome to avoid interference in the fluorescence intensity of the reagents used in each assay.

2.4. Phagocytosis Assays

The phagocytosis assay using Zymosan pHrodo Green™ (Thermo Fisher, Carlsbad, CA, USA) did not require wash steps and quencher dyes. So, after the incubation period, 10,000 cells were analyzed by flow cytometry using FACSCalibur™ (BD Biosciences, San Jose, CA, USA) with excitation/emission maxima of 509/533 nm. The results were expressed by the Phagocytosis Index (%). The experiments were performed in duplicate.

2.5. Tests for the Analysis of Free Radicals

The cells were incubated with Zymosan in the presence of 5 μ/mL of dihydrorhodamine 123 (DHR123) (Sigma, St Louis, MO, USA). The intensity of fluorescence emitted was proportional to the amount of reactive oxygen species released [33]. The Fluoroskan Ascent FL™ plate reader (Thermo Scientific, Vantaa, Finland) was used, with the 485-nm excitation and 538-nm emission filters. The results were expressed as the DHR123 mean fluorescence intensity. The experiments were performed in duplicate.

2.6. Intracellular Calcium Assay

The cells were incubated with Zymosan in the presence of the 5-μL Fluo-3 AM solution (Sigma, St Louis, MO, USA). The cells were washed and resuspended in HBSS (Hank's Balanced Salt Solution) containing bovine serum albumin (BSA). The fluorescence intensity was measured by a Fluoroskan Ascent FL™ Microplate reader (Thermo Scientific, Vantaa, Finland) using 485-nm excitation and 538-nm emission filters. The rate of intracellular Ca^{2+} release was expressed as the mean fluorescence intensity of Fluo-3. The experiments were performed in duplicate.

2.7. Apoptosis Assay

Cells undergoing apoptosis were detected using FITC Annexin V (BD Biosciences, Erembodegem, Belgium). The fluorescence intensity was obtained with the Fluoroskan Ascent FL™ Microplate reader (Thermo Scientific, Vantaa, Finland) using 485-nm excitation and 538-nm emission filters. The results were expressed as Apoptosis Index values (%). The experiments were performed in duplicate.

2.8. Statistical Analysis

Statistical analyses were performed with BioEstat® version 5.0 software (Mamirauá Institute, Belém, Brazil). The results are presented as mean (± standard deviation). The D'Agostino normality test and variance analysis (ANOVA) were used, followed by Tukey's test. Significant differences were considered when $p < 0.05$, and the power of the test for the sample size used was 80%.

3. Results

The women who had a high pre-pregnancy BMI also presented higher BMI values at the time of delivery. The other maternal/infant parameters did not differ between groups (Table 1).

Table 1. Maternal and neonate characteristics according to maternal pre-gestational body mass index (BMI) (normal or high BMI).

Maternal and Child Characteristics	Normal BMI (n = 40) (18.5–24.9 kg/m²)	High BMI (n = 40) (≥30 kg/m²)
Age (years)	26.02 (±5.43)	25.60 (±4.88)
Diabetes or gestational diabetes (%)	00 (0.00%)	00 (0.00%)
Maternal pre-gestational BMI (kg/m²)	21.51 (±2.32)	31.05 (±3.77) [#]
Delivery BMI (kg/m²)	25.72 (±2.32)	34.77 (±3.78) [#]
Gestational weight gain	10.52 (±4.08)	9.11 (±3.24)
Gestational age at delivery (weeks)	38.77 (±1.10)	38.70 (±1.04)
Infant sex—female (%)	23 (57.50%)	21 (52.50%)
Birth weight (g)	3263.00 (±430.31)	3367.375 (±495.50)
Birth height (cm)	47.31 (±2.84)	47.58 (±2.69)

Maternal and neonatal data are shown as mean (±SD) or number (%). They were assessed by ANOVA and Tukey's test. [#] Statistical difference among the normal and high BMI groups ($p < 0.05$).

Pre-gestational excess weight caused a reduction in the phagocytosis index of the human colostrum mononuclear cells ($p < 0.05$). However, the addition of adiponectin and leptin (100 ng/mL) increased the percentage of phagocytosis in the group with high BMI ($p < 0.05$) (Figure 1).

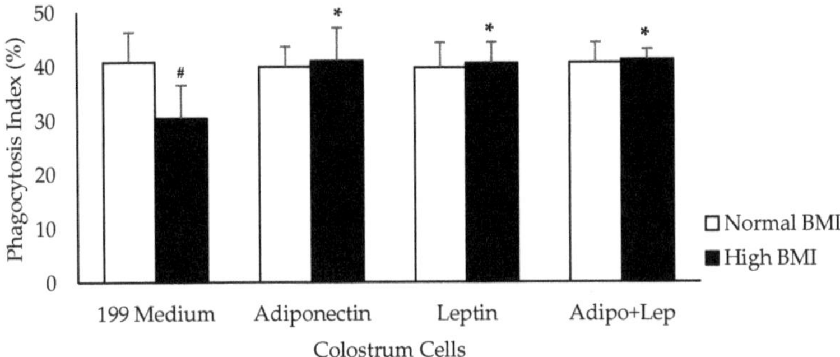

Figure 1. Effects of adiponectin and leptin on phagocytosis by colostrum mononuclear (MN) cells from women with normal BMI and high BMI values. Phagocytosis Index (%) values were determined by the assessment of pHRodo™ Green zymosan in MN cells from human colostrum. The results are presented as mean ± SD (n = 10 per group). They were assessed by ANOVA and Tukey's Test. * Statistical difference between colostrum cells incubated with 119 medium and hormones within groups ($p < 0.05$). # Statistical difference among groups with the same treatment and samples ($p < 0.05$).

The human colostrum mononuclear phagocytes from mothers with obesity decreased the release of reactive oxygen species. When the cells were treated with adiponectin, the reactive oxygen species release was similar between groups. Higher concentrations of reactive oxygen species were observed in cells treated with leptin from the group with high BMI values ($p < 0.05$). The association with adiponectin (adipo) and leptin (lep) increased the release of reactive oxygen species when compared with untreated cells in the high BMI group, and showed similar rates of release to the cells from normal BMI individuals (Figure 2).

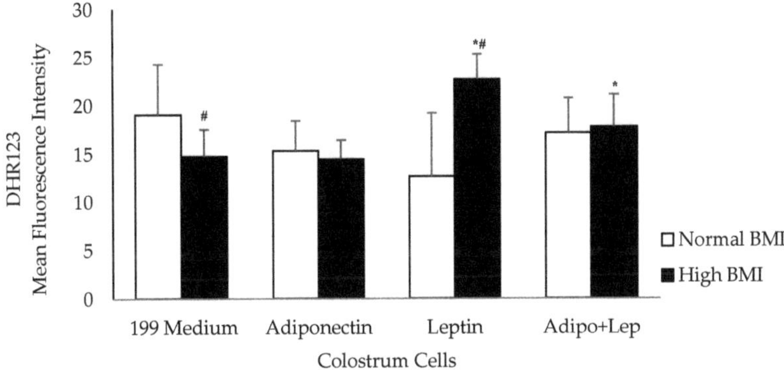

Figure 2. The effects of adiponectin and leptin on the oxidative burst of colostrum mononuclear cells from women with normal BMI and high BMI values incubated with Zymosan A from *Saccharomyces cerevisiae*. Reactive oxygen species were released by colostrum mononuclear phagocytes. The results are presented as mean ± SD (n = 10 per group) and assessed by ANOVA and Tukey's Test. * Statistical difference between colostrum cells incubated with 199 medium and hormones within groups ($p < 0.05$). # Statistical difference among groups with the same treatment and sample ($p < 0.05$).

Figure 3 shows intracellular Ca^{2+} release in the presence of adiponectin and leptin. Adiponectin increased the intracellular Ca^{2+} release by colostrum phagocytes in the obesity group, whereas this was decreased by leptin. Similar rates of release were observed between groups when the colostrum cells were treated with both hormones simultaneously.

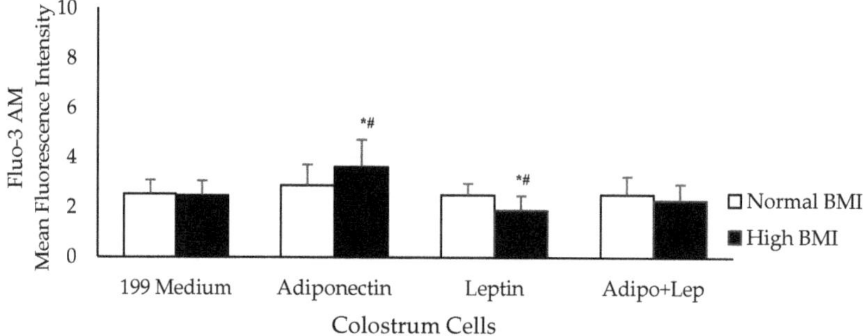

Figure 3. The effects of adiponectin and leptin on the release of intracellular calcium by colostrum mononuclear phagocytes from women with normal BMI and high BMI values incubated with Zymosan A from *Saccharomyces cerevisiae*. Intracellular calcium release was performed with Fluo-3AM (n = 10). The results are presented as the mean ± SD (n = 10 per group). They were assessed by ANOVA and Tukey's Test. * Statistical difference between colostrum cells incubated with 119 medium and hormones within groups ($p < 0.05$). # Statistical difference among groups with the same treatment and sample ($p < 0.05$).

Irrespective of the BMI status of the women tested, their colostrum cells showed similar apoptosis rates when treated with adiponectin or leptin compared to untreated cells. When the colostrum cells from higher BMI group were treated with both hormones, they exhibited higher apoptosis rates than in cells from normal BMI group (Figure 4).

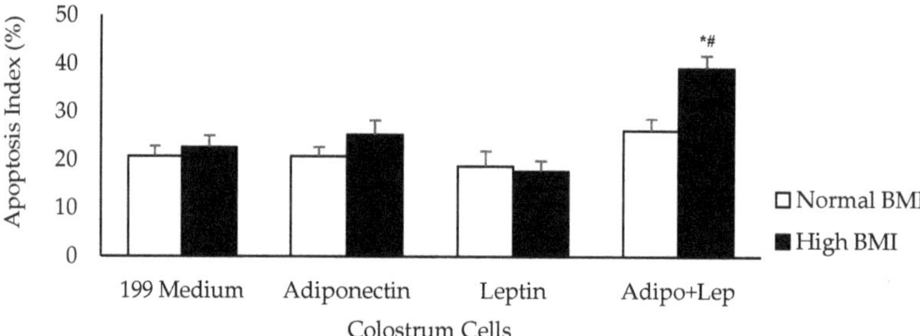

Figure 4. The effects of adiponectin and leptin on the apoptosis index of colostrum mononuclear phagocytes from women with normal BMI and high BMI values incubated with Zymosan A from *Saccharomyces cerevisiae*. The apoptosis assay was performed by Annexin V-fluorescein isothiocyanate (FITC) staining (n = 10). The results are presented as mean ± SD. They were assessed by ANOVA and Tukey's test. * Statistical difference between colostrum cells incubated with 119 medium and hormones within groups ($p < 0.05$). # Statistical difference among groups with the same treatment and sample ($p < 0.05$).

4. Discussion

Obesity is characterized by an increase in adipose tissue, which is considered a large organ with plastic and endocrine properties [34]. Adipocytes secrete inflammatory cytokines and hormones called adipokines that have complex interactions with each other and are altered in overweight and obese individuals. In obese individuals, an increase in proinflammatory adipokine levels and a reduction in anti-inflammatory adipokines occurs, promoting a state of chronic inflammation [35]. The most abundant adipokines secreted by adipocytes are adiponectin and leptin [36,37].

In obese mothers, a state of chronic inflammation can trigger an exaggerated inflammatory response in the placenta, leading to the accumulation of macrophages and the production of proinflammatory mediators [38]. However, a different mechanism that leads to the production and regulation of inflammatory mediators probably operates in colostrum from obese mothers, which exhibits immunological components that protect newborns without triggering inflammatory processes [15]. This suggests that breastfeeding may represent a strategy for enhance maternal and child health, as well as controlling weight in overweight and obese individuals [7,8,12].

The findings of the present study indicate that as an inflammatory agent, obesity alters phagocytosis and reactive oxygen species release in colostrum phagocytes. In addition, colostrum cells from obese mothers showed differences in functional activity in the presence of adiponectin and leptin.

Colostral macrophages produce reactive oxygen species (ROS) and show phagocytic activity [30–32]. Reactive oxygen species (ROS) release is one of the last events in the course of the innate immune response and plays an essential role in the death process of invasive microorganism [39]. In this study, colostrum phagocytes from women with high BMI values were associated with lower phagocytosis index values and ROS release in the presence of Zymosan.

Several studies have reported that maternal obesity causes alterations in the components of colostrum and human milk [15,26–29]. Some authors have shown that obesity is associated with a decreased macrophage phagocytic index [40,41], whereas others suggest an increase in phagocytes and oxidative stress in these cells [42].

Interestingly, the treatment of colostrum cells with adiponectin and leptin restored the phagocytic capacity of colostrum cells from obese mothers to similar phagocytic index values to normal weight mothers. However, ROS release only showed higher levels when the cells were treated with leptin.

Leptin also plays an important role in the action of phagocytes, and can be used as a potent phagocytosis-inducing agent [43]. Immunomodulatory effects have been attributed to this hormone, which represents a link between nutritional status and neuroendocrine and immunological functions [44] because of its dual action as both a hormone and cytokine [45] that can control macrophage activation [46]. Its mechanism of action involves the activation of the JAK/STAT signaling pathways, which induce phagocytosis, oxidative burst, and the increased secretion of proinflammatory cytokines [22].

Studies have shown a significant relationship between high levels of leptin and the inflammatory state present in obesity [47,48]. Increased ROS production modifies the response of intracellular Ca^{2+} during oxidative metabolism [49]. It is important to note that although leptin is related to ROS elevation, when macrophages are in a hyperinflammatory state, they may alter the oxidative burst [21]. In this study, it is possible that the alteration in ROS by colostrum cells treated with leptin may have been associated with the reduction of intracellular calcium released by the cells from the high BMI group.

In contrast to leptin, although the presence of adiponectin in colostrum cells from obese mothers increased phagocytosis and intracellular calcium release, it did not alter the ROS release. Adiponectin, despite being known for its anti-inflammatory action, is derived from its ability to stimulate the expression of macrophage markers of the M2 phenotype. This adipokine also causes stimuli in macrophages [50,51]. Adiponectin promotes macrophage activation, inducing a proinflammatory response that resembles M1 more than M2. It induces a limited program of inflammatory activation that likely desensitizes these cells to future proinflammatory stimuli [50]. Cot/tpl2 also play an important role in the production of inflammatory mediators upon the stimulation of macrophages with adiponectin. Furthermore, activation of this axis plays an important role in the M1 proinflammatory program induced by adiponectin in macrophages [52].

Through phosphatidyl inositol 3-kinase (PI3K)/Akt, adiponectin induces the activation of the enzyme IkB kinase. This enzyme is responsible for the degradation of inhibitory protein kB and thus the consequent activation of NF-kB, which will promote the release of proinflammatory cytokines in the nucleus and trigger the inflammatory response [51].

It must be highlighted that intracellular calcium levels are fundamental for the phagocytosis process and to control the subsequent steps involved in phagosome maturation [53]. Here, colostrum cells from mothers with high BMI values treated with adiponectin exhibited higher concentrations of intracellular calcium, suggesting that it plays an important role in its mechanism of action. Intracellular calcium is responsible for the control of various cellular processes including proliferation, differentiation, and cell death [54]. Alterations in the intracellular Ca^{2+} influx by human cells may cause cell damage and have been associated with apoptosis [55].

Considering that leptin was able to activate cellular oxidative mechanisms and that adiponectin increased the calcium influx, we suggest that treatment with adiponectin plus leptin induces a more efficient phagocytic response with an association mechanism that is dependent on ROS and intracellular calcium, which culminated in increased rates of apoptosis. Possibly, this association potentiated the microbicidal activity of these phagocytes, and thus led to the consequent increase in the levels of apoptosis in the high BMI group.

According to the scientific literature, adipokines can enhance the cellular response, possibly through activation of the classical NF-kB pathway, which regulates the genes responsible for the production of most ROS in the cells [56].

From the literature, it is already known that the ratio between leptin and adiponectin is an important parameter that can be used to indicate the risk of developing obesity and even metabolic syndrome [19,20].

However, in this study, the data suggest that the actions of adiponectin and leptin on the functional activity of colostrum cells occur by different mechanisms. Adiponectin is probably associated with microbicidal activity, which is essential for the immunological response. Leptin induces high levels of ROS, which, although important for microbicidal mechanisms, may be responsible for the inflammatory processes generated by the activation of macrophages. Thus, their association is a considerable defense strategy, because in addition to inducing apoptosis, it reduced oxidative stress to levels similar to those of the non-obese group, and consequently may have controlled inflammatory processes. The maintenance of the balance of adiponectin and leptin concentrations is essential for the adequate function of mononuclear phagocytes and is extremely important for the immunologic and metabolic programming that breastfeeding can confer to the child.

This study used a cross-sectional design that brings limitations to the research. It would be interesting to develop a prospective cohort study to evaluate the effect of adipokines on human colostrum phagocytes and on the health outcomes of infants during the childhood. Such research could add important contributions that may allow the development of new strategies against the obesity epidemic, for example, through the encouragement of maternal weight control or initiatives that use colostrum and breast milk as an intervention strategy. In this sense, breastfeeding could be used at strategic hours which take into account the serum fluctuations of adipokines. Thus, we could create new alternatives with potential impacts against possible inflammatory processes caused by obesity as well as reduce the prevalence of childhood infections.

5. Conclusions

Pre-gestational maternal overweightness and obesity alter the functional activity of human colostrum phagocytes. In this study, the action of adiponectin on colostrum phagocytes did not alter the level of reactive oxygen species and increased the release of intracellular calcium in obese women. In contrast, leptin increased the level of reactive oxygen species and reduced the intracellular calcium concentration. The association of adiponectin with leptin was shown to be essential to restore the level of reactive oxygen species and maintain the intracellular calcium level. It also contributed to better microbicidal activity, thus reflecting an increase in the apoptosis level.

These data reinforce the importance of maternal weight control from the pre-gestational period to ensure a balance among the endogenous concentrations of adipokines and to provide to the infant, through breastfeeding, colostrum mononuclear cells that are capable of eliciting a more effective response against childhood infection. Moreover, these data support the importance of breastfeeding in the context of obesity to control the inflammatory process and reduce childhood infections.

Author Contributions: Conceptualization, A.C.H.-F.; Data curation, T.C.M., O.B.d.Q., R.S.P. and M.F.; Formal analysis, T.C.M., B.E.G.D. and E.L.F.; Funding acquisition, L.C.d.A. and A.C.H.-F.; Investigation, R.S.P. and M.F.; Methodology, O.B.d.Q., M.F. and B.E.G.D.; Project administration, E.L.F.; Supervision, L.C.d.A., E.L.F. and A.C.H.-F.; Writing—original draft, T.C.M., E.L.F. and A.C.H.-F.; Writing—review & editing, L.C.d.A. and B.E.G.D.

Funding: Funding for this research was provided by Fundação de Amparo a Pesquisa de São Paulo (FAPESP N°: 2015/19922-0; N°: 2015/01051-3) and Conselho Nacional de DesenvolvimentoCientífico e Tecnológico (CNPq-N°: 303983/2016-7; N°: 403383/2016-1).

Conflicts of Interest: The authors declare no conflict of interest.

References

1. Manna, P.; Jain, S.K. Obesity, oxidative stress, adipose tissue dysfunction, and the associated health risks: causes and therapeutic strategies. *Metab. Syndr. Relat. Disord.* **2015**, *13*, 423–444. [CrossRef]

2. Bray, G.A.; Kim, K.K.; Wilding, J.P.H. World Obesity Federation. Obesity: a chronic relapsing progressive disease process. A position statement of the World Obesity Federation. *Obes. Rev.* **2017**, *18*, 715–723. [CrossRef]

3. Rogero, M.M.; Calder, P.C. Obesity, inflammation, toll-like receptor 4 and fatty acids. *Nutrients* **2018**, *10*, 432. [CrossRef]

4. WHO. Fact Sheet: Obesity and Overweight. Available online: http://www.who.int/news-room/fact-sheets/detail/obesity-and-overweight (accessed on 27 November 2018).

5. Ng, M.; Fleming, T.; Robinson, M.; Thomson, B.; Graetz, N.; Margono, C.; Mullany, E.C.; Biryukov, S.; Abbafati, C.; Abera, S.F.; et al. Global, regional, and national prevalence of overweight and obesity in children and adults during 1980-2013: a systematic analysis for the Global Burden of Disease Study 2013. *Lancet* **2014**, *384*, 766–781. [CrossRef]

6. Mitchell, S.; Shaw, D. The worldwide epidemic of female obesity. *Best. Pract. Res. Clin. Obstet. Gynaecol.* **2015**, *29*, 289–299. [CrossRef]

7. Mayer-Davis, E.J.; Rifas-Shiman, S.L.; Zhou, L.; Hu, F.B.; Colditz, G.A.; Gillman, M.W. Breast-feeding and risk for childhood obesity: does maternal diabetes or obesity status matter? *Diabetes Care* **2006**, *29*, 2231–2237. [CrossRef]

8. Dieterich, C.M.; Felice, J.P.; O'Sullivan, E.; Rasmussen, K.M. Breastfeeding and health outcomes for the mother-infant dyad. *Pediatr. Clin. N. Am.* **2013**, *60*, 31–48. [CrossRef]

9. Sharma, A.J.; Dee, D.L.; Harden, S.M. Adherence to breastfeeding guidelines and maternal weight 6 years after delivery. *Pediatrics* **2014**, *134* (Suppl. 1), S42–S49. [CrossRef]

10. Binns, C.; Lee, M.; Low, W.Y. The Long-Term Public Health Benefits of Breastfeeding. *Asia Pac. J. Public Health* **2016**, *28*, 7–14. [CrossRef]

11. Badillo-Suárez, P.A.; Rodríguez-Cruz, M.; Nieves-Morales, X. Impact of metabolic hormones secreted in human breast milk on nutritional programming in childhood obesity. *J. Mammary Gland. Biol. Neoplasia* **2017**, *22*, 171–191. [CrossRef]

12. Horta, B.L.; Victora, C.G.; França, G.V.A.; Hartwig, F.P.; Ong, K.K.; Rolfe, E.L.; Magalhães, E.I.S.; Lima, N.P.; Barros, F.C. Breastfeeding moderates FTO related adiposity: a birth cohort study with 30 years of follow-up. *Sci. Rep.* **2018**, *8*, 2530. [CrossRef]

13. Bronsky, J.; Mitrova, K.; Karpisek, M.; Mazoch, J.; Durilova, M.; Fisarkova, B.; Stechova, K.; Prusa, R.; Nevoral, J. Adiponectin, AFABP, and leptin in human breast milk during 12 months of lactation. *J. Pediatr. Gastroenterol. Nutr.* **2011**, *52*, 474–477. [CrossRef]

14. Andreas, N.J.; Hyde, M.J.; Gale, C.; Parkinson, J.R.; Jeffries, S.; Holmes, E.; Modi, N. Effect of maternal body mass index on hormones in breast milk: a systematic review. *PLoS ONE* **2014**, *9*, e115043. [CrossRef]

15. Fujimori, M.; França, E.L.; Morais, T.C.; Fiorin, V.; Abreu, L.C.; Honorio-França, A.C. Cytokine and adipokine are biofactors can act in blood and colostrum of obese mothers. *Biofactors* **2017**, *43*, 243–250. [CrossRef]

16. Gridneva, Z.; Kugananthan, S.; Rea, A.; Lai, C.T.; Ward, L.C.; Murray, K.; Hartmann, P.E.; Geddes, D.T. Human milk adiponectin and leptin and infant body composition over the first 12 months of lactation. *Nutrients* **2018**, *10*, 1125. [CrossRef]

17. Kratzsch, J.; Bae, Y.J.; Kiess, W. Adipokines in human breast milk. *Best Pract. Res. Clin. Endocrinol. Metab.* **2018**, *32*, 27–38. [CrossRef]

18. Sun, J.; Gao, Y.; Yao, T.; Huang, Y.; He, Z.; Kong, X.; Yu, K.; Wang, R.; Guo, H.; Yan, J.; et al. Adiponectin potentiates the acute effects of leptin in arcuate Pomc neurons. *Mol. Metab.* **2016**, *5*, 882–891. [CrossRef]

19. Konsoulova, P.S.; Nyagolova, P.V.; Orbetzova, M.M.; Simitchiev, K.K.; Terzieva, D.D.; Kaleva, N.N. Leptin, adiponectin, and leptin/adiponectin ratio in adolescents with metabolic syndrome. *Int. J. Pharm.* **2016**, *4*, 1–6.

20. van Zyl, S.; van der Merwe, L.J.; van Rooyen, F.C.; Joubert, G.; Walsh, C.M. The relationship between obesity, leptin, adiponectin and the components of metabolic syndrome in urban african women, Free State. *S. Afr. J. Clin. Nutr.* **2017**, *30*, 68–73. [CrossRef]

21. Sánchez-Pozo, C.; Rodriguez-Baño, J.; Domínguez-Castellano, A.; Muniain, M.A.; Goberna, R.; Sánchez-Margalet, V. Leptin stimulates the oxidative burst in control monocytes but attenuates the oxidative burst in monocytes from HIV-infected patients. *Clin. Exp. Immunol.* **2003**, *134*, 464–469. [CrossRef]

22. Shirshev, S.V.; Orlova, E.G. Molecular mechanisms of regulation of functional activity of mononuclear phagocytes by leptin. *Biochemistry (Mosc.)* **2005**, *70*, 841–847. [CrossRef]

23. Ohashi, K.; Parker, J.L.; Ouchi, N.; Higuchi, A.; Vita, J.A.; Gokce, N.; Pedersen, A.A.; Kalthoff, C.; Tullin, S.; Sams, A.; et al. Adiponectin promotes macrophage polarization toward an anti-inflammatory phenotype. *J. Biol. Chem.* **2010**, *285*, 6153–6160. [CrossRef]

24. Luo, Y.; Liu, M. Adiponectin: a versatile player of innate immunity. *J. Mol. Cell Biol.* **2016**, *8*, 120–128. [CrossRef]

25. Francisco, V.; Pino, J.; Campos-Cabaleiro, V.; Ruiz-Fernández, C.; Mera, A.; Gonzalez-Gay, M.A.; Gómez, R.; Gualillo, O. Obesity, fat mass and immune system: role for leptin. *Front. Physiol.* **2018**, *9*, 640. [CrossRef]

26. Fujimori, M.; França, E.L.; Fiorin, V.; Morais, T.C.; Honorio-França, A.C.; Abreu, L.C. Changes in the biochemical and immunological components of serum and colostrum of overweight and obese mothers. *BMC Pregnancy Childbirth* **2015**, *15*, 166. [CrossRef]

27. Andreas, N.J.; Hyde, M.J.; Herbert, B.R.; Jeffries, S.; Santhakumaran, S.; Mandalia, S.; Holmes, E.; Modi, N. Impact of maternal BMI and sampling strategy on the concentration of leptin, insulin, ghrelin and resistin in breast milk across a single feed: a longitudinal cohort study. *BMJ Open* **2016**, *6*, e010778. [CrossRef]

28. De Luca, A.; Frasquet-Darrieux, M.; Gaud, M.A.; Christin, P.; Boquien, C.Y.; Millet, C.; Herviou, M.; Darmaun, D.; Robins, R.J.; Ingrand, P.; et al. Higher leptin but not human milk macronutrient concentration distinguishes normal-weight from obese mothers at 1-month postpartum. *PLoS ONE* **2016**, *11*, e0168568. [CrossRef]

29. Fields, D.A.; George, B.; Williams, M.; Whitaker, K.; Allison, D.B.; Teague, A.; Demerath, E.W. Associations between human breast milk hormones and adipocytokines and infant growth and body composition in the first 6 months of life. *Pediatr. Obes.* **2017**, *12*, 78–85. [CrossRef]

30. França, E.L.; Bitencourt, R.V.; Fujimori, M.; Morais, T.C.; Calderon, I.M.; Honorio-França, A.C. Human colostral phagocytes eliminate enterotoxigenic Escherichia coli opsonized by colostrum supernatant. *J. Microbiol. Immunol. Infect.* **2011**, *44*, 1–7. [CrossRef]

31. Morceli, G.; Honorio-França, A.C.; Fagundes, D.L.; Calderon, I.M.; França, E.L. Antioxidant effect of melatonin on the functional activity of colostral phagocytes in diabetic women. *PLoS ONE* **2013**, *8*, e56915. [CrossRef]

32. Fagundes, D.L.G.; França, E.L.; Gonzatti, M.B.; Rugde, M.V.C.; Calderon, I.M.P.; Honorio-França, A.C. The modulatory role of cytokines IL-4 and IL-17 in the functional activity of phagocytes in diabetic pregnant women. *APMIS* **2018**, *126*, 56–64. [CrossRef]

33. Radogna, F.; Paternoster, L.; De Nicola, M.; Cerella, C.; Ammendola, S.; Bedini, A.; Tarzia, G.; Aquilano, K.; Ciriolo, M.; Ghibelli, L. Rapid and transient stimulation of intracellular reactive oxygen species by melatonin in normal and tumor leukocytes. *Toxicol. Appl. Pharmacol.* **2009**, *239*, 37–45. [CrossRef]

34. Cinti, S. Adipose organ development and remodeling. *Compr. Physiol.* **2018**, *8*, 1357–1431. [CrossRef] [PubMed]

35. Borges, M.D.; Franca, E.L.; Fujimori, M.; Silva, S.M.C.; de Marchi, P.G.F.; Deluque, A.L.; Honorio-Franca, A.C.; de Abreu, L.C. Relationship between proinflammatory cytokines/chemokines and adipokines in serum of young adults with obesity. *Endocr. Metab. Immune Disord. Drug Targets* **2018**, *18*, 260–267. [CrossRef]

36. Marseglia, L.; Manti, S.; D'Angelo, G.; Nicotera, A.; Parisi, E.; Di Rosa, G.; Gitto, E.; Arrigo, T. Oxidative stress in obesity: a critical component in human diseases. *Int. J. Mol. Sci.* **2015**, *16*, 378–400. [CrossRef]

37. Park, P.H. Autophagy induction: a critical event for the modulation of cell death/survival and inflammatory responses by adipokines. *Arch. Pharm. Res.* **2018**, *41*, 1062–1073. [CrossRef]

38. Challier, J.C.; Basu, S.; Bintein, T.; Minium, J.; Hotmire, K.; Catalano, P.M.; Hauguel-de-Mouzon, S. Obesity in pregnancy stimulates macrophage accumulation and inflammation in the placenta. *Placenta* **2008**, *29*, 274–281. [CrossRef]

39. Djordjević, V.B. Free radicals in cell biology. *Int. Rev. Cytol.* **2004**, *237*, 57–89. [CrossRef]

40. Loffreda, S.; Yang, S.Q.; Lin, H.Z.; Karp, C.L.; Brengman, M.L.; Wang, D.J.; Klein, A.S.; Bulkley, G.B.; Bao, C.; Noble, P.W.; et al. Leptin regulates proinflammatory immune responses. *FASEB J.* **1998**, *12*, 57–65. [CrossRef]

41. Hunsche, C.; Hernandez, O.; De la Fuente, M. Impaired immune response in old mice suffering from obesity and premature immunosenescence in adulthood. *J. Gerontol. A Biol. Sci. Med. Sci.* **2016**, *71*, 983–991. [CrossRef]

42. Nieman, D.C.; Henson, D.A.; Nehlsen-Cannarella, S.L.; Ekkens, M.; Utter, A.C.; Butterworth, D.E.; Fagoaga, O.R. Influence of obesity on immune function. *J. Am. Diet. Assoc.* **1999**, *99*, 294–299. [CrossRef]

43. Morshedi, A.; Zarkesh-Esfahani, S.H.; Behjati, M. Effect of leptin on neutrophils phagocytosis and lymphocytes apoptosis challenge by listeria monocytogenes and Escherichia coli. *Jundishapur J. Microbiol.* **2013**, *6*, 1–8. [CrossRef]

44. Ahima, R.S.; Flier, J.S. Leptin. *Annu. Rev. Physiol.* **2000**, *62*, 413–437. [CrossRef]

45. Procaccini, C.; Jirillo, E.; Matarese, G. Leptin as an immunomodulator. *Mol. Asp. Med.* **2012**, *33*, 35–45. [CrossRef]

46. van Dielen, F.M.; Van't Veer, C.; Schols, A.M.; Soeters, P.B.; Buurman, W.A.; Greve, J.W. Increased leptin concentrations correlate with increased concentrations of inflammatory markers in morbidly obese individuals. *Int. J. Obes. Relat. Metab. Disord.* **2001**, *25*, 1759–1766. [CrossRef]

47. Considine, R.V.; Sinha, M.K.; Heiman, M.L.; Kriauciunas, A.; Stephens, T.W.; Nyce, M.R.; Ohannesian, J.P.; Marco, C.C.; McKee, L.J.; Bauer, T.L.; et al. Serum immunoreactive-leptin concentrations in normal-weight and obese humans. *N. Engl. J. Med.* **1996**, *334*, 292–295. [CrossRef] [PubMed]

48. Morais, J.B.; Severo, J.S.; Santos, L.R.; de Sousa Melo, S.R.; de Oliveira Santos, R.; de Oliveira, A.R.; Cruz, K.J.; do Nascimento Marreiro, D. Role of magnesium in oxidative stress in individuals with obesity. *Biol. Trace Elem. Res.* **2017**, *176*, 20–26. [CrossRef]

49. Carrichon, L.; Picciocchi, A.; Debeurme, F.; Defendi, F.; Beaumel, S.; Jesaitis, A.J.; Dagher, M.C.; Stasia, M.J. Characterization of superoxide overproduction by the D-Loop(Nox4)-Nox2 cytochrome b(558) in phagocytes-Differential sensitivity to calcium and phosphorylation events. *Biochim. Biophys. Acta* **2011**, *1808*, 78–90. [CrossRef] [PubMed]

50. Cheng, X.; Folco, E.J.; Shimizu, K.; Libby, P. Adiponectin induces pro-inflammatory programs in human macrophages and CD4+ T cells. *J. Biol. Chem.* **2012**, *287*, 36896–36904. [CrossRef]

51. Lee, K.H.; Jeong, J.; Woo, J.; Lee, C.H.; Yoo, C.G. Globular adiponectin exerts a pro-inflammatory effect via IκB/NF-κB pathway activation and anti-inflammatory effect by IRAK-1 downregulation. *Mol. Cells* **2018**, *41*, 762–770. [CrossRef]

52. Sanz-Garcia, C.; Nagy, L.E.; Lasunción, M.A.; Fernandez, M.; Alemany, S. Cot/tpl2 participates in the activation of macrophages by adiponectin. *J. Leukoc. Biol.* **2014**, *95*, 917–930. [CrossRef]

53. Nunes, P.; Demaurex, N. The role of calcium signaling in phagocytosis. *J. Leukoc. Biol.* **2010**, *88*, 57–68. [CrossRef]

54. Ferrari, D.; Pinton, P.; Szabadkai, G.; Chami, M.; Campanella, M.; Pozzan, T.; Rizzuto, R. Endoplasmic reticulum, Bcl-2 and Ca(2+) handling in apoptosis. *Cell Calcium* **2002**, *32*, 413–420. [CrossRef]

55. Honorio-França, A.C.; Nunes, G.T.; Fagundes, D.L.; de Marchi, P.G.; Fernandes, R.T.; França, J.L.; França-Botelho, A.C.; Moraes, L.C.; Varotti, F.P.; França, E.L. Intracellular calcium is a target of modulation of apoptosis in MCF-7 cells in the presence of IgA adsorbed to polyethylene glycol. *Onco Targets Ther.* **2016**, *9*, 617–626. [CrossRef]
56. Morgan, M.J.; Liu, Z.G. Crosstalk of reactive oxygen species and NF-κB signaling. *Cell Res.* **2011**, *21*, 103–115. [CrossRef]

Review

Associations between Fatty Acid-Binding Protein 4–A Proinflammatory Adipokine and Insulin Resistance, Gestational and Type 2 Diabetes Mellitus

Marcin Trojnar [1], Jolanta Patro-Małysza [2], Żaneta Kimber-Trojnar [2,*], Bożena Leszczyńska-Gorzelak [2] and Jerzy Mosiewicz [1]

[1] Chair and Department of Internal Medicine, Medical University of Lublin, 20-081 Lublin, Poland; marcin.trojnar@umlub.pl (M.T.); jerzy.mosiewicz@umlub.pl (J.M.)
[2] Chair and Department of Obstetrics and Perinatology, Medical University of Lublin, 20-090 Lublin, Poland; jolapatro@wp.pl (J.P.-M.); b.leszczynska@umlub.pl (B.L.-G.)
* Correspondence: zkimber@poczta.onet.pl; Tel.: +48-81-7244-769 (Ż.K.-T.)

Received: 4 February 2019; Accepted: 3 March 2019; Published: 8 March 2019

Abstract: There is ample scientific evidence to suggest a link between the fatty acid-binding protein 4 (FABP4) and insulin resistance, gestational (GDM), and type 2 (T2DM) diabetes mellitus. This novel proinflammatory adipokine is engaged in the regulation of lipid metabolism at the cellular level. The molecule takes part in lipid oxidation, the regulation of transcription as well as the synthesis of membranes. An involvement of FABP4 in the pathogenesis of obesity and insulin resistance seems to be mediated via FABP4-dependent peroxisome proliferator-activated receptor γ (PPARγ) inhibition. A considerable number of studies have shown that plasma concentrations of FABP4 is increased in obesity and T2DM, and that circulating FABP4 levels are correlated with certain clinical parameters, such as body mass index, insulin resistance, and dyslipidemia. Since plasma-circulating FABP4 has the potential to modulate the function of several types of cells, it appears to be of extreme interest to try to develop potential therapeutic strategies targeting the pathogenesis of metabolic diseases in this respect. In this manuscript, representing a detailed review of the literature on FABP4 and the abovementioned metabolic disorders, various mechanisms of the interaction of FABP4 with insulin signaling pathways are thoroughly discussed. Clinical aspects of insulin resistance in diabetic patients, including women diagnosed with GDM, are analyzed as well.

Keywords: adipose tissue; fatty acid-binding protein 4; proinflammatory adipokine; insulin resistance; gestational diabetes mellitus; type 2 diabetes mellitus

1. Introduction

Type 2 diabetes mellitus (T2DM) represents a common metabolic disorder that is characterized by chronic hyperglycemia. For more than half a century, the link between insulin resistance and T2DM has been well recognized. Insulin resistance is not only the most powerful predictor of future development of T2DM, but it is also a therapeutic target. On the other hand, gestational diabetes mellitus (GDM) is one of the most common metabolic disorders of pregnancy and its incidence has considerably increased by 10–100% in the last 20 years [1]. It should be emphasized that women with a previous history of GDM have a significantly increased risk of developing T2DM, obesity, and cardiovascular diseases in the future [2–5]. Women who had prior GDM are nearly eight times more likely to develop future T2DM compared with those with normal glucose tolerance during their pregnancy [6]. Up to one-third of women with T2DM have been previously diagnosed with GDM [7,8]. The identification of women with GDM who are at high risk of developing subsequent diseases offers a remarkable opportunity to alter their future health [1,9].

There is ample evidence to suggest a link between fatty acid-binding protein 4 (FABP4) and insulin resistance, GDM, and T2DM.

2. Fatty Acid-Binding Protein 4

FABP4, also referred to in the literature as adipocyte fatty acid-binding protein (AFABP), is a relatively novel adipokine [10], which belongs to the calycin protein superfamily. This protein has also been termed adipocyte P2 (aP2) since there is high sequence similarity (67%) with the myelin P2 protein (M-FABP/FABP8) [11]. FABP4 is highly expressed in adipocytes and represents approximately 1% of all soluble proteins in adipose tissue [11]. FABP4 is able to reversibly bind to hydrophobic ligands, such as saturated and unsaturated long-chain fatty acids, eicosanoids, and other lipids. Accordingly, it takes part in the regulation of lipid trafficking and responses at the cellular level [12–15]. FABPs, a family of intracellular lipid chaperones, are engaged in the transport of fatty acids to specific organelles in the cell, including mitochondria, peroxisomes, the nucleus, and the endoplasmic reticulum [12,16]. Therefore, FABPs play a significant role in lipid oxidation, lipid-mediated transcriptional regulation, and the signaling, trafficking, and synthesis of membranes. In addition, FABPs are also engaged in the regulation of the enzymatic activity and storage of lipid droplets in the cytoplasm [17], the conversion of fatty acids to eicosanoids, and the stabilization of leukotrienes [18].

The human FABP4 consists of 132 amino acids. Its molecular mass has been assessed at 14.6 kDa. FABP4 expression markedly increases at the time of adipocyte differentiation [12]. Due to the abovementioned observation, this molecule has been suggested as an adipocyte differentiation marker [19]. FABP4 expression is also enhanced during differentiation from monocytes to macrophages. A wide spectrum of different proinflammatory factors modify and control the expression of FABP4 in these cells [20]. In macrophages, FABP4 stimulates the foam cell formation. Foam cell formation, which is believed to be mediated by modified low density lipoproteins (LDLs), often occurs in the presence of increased concentrations of insulin and glucose. These increased concentrations are characteristic of the insulin resistance associated with diabetes, obesity, and the metabolic syndrome [21]. Siersbaek et al. [22] found that peroxisome proliferator-activated receptor γ (PPARγ) and CCAAT/enhancer-binding protein (C/EBP) regulate the expression of most genes associated with adipogenesis. PPARγ not only promotes the proliferation and differentiation of adipocytes, but also confers insulin sensitivity to adipocytes [23]. An increase in insulin sensitivity can in turn promote the expression of the PPARγ gene in the adipose tissue, thereby positively accelerating the differentiation of adipocytes. FABP4 expression is controlled at the transcriptional level by PPARγ and C/EBP [12]. However, the aforementioned transcriptional regulators seem to be dysregulated in T2DM [24,25].

Other actions encompass modifications of the inflammatory response mediated by activation of the IKK-NF-κB and c-Jun N-terminal kinase (JNK)-activator protein-1 (AP-1) pathways [20]. In addition, FABP4 enhances the hydrolytic activity of hormone-sensitive lipase (HSL) [26].

FABP4 as an adipokine may influence insulin sensitivity. On the other hand, the expression of FABP4 is highly induced during adipocyte differentiation and transcriptionally controlled by PPARγ agonists, fatty acids, dexamethasone, and insulin [27]. Insulin downregulates only microvesicle-free-mediated and microvesicle-secreted FABP4. However, the release of FABP4 via adipocyte-derived microvesicles is a small fraction and conveys a minor activity [27].

Furthermore, FABP4$^{-/-}$ animals exhibit a defect in β-adrenergic stimulated insulin secretion, even under lean conditions, suggesting an effect on β cell function, another cell type that does not express FABP4. This is complemented by human studies showing that higher serum FABP4 levels correlate with a higher insulin response index in T2DM patients, and a higher insulinogenic index in non-diabetics [28]. In vivo, FABP4 levels are suppressed under refeeding or insulin [28].

Various biological effects of exogenous FABP4 have been demonstrated in different types of cells; FABP4 has been reported to enhance hepatic glucose production in vivo and in vitro and to increase the proliferation of vascular smooth muscle cells and glucose-stimulated insulin secretion in pancreatic β cells [29]. Furthermore, glucose oxidation and glycolysis are inhibited and glucose uptake as well as its utilization in the muscles and liver are considerably limited [30–32]. Moreover, FABP4 inhibits the activation of the insulin-signaling pathway, resulting in decreased activation of the endothelial nitric oxide synthase (eNOS) in vascular endothelial cells and nitric oxide production, inducing endothelial dysfunction [33].

3. Relationship between FABP4 and PPARγ

An involvement of FABP4 in the pathogenesis of obesity and insulin resistance might be mediated via FABP4-dependent PPARγ inhibition [34]. FABP4 triggers the ubiquitination and subsequent proteasomal degradation of PPARγ, a crucial regulator of adipogenesis and insulin responsiveness [35]. Importantly, FABP4-null mouse preadipocytes as well as macrophages exhibited increased expression of PPARγ, and complementation of FABP4 in the macrophages reversed the increase in PPARγ expression. The FABP4-null preadipocytes exhibited a remarkably enhanced adipogenesis compared with wild-type cells, indicating that FABP4 regulates adipogenesis by downregulating PPARγ [35].

The FABP4 level was higher and PPARγ level was lower in human visceral fat and mouse epididymal fat compared with their subcutaneous fat. Furthermore, FABP4 was higher in the adipose tissues of obese diabetic individuals compared with healthy ones. Suppression of PPARγ by FABP4 in visceral fat may explain the reported role of FABP4 in the development of obesity-related morbidities, including insulin resistance, diabetes, and atherosclerosis [35]. The most recent studies also suggest that the FABP4 inhibitor-BMS309403 significantly improves insulin sensitivity in the ob/ob mice, and the FABP4 inhibitor reduces plasma triacylglycerol levels [10].

PPARγ is activated by natural or synthetic agonists, such as the antidiabetic, thiazolidinedione (TZD) [36]. The disruption of PPARγ specifically in myeloid cells also predisposes mice to the development of diet-induced obesity, insulin resistance, and glucose intolerance, whereas activation of PPARγ within macrophages promotes lipid efflux, thereby stabilizing atherosclerotic lesions [37,38].

4. Relationship between FABP4 and Diseases of Civilization

It has been reported that increased circulating FABP4 levels are associated with obesity, insulin resistance, T2DM, cardiovascular disorders, arterial hypertension, cardiac dysfunction, kidney damage, fatty liver disease, and atherosclerosis [10,29,31,32,39–42]. Increased FABP4 concentrations were found in obese subjects and were positively correlated with waist circumference, blood pressure, and insulin resistance [40]. A 10-year prospective study also documented that high FABP4 levels independently predicted the development of T2DM [31]. An increased concentration of FABP4 was an independent biomarker of the development of metabolic syndrome in a five-year perspective in a Chinese population [41]. Also, Stejskal et al. [43] found that serum FABP4 levels might be a significant predictor of metabolic syndrome in a Caucasian population. Cabré et al. [44] reported that higher FABP4 plasma concentrations were associated with the early presence of metabolic syndrome components, along with inflammation and oxidation markers in T2D subjects.

Circulating FABP4 is not only a potent biomarker, but, as an adipokine, it also plays an important role in the development of metabolic syndrome and cardiovascular diseases [29]. Furthermore, FABP4 could be a treatment target in T2DM [45]. A small molecular ligand for FABP4 that blocks the binding of endogenous ligands may be developed into a drug for the treatment of T2DM [46].

5. Associations of FABP4 with Adipogenesis and Inflammation

FABP4 plays a crucial role in the regulation of lipid-mediated actions, such as the initialization of inflammation and oxidative stress processes [33]. The expression of this adipogenic protein can be induced by vascular endothelial growth factor (VEGF) signaling [47]. FABP4 increases the ciatriphosphate-binding cassette A1 pathway [32]. Emerging data suggests an essential involvement of FABP4 in endothelial dysfunction [47]. Exogenous FABP4 interferes with insulin stimulated production of nitric oxide in endothelial cells [48]. Cabré et al. [44] observed that increased FABP4 concentrations were associated with excessive oxidative stress and inflammatory markers in diabetes.

FABP4 negatively regulates PPARγ in macrophages and adipocytes, affecting adipocyte differentiation. Higher levels of FABP4 and lower levels of PPARγ in visceral adipose tissue, when compared with subcutaneous adipose tissue, suggest a causative link between FABP4 and the metabolic syndrome. It may also explain certain morphological and functional differences between the adipocytes found in these two types of fat tissue. Visceral preadipocytes are known to proliferate and differentiate into mature adipocytes less actively than those from subcutaneous fat. Visceral adipose tissue secretes more proinflammatory cytokines [35].

It has also been postulated that the transcription factor forkhead box protein O1 (FOXO1) is involved in lipid metabolism. The available scientific evidence suggests that metformin may have a protective effect against lipid accumulation in macrophages as it decreases FABP4 expression at the mRNA level. The exact mechanism is not yet fully understood; it may deal with decreases in transcription or by promotion of mRNA degradation. For this reason, some authors suggest that metformin should also be perceived as a therapeutic agent for the prevention and targeting of atherosclerosis in metabolic syndrome [49].

FABP4 contributes to the accumulation of short-chain free fatty acids and suppresses the activity of relevant proteins in the phosphatidylinositol 3′-kinase (PI3K)-AKT signal pathway. Accordingly, FABP4 inhibits the presence of glucose oxidation and glycolysis and decreases the uptake and utilization of glucose in human organs, such as in muscles and the liver [50]. It is known that FABP4 can bind various intracellular fatty acids and probably mediates intracellular lipid trafficking between cellular compartments [51]. It might also modulate the availability and composition of fatty acids in muscles and adipose tissues [52]. In the myocytes and adipose tissue in mice, improved glucose homeostasis after the ablation of FABP4 was documented [53].

Increased FABP4 production could contribute to macrophage activity, possibly by activation of inflammatory pathways, resulting in inflammation [42]. An alternative hypothesis, more in line with the published evidence, is that the elevated circulating FABP4 reflects the increased cellular production in both adipocytes and macrophages in response to greater lipid availability, with increased Kupffer cell production of FABP4 triggering an increased inflammatory response [42,54].

FABP4 plays a crucial role in mediating the endoplasmic reticulum stress observed in macrophages upon lipotoxic signal exposure, which contributes to atherosclerosis, inflammation, and perhaps plaque vulnerability [55]. High levels of FABP4 may contribute to adverse prognosis via the induction of endoplasmic reticulum stress in macrophages and upregulation of pro-inflammatory cytokine production. Indeed, FABP4 modulates inflammatory responses in macrophages through a positive feedback loop involving c-Jun NH2-terminal kinases and activator protein-1 [42,56]. FABP4 is also implicated in modulating the eicosanoid balance by affecting both cyclooxygenase 2 (COX2) activity and leukotriene A4 (LTA4) stability, and upregulates uncoupling protein 2 (UCP2); all these processes influence macrophage function and adipose tissue inflammation. FABP4 can also interact with HSL and Janus kinase 2 (JAK2) [57] (Figure 1). Two different mechanisms of the latter action have been proposed. The first one includes direct protein–protein interactions between FABP4 and HSL and JAK2 [58,59]. FABP4 and HSL may also interact indirectly via two protein kinases A and G pathways [29].

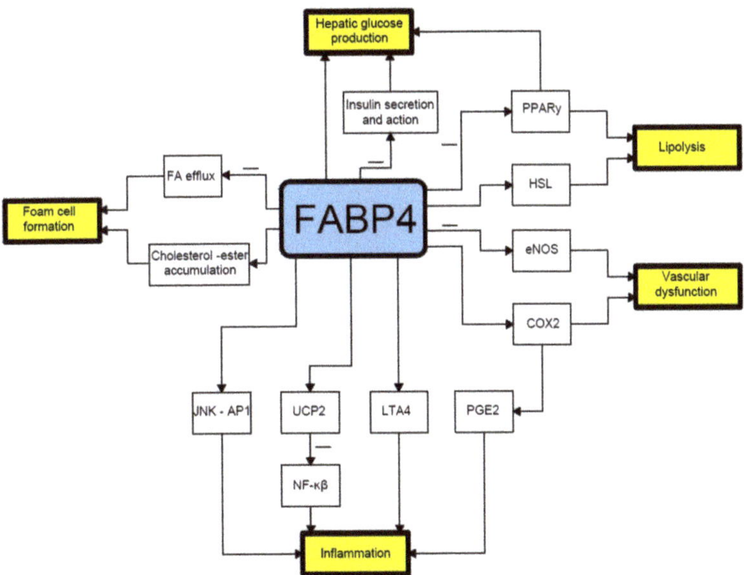

Figure 1. Relationship between fatty acid-binding protein 4 and pathophysiology of type 2 diabetes mellitus. COX2- cyclooxygenase-2; eNOS- endothelial nitric oxide synthase; FA- fatty acid; HSL-hormone sensitive lipase; JNK AP1-Jun N-terminal kinase-activator protein 1; LTA4-leukotriene A4; NF-κB-nuclear factor-kappa B; PGE2-prostaglandin E2; PPARγ–peroxisome proliferator-activated receptor γ; UCP2-uncoupling protein 2.

Several proinflammatory stimuli have been reported to upregulate FABP4 expression in macrophages, including oxidized low-density lipoproteins, toll-like receptor (TLR) agonists, and PPARγ agonists [60–63]. In this respect, lipopolysaccharide (LPS), a TLR4 ligand, stimulates FABP4 expression and the activated FABP4, reciprocally, enhances the LPS-TLR4 signaling-evoked JNK inflammatory pathways [56]. In addition, prolonged hyperglycemia has been shown to induce FABP4 expression in mesangial cells and trigger the release of proinflammatory cytokines [60,64] (Figure 2).

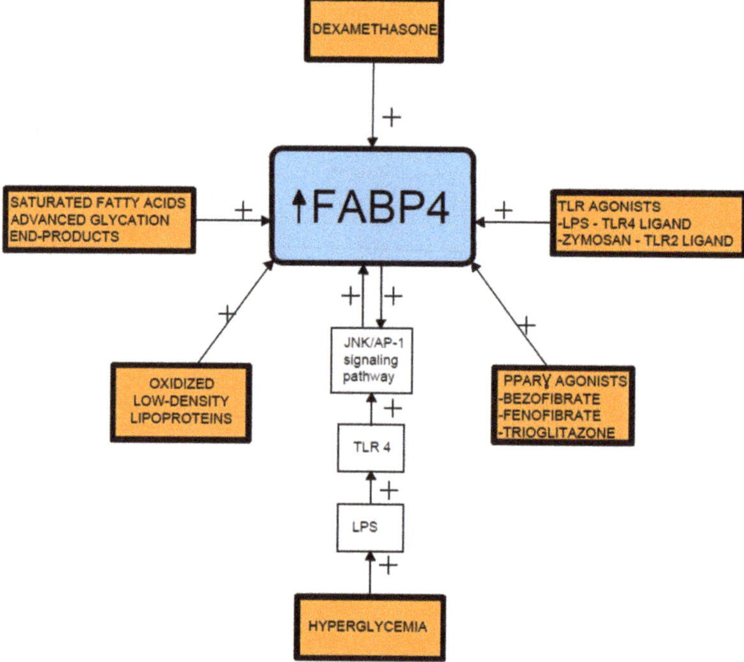

Figure 2. Fatty acid-binding protein 4 inducing factors. TLR- toll-like receptor; LPS- lipopolysaccharide; FABP4-fatty acid-binding protein 4.

6. Relationship between FABP4 and Insulin Resistance

FABP4 is a cytoplasmic fatty acid chaperone clearly engaged in the onset of insulin resistance [35]. Studies in animal models suggest that FABP4 is important for glucose homeostasis [50]. Deletion of the FABP4 gene protected mice against insulin resistance as well as hyperinsulinemia associated with both diet-induced obesity and genetic obesity [45,50,53,54,65,66]. A reduced ability of adipocytes to take up and retain free fatty acids, leading to ectopic lipid accumulation, and abnormalities in the release of adipokines by adipocytes are critical factors for insulin resistance and the development of T2DM [66].

FABP4 was also detected in apoptotic granulosa cells in atretic antral follicles of the mouse ovary, which suggests a potential link to polycystic ovary syndrome (PCOS), which is known to be frequently associated with insulin resistance [11]. Expression of FABP4 mRNA in isolated granulosa cells was found to be higher in patients diagnosed with PCOS than in controls [67].

It has been proven that FABP4 is negatively correlated with the glucose-disposal rate (GDR) [68]. In non-DM subjects, serum FABP concentrations were negatively correlated with the mean rate of glucose infusion during the last 30 min of the clamp test, which reflects insulin sensitivity [69]. Nakamura et al. [35] also showed that circulating FABP4 concentrations were negatively correlated with GDR, which is a marker of insulin resistance in skeletal muscles in individuals with T2DM. On the contrary, FABP4 concentration was positively related to the insulinogenic index in non-diabetic participants. Wu et al. [70] reported that circulating FABP4 concentrations were correlated with glucose-stimulated insulin secretion in healthy controls.

It has been suggested that insulinotropic potential of FABP4 is similar to the effects of glucagon-like peptide-1 (GLP-1) [35]. To maintain glucose homeostasis, FABP4 may stimulate β cells and alter insulin secretion. Moreover, FABP4 presented a positive relationship with insulin secretion at an early stage in the non-diabetic group, which may be due to the fact that insulin secretion is damaged relatively early in T2DM.

Nakamura et al. [35] found the most pronounced negative correlation between FABP4 and GDR when compared to some other markers of insulin resistance or body composition in T2DM. FABP4 has been reported to have a negative correlation with GDR in type 1 diabetes mellitus, T2DM, and the controls of Asian Americans [71]. FABP4 represents an important molecule dealing with insulin resistance in T2DM.

7. Relationship between FABP4 and Gestational Diabetes Mellitus

FABP4 and leptin are known to be involved in the pathophysiology of GDM and its long-term post-partum complications. Placental and non-placental origins of these adipokines are likely to contribute to insulin resistance and β-cell dysfunction [72].

Previous studies found that the serum FABP4 concentrations were significantly increased in women diagnosed with GDM when compared to the control group [65,73,74]. Zhang et al. [65] found that the fasting insulin and age adjusted FABP4 concentrations were significantly higher in the GDM group compared with the normal glucose tolerance participants in the mid and late stages of pregnancy. Furthermore, there was a significant increase in FABP4 from the second to third trimester in women with GDM [65]. Maternal FABP4 concentrations were obviously upregulated in the first trimester in women who later developed GDM [50]. Also, Li et al. [30] observed that FABP4 levels in the GDM group were significantly higher than those of controls, and noted that FABP4 was an independent risk factor for increased insulin resistance during pregnancy. Nevertheless, Ortega-Senovilla et al. [74] did not find any differences in FABP4 levels between the GDM and normal glucose tolerance groups when the FABP4 values were corrected with insulin. Meanwhile, the cited authors revealed significant correlations between maternal blood FABP4 levels and pre-pregnancy BMI values in both control and GDM pregnant patients [74].

Li et al. [32] explained the high circulating FABP4 levels in the maternal serum of pregnant GDM women by its additional release from placenta and adipocytes. Expression of FABP4 mRNA in the placenta and decidua of pregnant women with GDM is greater than that in normal organs [32]. Circulating FABP4 is associated with lipolysis and may aggravate insulin resistance compared to normal physiological insulin resistance during pregnancy [54].

Moreover, candidates for the placental hormones to induce FABP4 overexpression in the placenta and decidua in GDM include human placental lactogen and progesterone [32] as well as the combination of estrogen and progesterone [75]. Their concentrations increase continuously until term and may be associated with an increased insulin resistance along with the progressing gestational age [43,48]. Elevated placental hormones present in serum in GDM may increase the expression of FABP4 mRNA in adipocytes. The synergistic effects of FABP4 from the placenta and adipocytes can affect both metabolic and inflammatory pathways via adipocytes. These actions may play crucial roles in the development of insulin resistance and T2DM in the future lives of post-partum women [32].

On the other hand, FABP4 as expressed in human placental trophoblasts is a key regulator of trophoblastic lipid transport and accumulation during placental development. It has also been reported that maternal FABP4 levels are elevated in preeclampsia, even before the clinical onset of the disease [11]. The level of the second trimester plasma FABP4 in the preeclampsia GDM group was significantly higher than that of the patients with only GDM [76]. Besides, Wotherspoon et al. [77] revealed that increased second-trimester FABP4 levels independently predicted pre-eclampsia in women with type 1 diabetes mellitus.

The serum FABP4 levels were also associated with overweight in GDM patients. Ning et al. [11] concluded that serum FABP4 may be a potential biomarker in GDM diagnosis and is associated with overweight, insulin resistance, and TNF-α in GDM patients.

Our previous study showed that the serum FABP4 levels were significantly higher in the GDM group in the early puerperium in comparison with healthy mothers and women with excessive gestational weight gain [2]. Based on our findings and previous studies, it appears that increased circulating FABP4 concentrations can persist in GDM patients after delivery and might contribute to the increased risk of T2DM and metabolic syndrome. On the other hand, evaluation of FABP4 may be used as a predictive marker for mothers with a history of GDM [2].

In a follow-up study of women six years after GDM, Svensson et al. [78] identified three factors—all related to the adipose tissue and body composition—that, independently of body mass index (BMI) and ethnicity, may increase the risk of progression to T2DM following GDM. These factors included increased serum FABP4 levels, weight gain following index pregnancy, and a lower proportion of fat-free mass. High BMI and abdominal fat distribution were also associated with the development of T2DM after GDM [78].

8. Relationship between FABP4 and T2DM and its Complications

A 10-year prospective study proved that high levels of FABP4 at baseline independently predicted the development of T2DM [33]. Serum FABP4 concentrations have been reported to be associated with inadequate glucose control in T2DM [33]. Increased FABP4 levels are also linked to the early presence of metabolic syndrome components, as well as inflammation and oxidation markers in T2DM subjects [44]. Inflammation and oxidative stress are suggested to play a role in the pathogenesis of complications of T2DM [10,79,80].

A previous study showed that the concentrations of FABP4 were negatively associated with endothelial function in T2DM [81]. Another study identified a correlation between FABP4 and impaired endothelial function in diabetes, which lead to an increased cardiovascular risk [82]. Furthermore, the prognostic value of FABP4 for kidney damage [83] and cardiac contractile dysfunction [84] in patients with T2DM have also been proposed. In addition, previous studies suggest that increased FABP4 synthesis in atherosclerotic plaques is associated with disease severity [85,86]. Holm et al. [87] reported that FABP4 was linked to atherogenesis, plaque instability, and adverse outcomes in patients with carotid atherosclerosis and acute ischemic stroke. Two other studies have shown the association of enhanced FABP4 expression within human carotid atherosclerotic lesions with poor prognosis [85,86]. There is available data to suggest that higher levels of FABP4 are also associated with elevated cardiovascular diseases mortality among men with T2DM [42,88].

9. Relationship between FABP4 and Diabetic Retinopathy

FABP4 shows clinically relevant potential as a novel predictor of diabetic retinopathy (DR). According to some authors, strict glycemic control and more frequent retinal examination should be recommended for the T2DM patients with the detected highest quartile range of FABP4 [10].

The impact of FABP4 on atherosclerosis seems to be attributed to its action in macrophages [89]. Circulating FABP4 induces insulin resistance, which is an independent biomarker of proliferative retinopathy [90]. Lipopolysaccharides (LPS) stimulate FABP4 transcription through JNK, which in turn induces c-Jun recruitment to a highly conserved activator protein-1 recognition site within the proximal region of the FABP4 promoter [56]. LPS-binding protein is involved in the immune response triggered by inflammatory injury characteristic of DR. Moreover, FABP4 is an obligatory mediator coupling toxic lipids (i.e., saturated fatty acids) to endoplasmic reticulum stress in macrophages in vitro and in vivo [55]. Interestingly, endoplasmic reticulum stress represents an initial event in retina pathogenesis in diabetes [10,91].

10. Relationship between FABP4 and Diabetic Nephropathy

In the past decades, several biomarkers have emerged for the detection of early diabetic nephropathy (DN) besides the glomerular filtration rate (GFR) and urine albumin-to-creatinine ratio (UACR). Among them, FABP4 has attracted increased attention [92]. FABP4 was also reported at increased concentrations in nondiabetic as well as T2DM patients with end-stage renal disease [93]. Yeung et al. [94] reported that serum levels of FABP4 had a significantly inverse relationship with the estimated GFR (eGFR) and was independently associated with macrovascular complications and DN staging classified by albuminuria.

Cabré et al. [73] reported that FABP4 was independently associated with eGFR in T2DM patients with eGFR $\geq$ 60 mL/min/1.73 m^2. Ni X et al. [92] found that serum FABP4 along with UACR or a panel of biomarkers might be more sensitive for the detection of early DN. Serum FABP4 had an inverse correlation with GFR and could be an independent predictor for early DN [92].

The mechanisms behind the elevation of FABP4 in patients with diabetic kidney disease are not yet fully understood. It is known that FABP4 is abundantly expressed in adipocytes, macrophages, and endothelial cells [92]. Firstly, it is suggested that, during the early stage of DN, the accumulation of active macrophages is more evident in the kidney because of the increased oxidative stress and chronic inflammation, which consequently induces greater expression of serum FABP4 [44,94]. Secondly, damage to glomeruli and tubulointerstitium might result in both decreased glomerular filtration and increased tubular reabsorption, leading to an increase in FABP4 in the circulation [83]. Okazaki et al. [16] reported that urinary excretion of FABP4 was associated with the progression of proteinuria and renal dysfunction in healthy subjects. The cited authors suggested that the urinary FABP4 reflects damage of the glomerular with the hypothesis proposed by Tanaka et al. [13] that the main source of the urinary FABP4 is derived from ectopic expression of glomerular FABP4 rather than increased adiposity and that locally increased FABP4 in the glomerulus affects renal dysfunction.

Toruner et al. [93] found that serum FABP4 levels were independently and positively associated with the albumin excretion rate in patients with T2DM, suggesting an involvement of the increased serum FABP4 levels in the occurrence and development of microalbuminuria among patients with T2DM. Furthermore, researchers from Hong Kong [94] also documented that among patients with diabetes mellitus, serum FABP4 levels were shown to be independently associated with the severity of nephropathy. Their findings raised the possibility that FABP4 might be used as a serum biomarker for stratifying nephropathy stages in patients with T2DM. The serum FABP4 level can be used as an indicator of microalbuminuria not only in diabetic patients with early stage disease, but also for hyperglycemic individuals before the onset of diabetes [95].

11. Relationship between FABP4 and Non-Alcoholic Fatty Liver Disease

There are multiple reports available that clearly link metabolic disorders, including insulin resistance and T2DM, to non-alcoholic fatty liver disease (NAFLD), which nowadays represents the most frequent liver disease worldwide. Its incidence has been estimated at 20–30% in the populations of Western countries. Approximately 70% of T2DM and obese subjects present some extent of NAFLD. This abnormal lipid accumulation in the liver is considered a significant causative factor of cancerogenesis resulting in hepatocellular carcinoma (HCC) [96].

At present, HCC represents the third leading cause of cancer-related mortality. Interestingly, except for the well-known risk factors of malignancy, such as viral hepatitis and alcohol, a higher incidence of HCC was observed in T2DM patients. In the analysis, obese subjects and patients with metabolic syndrome were also at a higher risk of developing cancer [96]. Additionally, HCC treatment outcomes of all those patients appeared worse with T2DM, representing a prognostic predictor of the increased risk of mortality [96].

In a study by Thompson et al., the FABP4 expression was indeed proven to be significantly increased in animal models of obesity promoted HCC [97]. These findings are consistent with the previous report, which concluded that PPARγ is considered a tumor suppressor gene, whereas FABP4 plays a role in tumorigenesis [35].

12. Conclusions

FABP4, an intracellular lipid chaperone, is highly expressed in adipocytes and macrophages [98]. Abnormalities in the level of FABP4 have been correlated with the development of adiposity, oxidative stress, and atherosclerosis [40]. FABP4 is strongly involved in glucose and lipid metabolism, inflammation, and insulin resistance [76]. In one of the most recent systematic reviews by Bellos et al., targeting the potential association of 10 novel adipokines (i.e., apelin, chemerin, FABP4, fibroblast growth factor-1, monocyte chemoattractant protein-1, nesfatin-1, omentin-1, resistin, vaspin, and visfatin) with GDM, only FABP4 was found to be the most promising predictor of this metabolic pregnancy complication [99].

In fact, a large number of studies have shown that plasma levels of FABP4 are increased in obesity and T2DM, and that circulating FABP4 concentration correlated with clinical outcomes, such as body mass index, insulin resistance, and dyslipidemia [33]. Since the plasma-circulating FABP4 can modulate the function of several types of cells, it becomes extremely interesting to try to develop effective therapeutic strategies against the pathogenesis of metabolic and vascular diseases in this respect [33].

Furthermore, several studies have shown that serum FABP4 levels of GDM patients were higher than those found in women with normal pregnancies [30,65,73,76]. According to the current state of scientific knowledge, females with a previous history of GDM are much more prone to suffering from T2DM, obesity, and metabolic syndrome in the future. There is a huge need to use current research results regarding improved GDM management strategies, including primary prevention for mothers who are at risk of developing subsequent complications. Considering previous findings, it seems that FABP4 may be used as a predictive marker for mothers with a history of GDM. Further studies should focus on the evaluation of FABP4 in the pathogenesis of T2DM following GDM.

Author Contributions: Conceptualization, M.T.; Data curation, M.T., J.P.-M., Ż.K.-T.; Writing-review and editing, M.T., Ż.K.-T.; Visualization, J.P.-M.; Supervision, B.L.-G., J.M.

Funding: This research received no external funding.

Conflicts of Interest: The authors declare no conflict of interest.

References

1. Blumberg, J.; Ballares, V.; Durbin, J.L. Ethnic variations on gestational diabetes mellitus and evidence-based first-line interventions. *J. Matern. Fetal Neonatal Med.* **2018**, *31*, 2641–2647. [CrossRef] [PubMed]

2. Kimber-Trojnar, Ż.; Patro-Małysza, J.; Trojnar, M.; Skórzyńska-Dziduszko, K.E.; Bartosiewicz, J.; Oleszczuk, J.; Leszczyńska-Gorzelak, B. Fatty Acid-Binding Protein 4-An "Inauspicious" Adipokine-In Serum and Urine of Post-Partum Women with Excessive Gestational Weight Gain and Gestational Diabetes Mellitus. *J. Clin. Med.* **2018**, *7*, 505. [CrossRef] [PubMed]

3. Nikolic, D.; Al-Rasadi, K.; Al Busaidi, N.; Al-Waili, K.; Banerjee, Y.; Al-Hashmi, K.; Montalto, G.; Rizvi, A.A.; Rizzo, M.; Al-Dughaishi, T. Incretins, Pregnancy, and Gestational Diabetes. *Curr. Pharm. Biotechnol.* **2016**, *17*, 597–602. [CrossRef] [PubMed]

4. Skórzyńska-Dziduszko, K.E.; Kimber-Trojnar, Ż.; Patro-Małysza, J.; Olszewska, A.; Zaborowski, T.; Małecka-Massalska, T. An Interplay between obesity and inflammation in gestational diabetes mellitus. *Curr. Pharm. Biotechnol.* **2016**, *17*, 603–613. [CrossRef] [PubMed]

5. Marciniak, A.; Patro-Małysza, J.; Kimber-Trojnar, Ż.; Marciniak, B.; Oleszczuk, J.; Leszczyńska-Gorzelak, B. Fetal programming of the metabolic syndrome. *Taiwan. J. Obstet. Gynecol.* **2017**, *56*, 133–138. [CrossRef]

6. Gupta, Y.; Kapoor, D.; Josyula, L.K.; Praveen, D.; Naheed, A.; Desai, A.K.; Pathmeswaran, A.; de Silva, H.A.; Lombard, C.B.; Shamsul Alam, D.; et al. A lifestyle intervention programme for the prevention of Type 2 diabetes mellitus among South Asian women with gestational diabetes mellitus [LIVING study]: protocol for a randomized trial. *Diabet. Med.* **2019**, *36*, 243–251. [CrossRef]

7. Bellamy, L.; Casas, J.; Hingorani, A.; Williams, D. Type 2 diabetes mellitus after gestational diabetes: A systematic review and meta-analysis. *Lancet* **2009**, *373*, 1773–1779. [CrossRef]

8. Huopio, H.; Hakkarainen, H.; Pääkkönen, M.; Kuulasmaa, T.; Voutilainen, R.; Heinonen, S.; Cederberg, H. Long-Term changes in glucose metabolism after gestational diabetes: A double cohort study. *BMC Pregnancy Childbirth* **2014**, *14*, 296. [CrossRef]

9. Chiefari, E.; Arcidiacono, B.; Foti, D.; Brunetti, A. Gestational diabetes mellitus: an updated overview. *J. Endocrinol. Invest.* **2017**, *40*, 899–909. [CrossRef]

10. Zhang, X.Z.; Tu, W.J.; Wang, H.; Zhao, Q.; Liu, Q.; Sun, L.; Yu, L. Circulating Serum Fatty Acid-Binding Protein 4 Levels Predict the Development of Diabetic Retinopathy in Type 2 Diabetic Patients. *Am. J. Ophthalmol.* **2018**, *187*, 71–79. [CrossRef]

11. Ning, H.; Tao, H.; Weng, Z.; Zhao, X. Plasma fatty acid-binding protein 4 (FABP4) as a novel biomarker to predict gestational diabetes mellitus. *Acta Diabetol.* **2016**, *53*, 891–898. [CrossRef]

12. Rodríguez-Calvo, R.; Girona, J.; Alegret, J.M.; Bosquet, A.; Ibarretxe, D.; Masana, L. Role of the fatty acid-binding protein 4 in heart failure and cardiovascular disease. *J. Endocrinol.* **2017**, *233*, R173–R184. [CrossRef]

13. Tanaka, M.; Furuhashi, M.; Okazaki, Y.; Mita, T.; Fuseya, T.; Ohno, K.; Ishimura, S.; Yoshida, H.; Miura, T. Ectopic expression of fatty acid-binding protein 4 in the glomerulus is associated with proteinuria and renal dysfunction. *Nephron. Clin. Pract.* **2014**, *128*, 345–351. [CrossRef] [PubMed]

14. Furuhashi, M.; Ishimura, S.; Ota, H.; Miura, T. Lipid chaperones and metabolic inflammation. *Int. J. Inflam.* **2011**, *2011*, 642612. [CrossRef]

15. Kucharski, M.; Kaczor, U. Fatty Acid Binding Protein 4 (FABP4) and the Body Lipid Balance. *Folia Biologica (Kraków)* **2017**, *65*, 181–186. [CrossRef]

16. Okazaki, Y.; Furuhashi, M.; Tanaka, M.; Mita, T.; Fuseya, T.; Ishimura, S.; Watanabe, Y.; Hoshina, K.; Akasaka, H.; Ohnishi, H.; et al. Urinary excretion of fatty acid-binding protein 4 is associated with albuminuria and renal dysfunction. *PLoS ONE* **2014**, *9*, e115429. [CrossRef] [PubMed]

17. Furuhashi, M.; Hotamisligil, G.S. Fatty acid-binding proteins: role in metabolic diseases and potential as drug targets. *Nat. Rev. Drug Discov.* **2008**, *7*, 489–503. [CrossRef]

18. Zimmer, J.S.; Dyckes, D.F.; Bernlohr, D.A.; Murphy, R.C. Fatty acid binding proteins stabilize leukotriene A4: Competition with arachidonic acid but not other lipoxygenase products. *J. Lipid Res.* **2004**, *45*, 2138–2144. [CrossRef]

19. Smith, P.J.; Wise, L.S.; Berkowitz, R.; Wan, C.; Rubin, C.S. Insulin-like growth factor-I is an essential regulator of the differentiation of 3T3-L1 adipocytes. *J. Biol. Chem.* **1988**, *263*, 9402–9408.

20. Makowski, L.; Brittingham, K.C.; Reynolds, J.M.; Suttles, J.; Hotamisligil, G.S. The fatty acid-binding protein, aP2, coordinates macrophage cholesterol trafficking and inflammatory activity. Macrophage expression of aP2 impacts peroxisome proliferator-activated receptor gamma and IkappaB kinase activities. *J. Biol. Chem.* **2005**, *280*, 12888–12895. [CrossRef] [PubMed]

21. Shashkin, P.N.; Jain, N.; Miller, Y.I.; Rissing, B.A.; Huo, Y.; Keller, S.R.; Vandenhoff, G.E.; Nadler, J.L.; McIntyre, T.M. Insulin and glucose play a role in foam cell formation and function. *Cardiovasc. Diabetol.* **2006**, *5*, 13. [CrossRef] [PubMed]

22. Siersbaek, R.; Nielsen, R.; Mandrup, S. PPARgamma in adipocyte differentiation and metabolism—novel insights from genome-wide studies. *FEBS Lett.* **2010**, *584*, 3242–3249. [CrossRef] [PubMed]

23. Huang, Q.; Ma, C.; Chen, L.; Luo, D.; Chen, R.; Liang, F. Mechanistic Insights Into the Interaction Between Transcription Factors and Epigenetic Modifications and the Contribution to the Development of Obesity. *Front. Endocrinol. (Lausanne)* **2018**, *9*, 370. [CrossRef] [PubMed]

24. Mizukami, H.; Takahashi, K.; Inaba, W.; Tsuboi, K.; Osonoi, S.; Yoshida, T.; Yagihashi, S. Involvement of oxidative stress-induced DNA damage, endoplasmic reticulum stress, and autophagy deficits in the decline of β-cell mass in Japanese type 2 diabetic patients. *Diabetes Care* **2014**, *37*, 1966–1974. [CrossRef] [PubMed]

25. Lehrke, M.; Lazar, M.A. The many faces of PPARg. *Cell* **2005**, *123*, 993–999. [CrossRef]

26. Shen, W.J.; Liang, Y.; Hong, R.; Patel, S.; Natu, V.; Sridhar, K.; Jenkins, A.; Bernlohr, D.A.; Kraemer, F.B. Characterization of the functional interaction of adipocyte lipid-binding protein with hormone-sensitive lipase. *J. Biol. Chem.* **2001**, *276*, 49443–49448. [CrossRef]

27. Furuhashi, M. Fatty Acid-Binding Protein 4 in Cardiovascular and Metabolic Diseases. *J. Atheroscler. Thromb.* **2019**. [CrossRef]

28. Prentice, K.J.; Saksi, J.; Hotamisligil, G.S. Adipokine FABP4 integrates energy stores and counter regulatory metabolic responses. *J. Lipid Res.* **2019**, pii: jlr.S091793. [CrossRef]

29. Furuhashi, M.; Saitoh, S.; Shimamoto, K.; Miura, T. Fatty Acid-Binding Protein 4 (FABP4): Pathophysiological Insights and Potent Clinical Biomarker of Metabolic and Cardiovascular Diseases. *Clin. Med. Insights Cardiol.* **2015**, *8*, 23–33. [CrossRef]

30. Li, Y.Y.; Xiao, R.; Li, C.P.; Huangfu, J.; Mao, J.F. Increased plasma levels of FABP4 and PTEN is associated with more severe insulin resistance in women with gestational diabetes mellitus. *Med. Sci. Monit.* **2015**, *21*, 426–431.

31. Tso, A.W.; Xu, A.; Sham, P.C.; Wat, N.M.; Wang, Y.; Fong, C.H.; Cheung, B.M.; Janus, E.D.; Lam, K.S. Serum adipocyte fatty acid binding protein as a new biomarker predicting the development of type 2 diabetes: A 10-year prospective study in a Chinese cohort. *Diabetes Care* **2007**, *30*, 2667–2672. [CrossRef] [PubMed]

32. Li, L.; Lee, S.J.; Kook, S.Y.; Ahn, T.G.; Lee, J.Y.; Hwang, J.Y. Serum from pregnant women with gestational diabetes mellitus increases the expression of FABP4 mRNA in primary subcutaneous human pre-adipocytes. *Obstet. Gynecol. Sci.* **2017**, *60*, 274–282. [CrossRef] [PubMed]

33. Martínez-Micaelo, N.; Rodríguez-Calvo, R.; Guaita-Esteruelas, S.; Heras, M.; Girona, J.; Masana, L. Extracellular FABP4 uptake by endothelial cells is dependent on cytokeratin 1 expression. *Biochim. Biophys. Acta Mol. Cell. Biol. Lipids* **2018**, *1864*, 234–244. [CrossRef] [PubMed]

34. Cabia, B.; Andrade, S.; Carreira, M.C.; Casanueva, F.F.; Crujeiras, A.B. A role for novel adipose tissue-secreted factors in obesity-related carcinogenesis. *Obes. Rev.* **2016**, *17*, 361–376. [CrossRef] [PubMed]

35. Garin-Shkolnik, T.; Rudich, A.; Hotamisligil, G.S.; Rubinstein, M. FABP4 attenuates PPARγ and adipogenesis and is inversely correlated with PPARγ in adipose tissues. *Diabetes* **2014**, *63*, 900–911. [CrossRef] [PubMed]

36. Lehmann, J.M.; Moore, L.B.; Smith-Oliver, T.A.; Wilkison, W.O.; Kliewer, S.A. An antidiabetic thiazolidinedione is a high affinity ligand for peroxisome proliferator-activated receptor gamma (PPAR gamma). *J. Biol. Chem.* **1995**, *270*, 12953–12956. [CrossRef] [PubMed]

37. Odegaard, J.I.; Ricardo-Gonzalez, R.R.; Goforth, M.H.; Morel, C.R.; Subramanian, V.; Mukundan, L.; Red Eagle, A.; Vats, D.; Brombacher, F.; Ferrante, A.W.; et al. Macrophage-specific PPARgamma controls alternative activation and improves insulin resistance. *Nature* **2007**, *447*, 1116–1120. [CrossRef]

38. Chawla, A.; Boisvert, W.A.; Lee, C.H.; Laffitte, B.A.; Barak, Y.; Joseph, S.B.; Liao, D.; Nagy, L.; Edwards, P.A.; Curtiss, L.K.; et al. A PPAR gamma-LXR-ABCA1 pathway in macrophages is involved in cholesterol efflux and atherogenesis. *Mol. Cell* **2001**, *7*, 161–171. [CrossRef]

39. Kralisch, S.; Fasshauer, M. Adipocyte fatty acid binding protein: a novel adipokine involved in the pathogenesis of metabolic and vascular disease? *Diabetologia* **2013**, *56*, 10–21. [CrossRef]

40. Xu, A.; Wang, Y.; Xu, J.Y.; Stejskal, D.; Tam, S.; Zhang, J.; Wat, N.M.; Wong, W.K.; Lam, K.S. Adipocyte fatty acid-binding protein is a plasma biomarker closely associated with obesity and metabolic syndrome. *Clin. Chem.* **2006**, *52*, 405–413. [CrossRef]

41. Xu, A.; Tso, A.W.; Cheung, B.M.; Wang, Y.; Wat, N.M.; Fong, C.H.; Yeung, D.C.; Janus, E.D.; Sham, P.C.; Lam, K.S. Circulating adipocyte-fatty acid binding protein levels predict the development of the metabolic syndrome: a 5-year prospective study. *Circulation* **2007**, *115*, 1537–1543. [CrossRef] [PubMed]

42. Li, S.; Bi, P.; Zhao, W.; Lian, Y.; Zhu, H.; Xu, D.; Ding, J.; Wang, Q.; Yin, C. Prognostic Utility of Fatty Acid-Binding Protein 4 in Patients with Type 2 Diabetes and Acute Ischemic Stroke. *Neurotox. Res.* **2018**, *33*, 309–315. [CrossRef] [PubMed]

43. Stejskal, D.; Karpisek, M. Adipocyte fatty acid binding protein in a Caucasian population: A new marker of metabolic syndrome? *Eur. J. Clin. Invest.* **2006**, *36*, 621–625. [CrossRef] [PubMed]

44. Cabré, A.; Lázaro, I.; Girona, J.; Manzanares, J.M.; Marimón, F.; Plana, N.; Heras, M.; Masana, L. Fatty acid binding protein 4 is increased in metabolic syndrome and with thiazolidinedione treatment in diabetic patients. *Atherosclerosis* **2007**, *195*, e150–e158. [CrossRef]

45. Nakamura, R.; Okura, T.; Fujioka, Y.; Sumi, K.; Matsuzawa, K.; Izawa, S.; Ueta, E.; Kato, M.; Taniguchi, S.I.; Yamamoto, K. Serum fatty acid-binding protein 4 (FABP4) concentration is associated with insulin resistance in peripheral tissues, A clinical study. *PLoS ONE* **2017**, *12*, e0179737. [CrossRef] [PubMed]

46. Wang, Q.; Rizk, S.; Bernard, C.; Lai, M.P.; Kam, D.; Storch, J.; Stark, R.E. Protocols and pitfalls in obtaining fatty acid-binding proteins for biophysical studies of ligand-protein and protein-protein interactions. *Biochem. Biophys. Rep.* **2017**, *10*, 318–324. [PubMed]

47. Chen, R.A.; Sun, X.M.; Yan, C.Y.; Liu, L.; Hao, M.W.; Liu, Q.; Jiao, X.Y.; Liang, Y.M. Hyperglycemia-induced PATZ1 negatively modulates endothelial vasculogenesis via repression of FABP4 signaling. *Biochem. Biophys. Res. Commun.* **2016**, *477*, 548–555. [CrossRef]

48. Harjes, U.; Bridges, E.; McIntyre, A.; Fielding, B.A.; Harris, A.L. Fatty acid-binding protein 4, a point of convergence for angiogenic and metabolic signaling pathways in endothelial cells. *J. Biol. Chem.* **2014**, *289*, 23168–23176. [CrossRef]

49. Song, J.; Ren, P.; Zhang, L.; Wang, X.L.; Chen, L.; Shen, Y.H. Metformin reduces lipid accumulation in macrophages by inhibiting FOXO1-mediated transcription of fatty acid-binding protein 4. Biochem. *Biophys. Res. Commun.* **2010**, *393*, 89–94. [CrossRef]

50. Tu, W.J.; Guo, M.; Shi, X.D.; Cai, Y.; Liu, Q.; Fu, C.W. First-Trimester Serum Fatty Acid-Binding Protein 4 and Subsequent Gestational Diabetes Mellitus. *Obstet. Gynecol.* **2017**, *130*, 1011–1016. [CrossRef]

51. Coe, N.R.; Bernlohr, D.A. Physiological properties and functions of intracellular fatty acid-binding proteins. *Biochim. Biophys. Acta* **1998**, *1391*, 287–306. [CrossRef]

52. Maeda, K.; Cao, H.; Kono, K.; Gorgun, C.Z.; Furuhashi, M.; Uysal, K.T.; Cao, Q.; Atsumi, G.; Malone, H.; Krishnan, B.; et al. Adipocyte/macrophage fatty acid binding proteins control integrated metabolic responses in obesity and diabetes. *Cell. Metab.* **2005**, *1*, 107–119. [CrossRef] [PubMed]

53. Kralisch, S.; Klöting, N.; Ebert, T.; Kern, M.; Hoffmann, A.; Krause, K.; Jessnitzer, B.; Lossner, U.; Sommerer, I.; Stumvoll, M.; et al. Circulating adipocyte fatty acid-binding protein induces insulin resistance in mice in vivo. *Obesity (Silver Spring)* **2015**, *23*, 1007–1013. [CrossRef] [PubMed]

54. Furuhashi, M.; Tuncman, G.; Görgün, C.Z.; Makowski, L.; Atsumi, G.; Vaillancourt, E.; Kono, K.; Babaev, V.R.; Fazio, S.; Linton, M.F.; et al. Treatment of diabetes and atherosclerosis by inhibiting fatty-acid-binding protein aP2. *Nature* **2007**, *447*, 959–965. [CrossRef] [PubMed]

55. Erbay, E.; Babaev, V.R.; Mayers, J.R.; Makowski, L.; Charles, K.N.; Snitow, M.E.; Fazio, S.; Wiest, M.M.; Watkins, S.M.; Linton, M.F.; et al. Reducing endoplasmic reticulum stress through a macrophage lipid chaperone alleviates atherosclerosis. *Nat. Med.* **2009**, *15*, 1383–1391. [CrossRef] [PubMed]

56. Hui, X.; Li, H.; Zhou, Z.; Lam, K.S.; Xiao, Y.; Wu, D.; Ding, K.; Wang, Y.; Vanhoutte, P.M.; Xu, A. Adipocyte fatty acid-binding protein modulates inflammatory responses in macrophages through a positive feedback loop involving c-Jun NH2-terminal kinases and activator protein-1. *J. Biol. Chem.* **2010**, *285*, 10273–10280. [CrossRef] [PubMed]

57. Hotamisligil, G.S.; Bernlohr, D.A. Metabolic functions of FABPs–mechanisms and therapeutic implications. *Nat. Rev. Endocrinol.* **2015**, *11*, 592–605. [CrossRef] [PubMed]

58. Smith, A.J.; Thompson, B.R.; Sanders, M.A.; Bernlohr, D.A. Interaction of the adipocyte fatty acid-binding protein with the hormone-sensitive lipase: regulation by fatty acids and phosphorylation. *J. Biol. Chem.* **2007**, *282*, 32424–32432. [CrossRef]

59. Thompson, B.R.; Mazurkiewicz-Muñoz, A.M.; Suttles, J.; Carter-Su, C.; Bernlohr, D.A. Interaction of adipocyte fatty acid-binding protein (AFABP) and JAK2: AFABP/aP2 as a regulator of JAK2 signaling. *J. Biol. Chem.* **2009**, *284*, 13473–13480. [CrossRef]

60. Li, H.; Luo, H.Y.; Liu, Q.; Xiao, Y.; Tang, L.; Zhong, F.; Huang, G.; Xu, J.M.; Xu, A.M.; Zhou, Z.G.; et al. Intermittent High Glucose Exacerbates A-FABP Activation and Inflammatory Response through TLR4-JNK Signaling in THP-1 Cells. *J. Immunol. Res.* **2018**, *2018*, 1319272. [CrossRef]

61. Fu, Y.; Luo, N.; Lopes-Virella, M.F.; Garvey, W.T. The adipocyte lipid binding protein (ALBP/aP2) gene facilitates foam cell formation in human THP-1 macrophages. *Atherosclerosis* **2002**, *165*, 259–269. [CrossRef]

62. Kazemi, M.R.; McDonald, C.M.; Shigenaga, J.K.; Grunfeld, C.; Feingold, K.R. Adipocyte fatty acid-binding protein expression and lipid accumulation are increased during activation of murine macrophages by toll-like receptor agonists. *Arterioscler. Thromb. Vasc. Biol.* **2005**, *25*, 1220–1224. [CrossRef] [PubMed]

63. Pelton, P.D.; Zhou, L.; Demarest, K.T.; Burris, T.P. PPARgamma activation induces the expression of the adipocyte fatty acid binding protein gene in human monocytes. *Biochem. Biophys. Res. Commun.* **1999**, *261*, 456–458. [CrossRef] [PubMed]

64. Yao, F.; Li, Z.; Ehara, T.; Yang, L.; Wang, D.; Feng, L.; Zhang, Y.; Wang, K.; Shi, Y.; Duan, H.; et al. Fatty Acid-Binding Protein 4 mediates apoptosis via endoplasmic reticulum stress in mesangial cells of diabetic nephropathy. *Mol. Cell. Endocrinol.* **2015**, *411*, 232–242. [CrossRef] [PubMed]

65. Zhang, Y.; Zhang, H.H.; Lu, J.H.; Zheng, S.Y.; Long, T.; Li, Y.T.; Wu, W.Z.; Wang, F. Changes in serum adipocyte fatty acid-binding protein in women with gestational diabetes mellitus and normal pregnant women during mid- and late pregnancy. *J. Diabetes Investig.* **2016**, *7*, 797–804. [CrossRef] [PubMed]

66. Xie, X.; Yi, Z.; Sinha, S.; Madan, M.; Bowen, B.P.; Langlais, P.; Ma, D.; Mandarino, L.; Meyer, C. Proteomics analyses of subcutaneous adipocytes reveal novel abnormalities in human insulin resistance. *Obesity (Silver Spring)* **2016**, *24*, 1506–1514. [CrossRef]

67. Nourani, M.R.; Owada, Y.; Kitanaka, N.; Sakagami, H.; Hoshi, H.; Iwasa, H.; Spener, F.; Kondo, H. Occurrence of immunoreactivity for adipocyte-type fatty acid binding protein in degenerating granulosa cells in atretic antral follicles of mouse ovary. *J. Mol. Histol.* **2005**, *36*, 491–497. [CrossRef]

68. Hsu, W.C.; Okeke, E.; Cheung, S.; Keenan, H.; Tsui, T.; Cheng, K.; King, G.L. A cross-sectional characterization of insulin resistance by phenotype and insulin clamp in East Asian Americans with type 1 and type 2 diabetes. *PLoS ONE* **2011**, *6*, e28311. [CrossRef]

69. Ota, H.; Furuhashi, M.; Ishimura, S.; Koyama, M.; Okazaki, Y.; Mita, T.; Fuseya, T.; Yamashita, T.; Tanaka, M.; Yoshida, H.; et al. Elevation of fatty acid-binding protein 4 is predisposed by family history of hypertension and contributes to blood pressure elevation. *Am. J. Hypertens.* **2012**, *25*, 1124–1130. [CrossRef]

70. Wu, L.E.; Samocha-Bonet, D.; Whitworth, P.T.; Fazakerley, D.J.; Turner, N.; Biden, T.J.; James, D.E.; Cantley, J. Identification of fatty acid binding protein 4 as an adipokine that regulates insulin secretion during obesity. *Mol. Metab.* **2014**, *3*, 465–473. [CrossRef]

71. Greco, E.A.; Francomano, D.; Fornari, R.; Marocco, C.; Lubrano, C.; Papa, V.; Wannenes, F.; di Luigi, L.; Donini, L.M.; Lenzi, A.; et al. Negative association between trunk fat, insulin resistance and skeleton in obese women. *World J. Diabetes* **2013**, *4*, 31–39. [CrossRef] [PubMed]

72. Fasshauer, M.; Blüher, M.; Stumvoll, M. Adipokines in gestational diabetes. *Lancet Diabetes Endocrinol.* **2014**, *2*, 488–499. [CrossRef]

73. Kralisch, S.; Stepan, H.; Kratzsch, J.; Verlohren, M.; Verlohren, H.J.; Drynda, K.; Lössner, U.; Blüher, M.; Stumvoll, M.; Fasshauer, M. Serum levels of adipocyte fatty acid binding protein are increased in gestational diabetes mellitus. *Eur. J. Endocrinol.* **2009**, *160*, 33–38. [CrossRef] [PubMed]

74. Ortega-Senovilla, H.; Schaefer-Graf, U.; Meitzner, K.; Abou-Dakn, M.; Graf, K.; Kintscher, U.; Herrera, E. Gestational diabetes mellitus causes changes in the concentrations of adipocyte fatty acid-binding protein and other adipocytokines in cord blood. *Diabetes Care* **2011**, *34*, 2061–2066. [CrossRef] [PubMed]

75. Wang, P.; Zhu, Q.; Peng, H.; Du, M.; Dong, M.; Wang, H. Fatty Acid-Binding Protein 4 in Endometrial Epithelium Is Involved in Embryonic Implantation. *Cell. Physiol. Biochem.* **2017**, *41*, 501–509. [CrossRef]

76. Li, B.; Yang, H.; Zhang, W.; Shi, Y.; Qin, S.; Wei, Y.; He, Y.; Yang, W.; Jiang, S.; Jin, H. Fatty acid-binding protein 4 predicts gestational hypertension and preeclampsia in women with gestational diabetes mellitus. *PLoS ONE* **2018**, *13*, e0192347. [CrossRef] [PubMed]

77. Wotherspoon, A.C.; Young, I.S.; McCance, D.R.; Patterson, C.C.; Maresh, M.J.; Pearson, D.W.; Walker, J.D.; Holmes, V.A. Diabetes and Pre-eclampsia Intervention Trial (DAPIT) Study Group. Serum Fatty Acid Binding Protein 4 (FABP4) Predicts Pre-eclampsia in Women With Type 1 Diabetes. *Diabetes Care* **2016**, *39*, 1827–1829. [CrossRef]

78. Svensson, H.; Wetterling, L.; Andersson-Hall, U.; Jennische, E.; Edén, S.; Holmäng, A.; Lönn, M. Adipose tissue and body composition in women six years after gestational diabetes: Factors associated with development of type 2 diabetes. *Adipocyte* **2018**, *7*, 229–237. [CrossRef]

79. Joussen, A.M.; Poulaki, V.; Le, M.L.; Koizumi, K.; Esser, C.; Janicki, H.; Schraermeyer, U.; Kociok, N.; Fauser, S.; Kirchhof, B.; et al. A central role for inflammation in the pathogenesis of diabetic retinopathy. *FASEB J.* **2004**, *18*, 1450–1452. [CrossRef]

80. Behl, T.; Kaur, I.; Kotwani, A. Implication of oxidative stress in progression of diabetic retinopathy. *Surv. Ophthalmol.* **2016**, *61*, 187–196. [CrossRef]

81. Aragonès, G.; Ferré, R.; Lázaro, I.; Cabré, A.; Plana, N.; Merino, J.; Heras, M.; Girona, J.; Masana, L. Fatty acid-binding protein 4 is associated with endothelial dysfunction in patients with type 2 diabetes. *Atherosclerosis* **2010**, *213*, 329–331. [CrossRef] [PubMed]

82. Aragonès, G.; Saavedra, P.; Heras, M.; Cabré, A.; Girona, J.; Masana, L. Fatty acid-binding protein 4 impairs the insulin-dependent nitric oxide pathway in vascular endothelial cells. *Cardiovasc. Diabetol.* **2012**, *11*, 72. [CrossRef] [PubMed]

83. Cabré, A.; Lázaro, I.; Girona, J.; Manzanares, J.M.; Marimón, F.; Plana, N.; Heras, M.; Masana, L. Plasma fatty acid binding protein 4 is associated with atherogenic dyslipidemia in diabetes. *J. Lipid Res.* **2008**, *49*, 1746–1751. [CrossRef] [PubMed]

84. Lamounier-Zepter, V.; Look, C.; Alvarez, J.; Christ, T.; Ravens, U.; Schunck, W.H.; Ehrhart-Bornstein, M.; Bornstein, S.R.; Morano, I. Adipocyte fatty acid-binding protein suppresses cardiomyocyte contraction: a new link between obesity and heart disease. *Circ. Res.* **2009**, *105*, 326–334. [CrossRef] [PubMed]

85. Peeters, W.; de Kleijn, D.P.; Vink, A.; van de Weg, S.; Schoneveld, A.H.; Sze, S.K.; van der Spek, P.J.; de Vries, J.P.; Moll, F.L.; Pasterkamp, G. Adipocyte fatty acid binding protein in atherosclerotic plaques is associated with local vulnerability and is predictive for the occurrence of adverse cardiovascular events. *Eur. Heart J.* **2011**, *32*, 1758–1768. [CrossRef] [PubMed]

86. Agardh, H.E.; Folkersen, L.; Ekstrand, J.; Marcus, D.; Swedenborg, J.; Hedin, U.; Gabrielsen, A.; Paulsson-Berne, G. Expression of fatty acid-binding protein 4/aP2 is correlated with plaque instability in carotid atherosclerosis. *J. Intern. Med.* **2011**, *269*, 200–210. [CrossRef] [PubMed]

87. Holm, S.; Ueland, T.; Dahl, T.B.; Michelsen, A.E.; Skjelland, M.; Russell, D.; Nymo, S.H.; Krohg-Sørensen, K.; Clausen, O.P.; Atar, D.; et al. Fatty Acid binding protein 4 is associated with carotid atherosclerosis and outcome in patients with acute ischemic stroke. *PLoS ONE* **2011**, *6*, e28785. [CrossRef]

88. Liu, G.; Ding, M.; Chiuve, S.E.; Rimm, E.B.; Franks, P.W.; Meigs, J.B.; Hu, F.B.; Sun, Q. Plasma Levels of Fatty Acid-Binding Protein 4, Retinol-Binding Protein 4, High-Molecular-Weight Adiponectin, and Cardiovascular Mortality Among Men With Type 2 Diabetes: A 22-Year Prospective Study. *Arterioscler. Thromb. Vasc. Biol.* **2016**, *36*, 2259–2267. [CrossRef]

89. Makowski, L.; Boord, J.B.; Maeda, K.; Babaev, V.R.; Uysal, K.T.; Morgan, M.A.; Parker, R.A.; Suttles, J.; Fazio, S.; Hotamisligil, G.S.; et al. Lack of macrophage fatty-acid-binding protein aP2 protects mice deficient in apolipoprotein E against atherosclerosis. *Nat. Med.* **2001**, *7*, 699–705. [CrossRef]

90. Parvanova, A.I.; Trevisan, R.; Iliev, I.P.; Dimitrov, B.D.; Vedovato, M.; Tiengo, A.; Remuzzi, G.; Ruggenenti, P. Insulin resistance and microalbuminuria: a cross-sectional, case-control study of 158 patients with type 2 diabetes and different degrees of urinary albumin excretion. *Diabetes* **2006**, *55*, 1456–1462. [CrossRef]

91. Sanchez-Chavez, G.; Hernandez-Ramırez, E.; Osorio-Paz, I.; Hernandez-Espinosa, C.; Salceda, R. Potential role of endoplasmic reticulum stress in pathogenesis of diabetic retinopathy. *Neurochem. Res.* **2016**, *41*, 1098–1106. [CrossRef] [PubMed]

92. Ni, X.; Gu, Y.; Yu, H.; Wang, S.; Chen, Y.; Wang, X.; Yuan, X.; Jia, W. Serum Adipocyte Fatty Acid-Binding Protein 4 Levels Are Independently Associated with Radioisotope Glomerular Filtration Rate in Type 2 Diabetic Patients with Early Diabetic Nephropathy. *Biomed. Res. Int.* **2018**, *2018*, 4578140. [CrossRef] [PubMed]

93. Toruner, F.; Altinova, A.E.; Akturk, M.; Kaya, M.; Arslan, E.; Bukan, N.; Kan, E.; Yetkin, I.; Arslan, M. The relationship between adipocyte fatty acid binding protein-4, retinol binding protein-4 levels and early diabetic nephropathy in patients with type 2 diabetes. *Diabetes Res. Clin. Pract.* **2011**, *91*, 203–207. [CrossRef] [PubMed]

94. Yeung, D.C.; Xu, A.; Tso, A.W.; Chow, W.S.; Wat, N.M.; Fong, C.H.; Tam, S.; Sham, P.C.; Lam, K.S. Circulating levels of adipocyte and epidermal fatty acid-binding proteins in relation to nephropathy staging and macrovascular complications in type 2 diabetic patients. *Diabetes Care* **2009**, *32*, 132–134. [CrossRef] [PubMed]

95. Hu, X.; Ma, X.; Luo, Y.; Xu, Y.; Xiong, Q.; Pan, X.; Bao, Y.; Jia, W. Contribution of serum adipocyte fatty acid-binding protein levels to the presence of microalbuminuria in a Chinese hyperglycemic population. *J. Diabetes Investig.* **2017**, *8*, 582–589. [CrossRef] [PubMed]

96. Scalera, A.; Tarantino, G. Could metabolic syndrome lead to hepatocarcinoma via non-alcoholic fatty liver disease? *World J. Gastroenterol.* **2014**, *20*, 9217–9228.

97. Thompson, K.J.; Austin, R.G.; Nazari, S.S.; Gersin, K.S.; Iannitti, D.A.; McKillop, I.H. Altered fatty acid-binding protein 4 (FABP4) expression and function in human and animal models of hepatocellular carcinoma. *Liver Int.* **2018**, *38*, 1074–1083. [CrossRef]

98. Tu, W.J.; Zeng, X.W.; Deng, A.; Zhao, S.J.; Luo, D.Z.; Ma, G.Z.; Wang, H.; Liu, Q. Circulating FABP4 (Fatty Acid-Binding Protein 4) Is a Novel Prognostic Biomarker in Patients With Acute Ischemic Stroke. *Stroke* **2017**, *48*, 1531–1538. [CrossRef]

99. Bellos, I.; Fitrou, G.; Pergialiotis, V.; Perrea, D.N.; Daskalakis, G. Serum levels of adipokines in gestational diabetes: A systematic review. *J. Endocrinol. Invest.* **2018**. [CrossRef]

Review

Obesity-Induced Adipose Tissue Inflammation as a Strong Promotional Factor for Pancreatic Ductal Adenocarcinoma

Hui-Hua Chang and Guido Eibl *

Department of Surgery, David Geffen School of Medicine at UCLA, Los Angeles, CA 90095, USA
* Correspondence: GEibl@mednet.ucla.edu; Tel.: +1-310-825-5743

Received: 14 June 2019; Accepted: 2 July 2019; Published: 3 July 2019

Abstract: Pancreatic ductal adenocarcinoma (PDAC) is expected to soon become the second leading cause of cancer related deaths in the United States. This may be due to the rising obesity prevalence, which is a recognized risk factor for PDAC. There is great interest in deciphering the underlying driving mechanisms of the obesity–PDAC link. Visceral adiposity has a strong correlation to certain metabolic diseases and gastrointestinal cancers, including PDAC. In fact, our own data strongly suggest that visceral adipose tissue inflammation is a strong promoter for PDAC growth and progression in a genetically engineered mouse model of PDAC and diet-induced obesity. In this review, we will discuss the relationship between obesity-associated adipose tissue inflammation and PDAC development, with a focus on the key molecular and cellular components in the dysfunctional visceral adipose tissue, which provides a tumor permissive environment.

Keywords: obesity; adipose tissue inflammation; pancreatic ductal adenocarcinoma

---

1. Obesity and Pancreatic Ductal Adenocarcinoma

Pancreatic cancer, of which pancreatic ductal adenocarcinoma (PDAC) accounts for the vast majority, is an aggressive disease with an overall 5-year survival rate of ~8% [1]. Currently, as the third leading cause of cancer mortality in both men and women, pancreatic cancer is expected to soon become the second leading cause of cancer-related deaths in the United States [2]. One of the recognized risk factors for PDAC and several other gastrointestinal cancers is obesity [3–8]. A recent analysis by the National Institutes of Health reported that 16.9% of all PDAC cases in the Unites States are estimated to be attributable to excess body weight (in comparison, the proportion of PDAC cases attributable to cigarette smoking was 10.2%) [9]. Besides promoting tumorigenesis, obesity might also complicate the treatment of established cancers [10], not only by altering the pharmacokinetics and pharmacodynamics of administered anti-cancer drugs [11], but also by modulating the tumor microenvironment (TME) [12]. Notably, worse prognosis is reported in patients who are either overweight or obese at the time of PDAC diagnosis [13,14]. Considering the current obesity pandemic, there is great interest in deciphering the underlying biological basis of this compelling epidemiological link.

The mechanisms underlying the obesity–cancer link are likely multifactorial, with metabolic perturbations and inflammation being the current research focus. Other mechanisms, such as alterations in the gut microbiome, changes in gastrointestinal peptides, and sex hormones, certainly play an important role as well. For example, obesity-associated insulin resistance and hormonal changes (e.g., compensatory hyperinsulinemia and elevated insulin-like growth factors (IGFs)) are among the various driving mechanisms postulated to explain the obesity–cancer link [15]. In accordance with this notion, type 2 diabetes, hyperinsulinemia, and increased circulating IGF-1 are established risk factors for pancreatic and other types of cancers [16–20]. In addition to the insulin/IGF axis, increasing attention has focused on adipose tissue (AT) dysfunction characterized by a pro-inflammatory state, which can contribute not only to the pathogenesis of insulin resistance [21] but also to the development and promotion of cancer. Utilizing a genetically engineered mouse model of PDAC and diet-induced obesity, our own data suggest that inflammatory responses in the AT surrounding the pancreas is a strong promotional factor for PDAC growth and progression [22]. AT inflammation associated with increased abdominal adiposity, which involves various cellular and extracellular components (e.g., infiltrated macrophages and cytokines), can shape the peri-tumoral micro-environment favorable for PDAC development, both locally and systemically. More detailed cellular and molecular characteristics of dysfunctional and inflamed AT associated with the obese state, and their roles in tumor development, will be discussed in the later sections.

2. The Role of Adipose Tissue Inflammation in Obesity-Promoted Tumor Development

2.1. Classification and Physiological Roles of AT

White adipose tissue (WAT), the most abundant type of AT in adult humans, stores excess energy in the form of triglycerides, which can be mobilized via lipolysis to release fatty acids for usage by other tissues. Based on anatomical locations, AT can also be organized into different types of depots: Subcutaneous and visceral. These depots are heterogeneous with regard to their morphological, molecular, and metabolic profiles [23], and exhibit major gender differences. Visceral adipose tissue (VAT), which encompasses omental, mesenteric, and other intra-abdominal fat pads, is considered of great relevance in obesity-related comorbidities, as increased VAT area is more closely associated with metabolic dysfunction and cancer than subcutaneous fat [7,23–25]. Surgical removal of VAT prior to high fat diet feeding is shown to prevent obesity-induced insulin resistance in mice [26]. Since VAT is in close proximity to digestive organs, including the pancreas, and is the primary source of systemic and local chronic inflammation during obesity [27], it is particularly pertinent to the topics covered in this review. For example, VAT (but not subcutaneous fat) and pancreatic fatty infiltration are significantly correlated with pancreatic intraepithelial neoplasia (PanIN), the precursor lesions of PDAC [28]. Interestingly, the increased PDAC incidence in obese conditional Kras^{G12D} (KC) mice was largely observed in male animals [29], which correlated with a greater expansion of VAT in obese male mice, as compared to obese female mice that preferentially gained subcutaneous fat [22].

Besides being a key regulator of energy homeostasis, AT is regarded as an important endocrine organ secreting various factors collectively known as "adipokines", and growingly recognized as a prominent immune site harboring macrophages and other types of immune cells. Overall, this dynamic organ is composed of mature adipocytes and the "stromal-vascular fraction" comprising a mixture of mesenchymal, hematopoietic, and endothelial cell types. Under homeostatic conditions, the cells with diverse functional roles act cooperatively to support the metabolic and physiological functions of the organ. Specifically, healthy AT maintains an anti-inflammatory phenotype characterized by the secretion of anti-inflammatory adipocytokines (e.g., adiponectin) and enrichment of regulatory immune cells, such as M2 macrophages, eosinophils, Th2 leukocytes, and myeloid-derived suppressor cells that limit AT inflammation. In obesity, on the other hand, the regulatory anti-inflammatory network in AT is disrupted and skewed to a pro-inflammatory state, with substantial changes in adipokines and the number and function of AT immune cells (see next section), leading to sustained chronic inflammation.

2.2. Obesity-Associated AT Inflammation

Chronic over-nutrition is manifested as adipose tissue expansion via hyperplasia (increase in adipocyte numbers) and/or hypertrophy (enlargement of existing adipocytes). Adipocyte hyperplasia, which occurs more readily in subcutaneous depots than in VAT, is normally associated with no/weak inflammation and improved insulin sensitivity and therefore regarded as a "metabolically healthy" expansion of AT. Conversely, visceral adipocyte hypertrophy, which is often associated with cellular stress, pro-inflammatory responses, and impaired insulin sensitivity, is recognized as unhealthy pathological expansion and an important contributor to the obesity-related metabolic disorders [30]. When hypertrophic adipocytes are overloaded with lipids, they become hypoxic and undergo stress responses and cell death, which are correlated with a pro-inflammatory secretome, increased pro-inflammatory immune cell infiltration, and ectopic lipid deposition in other metabolically active organs [31], each of which may contribute to progressive insulin resistance and the pathogenesis of type 2 diabetes.

Adipocyte hypertrophy is associated with a shift in the adipokine secretion profile [32]. Specifically, there is a significant increase in the production of pro-inflammatory factors, such as leptin, tumor necrosis factor-α (TNF-α), interleukin 6 (IL-6), monocyte chemoattractive protein-1 (MCP-1), and plasminogen activator inhibitor-1 (PAI-1). On the contrary, the release of adiponectin, a major anti-inflammatory adipokine, is reduced with increased adiposity in response to AT inflammation [33]. These adipocyte-derived mediators, which link adipocytes with adipose-resident immune cells, lead to an altered composition of the adipose stromal-vascular fraction characterized by reduced numbers of anti-inflammatory cells (e.g., eosinophils, T-regs, Th2 cells) and an increased recruitment of pro-inflammatory Th1 cells, neutrophils, IFNγ-producing natural killer (NK) cells, and monocytes/macrophages [34,35].

The detailed landscape of immune cell changes in AT during obesity is extremely complex and virtually every immune cell type plays a role in obesity-associated metabolic diseases [36]. However, the most prominent feature is monocyte infiltration and differentiation into pro-inflammatory macrophages, which surround necrotic adipocytes and form the histological hallmark of AT inflammation called "crown-like structures" [37]. AT macrophages (ATM), which can account for up to 50% of the cellular content in the obese AT, are the primary source of pro-inflammatory cytokines (e.g., IL-6, TNF-α, IL1-β), and have been shown to be centrally important in obesity-related metabolic dysfunction [38–41]. The increase in ATM abundance during obesity is crucial for exacerbating the inflammatory response and is mainly caused by the recruitment of circulating bone marrow-derived monocytes through an MCP-1 (CCL2)/CCR2- dependent mechanism [42–44]. While resident ATMs in the lean state display an anti-inflammatory M2-polarized phenotype, obesity-associated ATM polarize towards the pro-inflammatory M1 state in response to the disturbed immune and cytokine profile and other environmental cues [35]. Interestingly, the oversimplified M1/M2 dichotomy is revised by the recent discovery of the "metabolically-activated" type ATM, which is metabolically activated by excessive saturated free fatty acids (FFAs), hyperinsulinemia, and hyperglycemia and is functionally distinct from the classical M1 macrophage activated by bacterial challenge [45]. As mentioned above, it is well established that AT inflammation and ATMs play a key role in the pathogenesis of metabolic syndromes, mainly via secreted mediators. However, much less is known about their role in cancer, especially PDAC.

Besides intrinsic danger signals, exogenous factors, such as gut microbiota-derived pathogen-associated molecular patterns (PAMPs), are also implicated in the reprogramming of immune cells toward pro-inflammatory subtypes in obese AT. Diet-induced obesity is associated with a robust change in gut microbiota composition [46,47] and impaired intestinal barrier integrity, correlating with an increased expression of pro-inflammatory cytokines in the AT [48,49]. Induced by gut dysbiosis and leaky intestinal tight junctions, circulating levels of lipopolysaccharide (LPS; a cell wall component of gram-negative bacteria) are found to be 2 to 3-fold higher in obese humans and rodents as compared to their lean counterparts, a phenomenon known as "metabolic endotoxemia" [50,51]. Elevated LPS levels in tissues, e.g., AT, can stimulate the innate immune response by activating Toll-like receptor 4 (TLR4) on macrophages, leading to the induction of NF-κB and pro-inflammatory cytokine production. Using germ-free mouse models, microbiota-derived LPS has been shown to promote macrophage infiltration and M1 polarization in WAT, independent of its role in regulating fat mass and glucose metabolism [52]. Additionally, mice fed a diet rich in saturated lipids (lard) exhibited reduced insulin sensitivity, enhanced WAT inflammation, and increased activation of TLRs compared with mice fed a fish oil diet, and such phenotypic differences were in part attributable to differences in gut microbiota composition [53]. These authors further demonstrated that lard-induced ATM accumulation was mediated by adipocyte-produced MCP-1, which was induced by microbial-derived factors through activation of TLR4 and the downstream adaptor molecules, MyD88 and TIR-domain-containing adapter-inducing interferon-β (TRIF) [53]. Together, these data strongly suggest that intestinal microbiota contribute to changes in immune cell function and AT inflammation during diet-induced obesity.

2.3. AT Inflammation in PDAC Development: The Molecular Links

Inflammation has long been associated with the development of cancer, including PDAC [54,55]. In a mouse model of obesity-promoted Kras^{G12D}-driven pancreatic neoplasia, with obesity induced by a high-fat high-calorie diet (HFCD), we observed a higher percentage of advanced PanIN lesions [56] and increased PDAC incidence [29], accompanied by prominent signs of inflammatory immune cell infiltration and elevation of pro-inflammatory cytokines in the pancreas [56] as well as the peri-pancreatic AT [22]. In addition, deficiency in hormone-sensitive lipase (HSL) was associated with AT inflammation and an acceleration of pancreatic cancer development in conditional Kras^{G12D} mice [57]. Cumulating evidence suggest a model in which diet-induced obesity leads to a robust inflammatory response in visceral peri-pancreatic (or intra-pancreatic [28]) adipose tissue, which may in turn promote inflammation and neoplastic progression in the neighboring pancreas, via soluble factors secreted from adipocytes and/or adipose-infiltrating immune cells.

2.3.1. Leptin, Adiponectin, and Other Adipokines

Obese individuals are characterized by dramatic changes in adipokine production, and several of these adipokine changes are associated with cancer development. The most prominent examples are the increased levels of the pro-inflammatory leptin [58,59] and the reduced levels of the anti-inflammatory adiponectin [60,61]. In several cancer models, leptin has been shown to promote cell proliferation, migration, and invasion through activation of the Janus Kinase/Signal Transducer and Activator of Transcription (JAK/STAT) pathway and the subsequent oncogenic phosphatidylinositol 3-kinase (PI3K)/Akt and extracellular signal-regulated kinase (ERK) signaling, leading to increased expression of inflammatory cytokines, angiogenic factors, and apoptotic proteins [61–64]. Adiponectin, on the other hand, seems to exert antagonistic effects on tumor growth and progression through activation of AMPK and protein tyrosine phosphatase 1B (PTP1B) [62,65]. However, the roles of circulating leptin and adiponectin in PDAC development remain debatable [66], as some studies found higher adiponectin/leptin ratios associated with pancreatic cancer [67,68]. Also, adiponectin transcription is activated by nuclear receptor 5A2 (NR5A2), an important risk factor identified by a genome-wide association study (GWAS) for pancreatic cancer [69]. Further, in our recent study using HSL deficient conditional Kras^{G12D} mice, decreased plasma leptin levels were observed in animals with advanced PDAC [57], suggesting that leptin may not play a central mechanistic role in PDAC promoted by AT inflammation. Besides systemic effects, adipokines generated by adjacent adipose tissue might act locally directly on tumor cells [70,71] or modulate inflammation and immunity in the pancreatic TME [72,73]. In that regard, it is noteworthy that lipocalin 2, a novel adipokine elevated in obesity subjects [74], has been shown to promote obesity-associated PDAC by stimulating the pro-inflammatory response in the TME [75].

2.3.2. Pro-Inflammatory Cytokines and Chemokines

Although the events initiating obesity-associated AT inflammation are not completely understood, increased pro-inflammatory factors produced by adipocytes and infiltrating ATMs or other immune cells in the dysfunctional AT create a systemic and local inflammatory environment. Some of these factors elevated in obese subjects, such as tumor-promoting cytokines (e.g., TNF-α and IL-6) and monocyte chemo-attractants (e.g., MCP-1 [76,77]), are implicated in PDAC development and progression.

Increased expression of TNF-α is observed in obese AT [40,78], where this key pro-inflammatory cytokine is thought to act locally as an autocrine or paracrine cytokine, since systemic TNF-α levels do not necessarily reflect local changes in TNF-α concentrations [78,79]. A central role for TNF-α in obesity-associated early pancreatic tumor promotion has been demonstrated by genetic deletion of tumor necrosis factor receptor 1A (*TNFR1*) [80]. Functionally, TNF-α activates JNK and NF-κB pathways, which are known to promote cell proliferation, invasion, and metastasis in various cancers, including human PDAC, where NF-κB is constitutively activated [81–83]. NF-κB activation by TNF-α can crosstalk with Notch signaling, which cooperates with oncogenic KRAS to promote PDAC progression [84]. In addition, activated NF-κB can stimulate the production of chemokines (such as MCP-1), by adipocytes and pre-adipocytes, leading to increased infiltration of pro-inflammatory macrophages [85,86]. Further, NF-κB is a critical transcription factor of M1 macrophages, regulating the expression of a variety of inflammatory genes, including those encoding IL-6, IL-1β, MCP-1, and cyclooxygenase-2 (COX-2), shown to play important roles in obesity-related PDAC promotion [55]. Adipose expression of IL-6, another key modulator in inflammation-associated tumorigenesis [87,88], is also upregulated in obese individuals [40,79]. IL-6 exerts its pro-tumorigenic and pro-invasive activities through the activation of the STAT3 signaling pathway, which is essential for PDAC development in the Kras-driven mouse model [89–91] and observed in human PDAC [92,93]. The effects of IL-1 are similar to those of TNF-α, and IL1β gene promoter single nucleotide polymorphisms (SNPs) are linked with pancreatic cancer risk [94]. These key cytokines, whose production is increased with obesity, also play critical roles in insulin resistance [78,95], adipokine regulation [96], and tumor stroma modulation [12,97,98], indirectly influencing tumor cell growth. Importantly, these inflammatory cytokines are associated with a poor prognosis in PDAC [99–101]. Overall, PDAC progression is characterized by convergent activation of inflammatory transcriptional factors (i.e., NF-κB and STAT3) [102], which can be stimulated by the pro-inflammatory cytokines whose secretion is markedly enhanced in dysfunctional AT with increased adiposity (especially visceral adiposity).

2.3.3. Microbiota and Obesity-Promoted PDAC

The gut microbiome could be closely related to obesity-associated chronic inflammation though TLR activation in the immune cell compartment, amplifying inflammatory responses and cytokine production that can fuel tumor growth. Although the effects of gut microbiota on tumorigenesis are mostly highlighted in other inflammation-driven, obesity-related gastrointestinal malignancies (e.g., colon and liver cancer) [103], it is well-documented that dysbiosis of oral bacteria is associated with increased risk for pancreatic cancer [104–106]. In experimental mouse models, LPS and TLR4 activation could enhance the severity of acute pancreatitis [107,108]. Also, ligation of TLRs, possibly through NF-κB and MAP kinase pathways, are shown to exacerbate pancreatic fibro-inflammation and accelerate Kras-driven pancreatic tumorigenesis [109–111]. However, less is known regarding how obesity-induced gut dysbiosis and AT inflammation contribute to PDAC development. Based on evidence linking the microbiome and inflammation-associated PDAC, and the notion that metabolic endotoxemia can be a strong promoter of ATM infiltration and M1 polarization, gut bacteria are likely to play a role in the interface of obesity, AT inflammation, and PDAC. Regardless, these interactions need to be further tested in the context of obesity.

2.3.4. Other Factors Associated with AT Dysfunction in Obesity

Obesity and the inflammatory environment are typically associated with oxidative stress, another important feature of dysfunctional AT, conferring genetic instability that can promote the acquisition of oncogenic mutations in neighboring pancreatic cells [112,113]. The elevated oxidative stress can result from enriched reactive oxygen species (ROS) and reactive nitrogen intermediates (RNIs) generated by inflammatory cells, as well as mitochondrial dysfunction and lipid oxidation related to increased release of surplus free fatty acids (FFA) and ectopic fat deposition. Interestingly, in our previous study involving a mouse model of obesity-promoted PDAC, exome sequencing of advanced pancreatic intraepithelial neoplasia (PanIN) lesions identified numerous genetic variants unique to the HFCD (and diet-induced obesity) group [29]. These genetic alterations are found in genes involved in oncogenic pathways that are commonly implied in PDAC, including the insulin and PI3K/Akt pathway.

3. Interventional Perspectives

Based on the increasingly recognized link between obesity-induced AT inflammation and PDAC development, AT inflammation has become an intriguing target for PDAC interception. There are several possibilities to disrupt the promoting effects of (obesity-induced) AT inflammation on PDAC growth and progression. Besides general health-promoting strategies aimed at reducing obesity (e.g., weight reduction), which certainly will positively impact AT inflammation, a detailed understanding of the mechanistic link between AT inflammation and PDAC will lead to targeted approaches. Inhibiting major pro-inflammatory mediators secreted by inflamed AT or blocking cytokine receptors in the pancreas might be a promising approach. For example, anti-TNF-α antibodies and antagonists of IL-1 and IL6 receptors exhibit anti-tumor effects in pre-clinical models of PDAC [101,114–117]. However, the blockade of these cytokines, although proven to be effective in treating other inflammatory diseases, has had limited success in patients of metabolic diseases [118,119] or PDAC [120,121]. Given the complexity of the immune cell changes and inflammatory responses in the AT during obesity, with the multitude of inflammatory mediators, targeting single or even few cytokines simultaneously might not be the best strategy. Blocking major pathways driving the inflammatory response in obese AT seems to be preferable. An intriguing approach with clear and rapid translational potential is the use of Food and Drug Administration (FDA)-approved drugs, e.g., statins.

Statins are lipid-lowering drugs that inhibit HMG-CoA reductase, which plays a central role in the production of cholesterol. High cholesterol levels have been associated with cardiovascular disease (CVD). Statins have been found to reduce CVD and mortality in those who are at high risk [122]. The evidence is strong that statins are effective for treating CVD in the early stages of a disease (secondary prevention) and in those at elevated risk but without CVD (primary prevention). There is intense interest in repurposing statins in cancer [123]. A recent meta-analysis of 95 cohorts including 1,111,407 individuals concluded that statin therapy has potential survival benefit for patients with malignancy [124]. Several large studies support the preventive effect of statins in selected cancers. Although some early epidemiologic studies reported no beneficial effects of statins on PDAC risk [125–128], several recent studies associated statins with a significantly lower risk of cancer, including PDAC [129–134]. In a large case-control study, statin use was associated with a 34% reduced PDAC risk, with a stronger association in male subjects [133]. In a case control study of 408 patients with PDAC and 816 matched controls, statin use was associated with reduced PDAC risk (odds ratio: 0.61; 95% CI: 0.43–0.88) [135]. In this study, the protective effect of statin was dose-dependent and stronger in obese patients. Very recently, a meta-analysis of 26 studies containing more than 3 million participants and 170,000 PDAC patients found a significant decrease in PDAC risk with statin use (RR: 0.84; 95% CI: 0.73–0.97) [136]. In patients with chronic pancreatitis, statins lowered the risk of progression and pancreatic cancer [137]. In addition, statins have been shown to improve survival after resection of early PDAC, indicating a potential benefit for secondary prevention [138–140].

Besides their cholesterol lowering effects, statins are known to have anti-inflammatory properties [141]. Statins attenuate AT inflammation in various human conditions and experimental models [142–145]. The improvement of AT inflammation by statins correlated with a reduction of inflammatory cell infiltration. In addition, statins have been found to decrease circulating MCP-1 levels, which are elevated in obese patients [76,146]. Importantly, statins elicited an anti-inflammatory M_2-polarization of macrophages [145,147]. In our own study using a genetically engineered mouse model of pancreatic cancer promoted by diet-induced obesity, we found that oral administration of simvastatin attenuated early pancreatic neoplastic progression [148], which was accompanied by a marked reduction of inflammation in the VAT (unpublished; Figure 1).

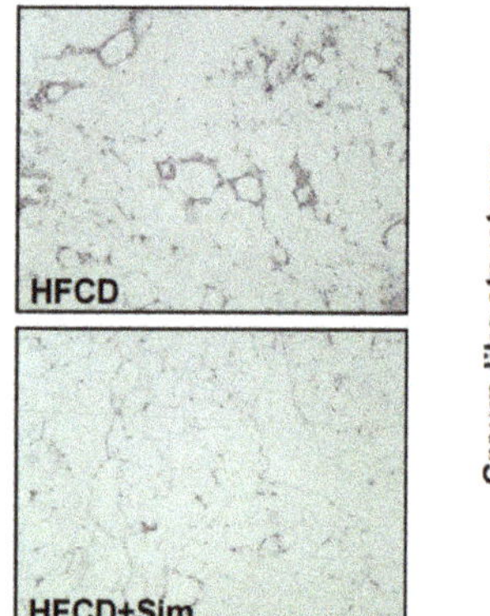

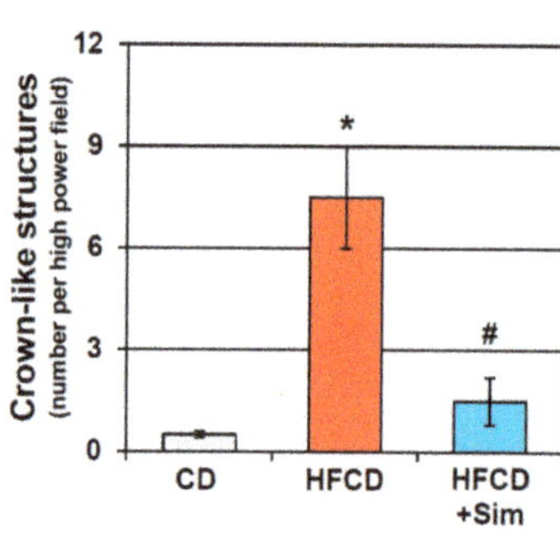

Figure 1. Representative histology of mesenteric adipose tissue of conditional KrasG12D mice fed an obesogenic high fat high calorie diet (HFCD) for 3 months supplemented without (upper panel) or with simvastatin (sim; lower panel). Quantification of crown-like structures demonstrates significant elevation of adipose tissue inflammation in obese HFCD-fed mice, which was abrogated by simvastatin (sim). *: $p < 0.01$ vs. CD; #: $p < 0.01$ vs. HFCD

Mechanistically, statins inhibit 3-hydroxy-3-methlyglutaryl coenzyme (HMG-CoA) reductase, the rate-limiting enzyme that controls the conversion of HMG-CoA to mevalonic acid. Mevalonic acid is a precursor of isoprenoids, which are essential for the post-translational modification of small G-proteins, e.g., Ras homolog gene family, member A (RhoA), cell division control protein 42 homolog (Cdc42), and Ras-related C3 botulinum toxin substrate 1 (Rac1), to attach to lipid membranes. Activation of Rho family members GTPases (guanosine triphosphate hydrolases) modulate the actin cytoskeleton, which has been shown to be an important regulator for macrophage polarization [149]. Our own unpublished data showed that lipophilic statins significantly attenuate pro-inflammatory signaling in macrophage cell lines through inhibition of the mevalonic acid/protein prenylation/actin cytoskeleton pathway. Besides their beneficial effects on AT inflammation, statins also have significant effects on PDAC cells [150–152]. Overall, statins show great promise for PDAC prevention/interception, in particular in the setting of obesity.

4. Conclusions

There is clear evidence that obesity increases the risk of pancreatic cancer. Obesity-induced AT inflammation is increasingly recognized as an important driver of the pathophysiologic process underlying the obesity–PDAC link. Dramatic changes in immune cell number and function in obese AT, especially VAT, with accompanying increased production of various pro-inflammatory cytokines may thereby promote the proliferation and survival of transformed cells in the adjacent pancreas. Although numerous immune cells have been shown to be involved in maintaining sustained AT inflammation in obesity, AT macrophages clearly play a central and critical role in orchestrating the inflammatory response. Given the importance of AT inflammation as a strong promotional driver of obesity-associated PDAC, strategies to inhibit obesity-induced AT inflammation are an intriguing approach. In this regard, the re-purposing of FDA-approved drugs aimed at reducing obesity-induced AT inflammation holds great translational significance. Recent epidemiologic data and preclinical evidence strongly suggest a role of statins, especially lipophilic statins, as promising drugs for PDAC prevention/interception. Besides their direct effects on PDAC cells, current data suggest that the anti-cancer properties of statins might at least partially include their anti-inflammatory effects on obese AT.

Author Contributions: Writing—original draft preparation, H.-H.C.; writing—review and editing, G.E.

Funding: This research was funded by the National Institutes of Health, National Cancer Institute, P01 CA163200, and the Hirshberg Foundation for Pancreatic Cancer Research.

Conflicts of Interest: The authors declare no conflict of interest.

References

1. Siegel, R.L.; Miller, K.D.; Jemal, A. Cancer statistics, 2019. *Ca: A Cancer J. Clin.* **2019**, *69*, 7–34. [CrossRef] [PubMed]
2. Rahib, L.; Smith, B.D.; Aizenberg, R.; Rosenzweig, A.B.; Fleshman, J.M.; Matrisian, L.M. Projecting Cancer Incidence and Deaths to 2030: The Unexpected Burden of Thyroid, Liver, and Pancreas Cancers in the United States. *Cancer Res.* **2014**, *74*, 2913–2921. [CrossRef] [PubMed]
3. Arslan, A.A.; Helzlsouer, K.J.; Kooperberg, C.; Shu, X.O.; Steplowski, E.; Bueno-de-Mesquita, H.B.; Fuchs, C.S.; Gross, M.D.; Jacobs, E.J.; Lacroix, A.Z.; et al. Anthropometric measures, body mass index, and pancreatic cancer: A pooled analysis from the Pancreatic Cancer Cohort Consortium (PanScan). *Arch. Intern. Med.* **2010**, *170*, 791–802. [CrossRef] [PubMed]
4. Bracci, P.M. Obesity and pancreatic cancer: Overview of epidemiologic evidence and biologic mechanisms. *Mol. Carcinog.* **2012**, *51*, 53–63. [CrossRef] [PubMed]
5. Lauby-Secretan, B.; Scoccianti, C.; Loomis, D.; Grosse, Y.; Bianchini, F.; Straif, K. Body Fatness and Cancer — Viewpoint of the IARC Working Group. *N. Engl. J. Med.* **2016**, *375*, 794–798. [CrossRef] [PubMed]
6. Kyrgiou, M.; Kalliala, I.; Markozannes, G.; Gunter, M.J.; Paraskevaidis, E.; Gabra, H.; Martin-Hirsch, P.; Tsilidis, K.K. Adiposity and cancer at major anatomical sites: Umbrella review of the literature. *BMJ* **2017**, *356*, j477. [CrossRef]
7. Aune, D.; Greenwood, D.C.; Chan, D.S.M.; Vieira, R.; Vieira, A.R.; Navarro Rosenblatt, D.A.; Cade, J.E.; Burley, V.J.; Norat, T. Body mass index, abdominal fatness and pancreatic cancer risk: A systematic review and non-linear dose–response meta-analysis of prospective studies. *Ann. Oncol.* **2011**, *23*, 843–852. [CrossRef]
8. Juo, Y.Y.; Gibbons, M.A.M.; Dutson, E.; Lin, A.Y.; Yanagawa, J.; Hines, O.J.; Eibl, G.; Chen, Y. Obesity Is Associated with Early Onset of Gastrointestinal Cancers in California. *J. Obes.* **2018**, *2018*, 7014073. [CrossRef]
9. Islami, F.; Goding Sauer, A.; Miller, K.D.; Siegel, R.L.; Fedewa, S.A.; Jacobs, E.J.; McCullough, M.L.; Patel, A.V.; Ma, J.; Soerjomataram, I.; et al. Proportion and number of cancer cases and deaths attributable to potentially modifiable risk factors in the United States. *Ca: A Cancer J. Clin.* **2018**, *68*, 31–54. [CrossRef]
10. Cascetta, P.; Cavaliere, A.; Piro, G.; Torroni, L.; Santoro, R.; Tortora, G.; Melisi, D.; Carbone, C. Pancreatic Cancer and Obesity: Molecular Mechanisms of Cell Transformation and Chemoresistance. *Int. J. Mol. Sci.* **2018**, *19*, 3331. [CrossRef]
11. Griggs, J.J.; Mangu, P.B.; Anderson, H.; Balaban, E.P.; Dignam, J.J.; Hryniuk, W.M.; Morrison, V.A.; Pini, T.M.; Runowicz, C.D.; Rosner, G.L.; et al. Appropriate Chemotherapy Dosing for Obese Adult Patients With Cancer: American Society of Clinical Oncology Clinical Practice Guideline. *J. Clin. Oncol.* **2012**, *30*, 1553–1561. [CrossRef] [PubMed]
12. Incio, J.; Liu, H.; Suboj, P.; Chin, S.M.; Chen, I.X.; Pinter, M.; Ng, M.R.; Nia, H.T.; Grahovac, J.; Kao, S.; et al. Obesity-Induced Inflammation and Desmoplasia Promote Pancreatic Cancer Progression and Resistance to Chemotherapy. *Cancer Discov.* **2016**, *6*, 852–869. [CrossRef] [PubMed]
13. Liu, J.; Divoux, A.; Sun, J.; Zhang, J.; Clément, K.; Glickman, J.N.; Sukhova, G.K.; Wolters, P.J.; Du, J.; Gorgun, C.Z.; et al. Genetic deficiency and pharmacological stabilization of mast cells reduce diet-induced obesity and diabetes in mice. *Nat. Med.* **2009**, *15*, 940. [CrossRef] [PubMed]
14. Yuan, C.; Bao, Y.; Wu, C.; Kraft, P.; Ogino, S.; Ng, K.; Qian, Z.R.; Rubinson, D.A.; Stampfer, M.J.; Giovannucci, E.L.; et al. Prediagnostic Body Mass Index and Pancreatic Cancer Survival. *J. Clin. Oncol.* **2013**, *31*, 4229–4234. [CrossRef] [PubMed]
15. Murphy, N.; Jenab, M.; Gunter, M.J. Adiposity and gastrointestinal cancers: Epidemiology, mechanisms and future directions. *Nat. Rev. Gastroenterol. Hepatol.* **2018**, *15*, 659–670. [CrossRef] [PubMed]
16. Huxley, R.; Ansary-Moghaddam, A.; Berrington de Gonzalez, A.; Barzi, F.; Woodward, M. Type-II diabetes and pancreatic cancer: A meta-analysis of 36 studies. *Br. J. Cancer* **2005**, *92*, 2076–2083. [CrossRef] [PubMed]
17. Stolzenberg-Solomon, R.Z.; Graubard, B.I.; Chari, S.; Limburg, P.; Taylor, P.R.; Virtamo, J.; Albanes, D. Insulin, glucose, insulin resistance, and pancreatic cancer in male smokers. *JAMA* **2005**, *294*, 2872–2878. [CrossRef] [PubMed]

18. Elena, J.W.; Steplowski, E.; Yu, K.; Hartge, P.; Tobias, G.S.; Brotzman, M.J.; Chanock, S.J.; Stolzenberg-Solomon, R.Z.; Arslan, A.A.; Bueno-de-Mesquita, H.B.; et al. Diabetes and risk of pancreatic cancer: A pooled analysis from the pancreatic cancer cohort consortium. *Cancer Causes Control* **2013**, *24*, 13–25. [CrossRef]

19. Abbruzzese, J.L.; Andersen, D.K.; Borrebaeck, C.A.K.; Chari, S.T.; Costello, E.; Cruz-Monserrate, Z.; Eibl, G.; Engleman, E.G.; Fisher, W.E.; Habtezion, A.; et al. The Interface of Pancreatic Cancer With Diabetes, Obesity, and Inflammation: Research Gaps and Opportunities: Summary of a National Institute of Diabetes and Digestive and Kidney Diseases Workshop. *Pancreas* **2018**, *47*, 516–525. [CrossRef]

20. Andersen, D.K.; Korc, M.; Petersen, G.M.; Eibl, G.; Li, D.; Rickels, M.R.; Chari, S.T.; Abbruzzese, J.L. Diabetes, Pancreatogenic Diabetes, and Pancreatic Cancer. *Diabetes* **2017**, *66*, 1103–1110. [CrossRef]

21. Blüher, M. Adipose tissue inflammation: A cause or consequence of obesity-related insulin resistance? *Clin. Sci.* **2016**, *130*, 1603–1614. [CrossRef] [PubMed]

22. Hertzer, K.M.; Xu, M.; Moro, A.; Dawson, D.W.; Du, L.; Li, G.; Chang, H.-H.; Stark, A.P.; Jung, X.; Hines, O.J.; et al. Robust Early Inflammation of the Peripancreatic Visceral Adipose Tissue During Diet-Induced Obesity in the KrasG12D Model of Pancreatic Cancer. *Pancreas* **2016**, *45*, 458–465. [CrossRef] [PubMed]

23. Tchkonia, T.; Thomou, T.; Zhu, Y.; Karagiannides, I.; Pothoulakis, C.; Jensen, M.D.; Kirkland, J.L. Mechanisms and metabolic implications of regional differences among fat depots. *Cell Metab.* **2013**, *17*, 644–656. [CrossRef] [PubMed]

24. Kwon, H.; Kim, D.; Kim, J.S. Body Fat Distribution and the Risk of Incident Metabolic Syndrome: A Longitudinal Cohort Study. *Sci. Rep.* **2017**, *7*, 10955. [CrossRef] [PubMed]

25. Kwee, T.C.; Kwee, R.M. Abdominal Adiposity and Risk of Pancreatic Cancer. *Pancreas* **2007**, *35*, 285–286. [CrossRef] [PubMed]

26. Wensveen, F.M.; Jelenčić, V.; Valentić, S.; Šestan, M.; Wensveen, T.T.; Theurich, S.; Glasner, A.; Mendrila, D.; Štimac, D.; Wunderlich, F.T.; et al. NK cells link obesity-induced adipose stress to inflammation and insulin resistance. *Nat. Immunol.* **2015**, *16*, 376. [CrossRef] [PubMed]

27. Lysaght, J.; van der Stok, E.P.; Allott, E.H.; Casey, R.; Donohoe, C.L.; Howard, J.M.; McGarrigle, S.A.; Ravi, N.; Reynolds, J.V.; Pidgeon, G.P. Pro-inflammatory and tumour proliferative properties of excess visceral adipose tissue. *Cancer Lett.* **2011**, *312*, 62–72. [CrossRef] [PubMed]

28. Rebours, V.; Gaujoux, S.; d'Assignies, G.; Sauvanet, A.; Ruszniewski, P.; Lévy, P.; Paradis, V.; Bedossa, P.; Couvelard, A. Obesity and Fatty Pancreatic Infiltration Are Risk Factors for Pancreatic Precancerous Lesions (PanIN). *Clin. Cancer Res.* **2015**, *21*, 3522–3528. [CrossRef]

29. Chang, H.-H.; Moro, A.; Takakura, K.; Su, H.-Y.; Mo, A.; Nakanishi, M.; Waldron, R.T.; French, S.W.; Dawson, D.W.; Hines, O.J.; et al. Incidence of pancreatic cancer is dramatically increased by a high fat, high calorie diet in KrasG12D mice. *PLoS ONE* **2017**, *12*, e0184455. [CrossRef]

30. Kusminski, C.M.; Bickel, P.E.; Scherer, P.E. Targeting adipose tissue in the treatment of obesity-associated diabetes. *Nat. Rev. Drug Discov.* **2016**, *15*, 639. [CrossRef]

31. Singh, R.G.; Yoon, H.D.; Wu, L.M.; Lu, J.; Plank, L.D.; Petrov, M.S. Ectopic fat accumulation in the pancreas and its clinical relevance: A systematic review, meta-analysis, and meta-regression. *Metabolism* **2017**, *69*, 1–13. [CrossRef] [PubMed]

32. Alberti-Huber, C.; Hauner, H.; Skurk, T.; Herder, C. Relationship between Adipocyte Size and Adipokine Expression and Secretion. *J. Clin. Endocrinol. Metab.* **2007**, *92*, 1023–1033. [CrossRef]

33. Ouchi, N.; Parker, J.L.; Lugus, J.J.; Walsh, K. Adipokines in inflammation and metabolic disease. *Nat. Rev. Immunol.* **2011**, *11*, 85. [CrossRef] [PubMed]

34. Deng, T.; Lyon, C.J.; Bergin, S.; Caligiuri, M.A.; Hsueh, W.A. Obesity, Inflammation, and Cancer. *Annu. Rev. Pathol.: Mech. Dis.* **2016**, *11*, 421–449. [CrossRef] [PubMed]

35. Mathis, D. Immunological Goings-on in Visceral Adipose Tissue. *Cell Metab.* **2013**, *17*, 851–859. [CrossRef]

36. Wensveen, F.M.; Valentic, S.; Sestan, M.; Turk Wensveen, T.; Polic, B. The "Big Bang" in obese fat: Events initiating obesity-induced adipose tissue inflammation. *Eur. J. Immunol.* **2015**, *45*, 2446–2456. [CrossRef]

37. Mohanan, S.; Horibata, S.; McElwee, J.L.; Dannenberg, A.J.; Coonrod, S.A. Identification of macrophage extracellular trap-like structures in mammary gland adipose tissue: A preliminary study. *Front Immunol.* **2013**, *4*, 67. [CrossRef]

38. Kim, J.; Chung, K.; Choi, C.; Beloor, J.; Ullah, I.; Kim, N.; Lee, K.Y.; Lee, S.-K.; Kumar, P. Silencing CCR2 in Macrophages Alleviates Adipose Tissue Inflammation and the Associated Metabolic Syndrome in Dietary Obese Mice. *Mol. Ther. - Nucleic Acids* **2016**, *5*, e280. [CrossRef]

39. Romeo Giulio, R.; Lee, J.; Shoelson Steven, E. Metabolic Syndrome, Insulin Resistance, and Roles of Inflammation – Mechanisms and Therapeutic Targets. *Arterioscler. Thromb. Vasc. Biol.* **2012**, *32*, 1771–1776. [CrossRef]

40. Weisberg, S.P.; McCann, D.; Desai, M.; Rosenbaum, M.; Leibel, R.L.; Ferrante, A.W., Jr. Obesity is associated with macrophage accumulation in adipose tissue. *J. Clin. Investig.* **2003**, *112*, 1796–1808. [CrossRef]

41. Patsouris, D.; Li, P.-P.; Thapar, D.; Chapman, J.; Olefsky, J.M.; Neels, J.G. Ablation of CD11c-Positive Cells Normalizes Insulin Sensitivity in Obese Insulin Resistant Animals. *Cell Metab.* **2008**, *8*, 301–309. [CrossRef] [PubMed]

42. Kanda, H.; Tateya, S.; Tamori, Y.; Kotani, K.; Hiasa, K.-I.; Kitazawa, R.; Kitazawa, S.; Miyachi, H.; Maeda, S.; Egashira, K.; et al. MCP-1 contributes to macrophage infiltration into adipose tissue, insulin resistance, and hepatic steatosis in obesity. *J. Clin. Investig.* **2006**, *116*, 1494–1505. [CrossRef] [PubMed]

43. Weisberg, S.P.; Hunter, D.; Huber, R.; Lemieux, J.; Slaymaker, S.; Vaddi, K.; Charo, I.; Leibel, R.L.; Ferrante, A.W., Jr. CCR2 modulates inflammatory and metabolic effects of high-fat feeding. *J. Clin. Investig.* **2006**, *116*, 115–124. [CrossRef] [PubMed]

44. Kamei, N.; Tobe, K.; Suzuki, R.; Ohsugi, M.; Watanabe, T.; Kubota, N.; Ohtsuka-Kowatari, N.; Kumagai, K.; Sakamoto, K.; Kobayashi, M.; et al. Overexpression of Monocyte Chemoattractant Protein-1 in Adipose Tissues Causes Macrophage Recruitment and Insulin Resistance. *J. Biol. Chem.* **2006**, *281*, 26602–26614. [CrossRef] [PubMed]

45. Coats, B.R.; Schoenfelt, K.Q.; Barbosa-Lorenzi, V.C.; Peris, E.; Cui, C.; Hoffman, A.; Zhou, G.; Fernandez, S.; Zhai, L.; Hall, B.A.; et al. Metabolically Activated Adipose Tissue Macrophages Perform Detrimental and Beneficial Functions during Diet-Induced Obesity. *Cell Rep.* **2017**, *20*, 3149–3161. [CrossRef]

46. Parks, B.W.; Nam, E.; Org, E.; Kostem, E.; Norheim, F.; Hui, S.T.; Pan, C.; Civelek, M.; Rau, C.D.; Bennett, B.J.; et al. Genetic Control of Obesity and Gut Microbiota Composition in Response to High-Fat, High-Sucrose Diet in Mice. *Cell Metab.* **2013**, *17*, 141–152. [CrossRef] [PubMed]

47. Turnbaugh, P.J.; Bäckhed, F.; Fulton, L.; Gordon, J.I. Diet-Induced Obesity Is Linked to Marked but Reversible Alterations in the Mouse Distal Gut Microbiome. *Cell Host Microbe* **2008**, *3*, 213–223. [CrossRef] [PubMed]

48. Amar, J.; Chabo, C.; Waget, A.; Klopp, P.; Vachoux, C.; Bermúdez-Humarán, L.G.; Smirnova, N.; Bergé, M.; Sulpice, T.; Lahtinen, S.; et al. Intestinal mucosal adherence and translocation of commensal bacteria at the early onset of type 2 diabetes: Molecular mechanisms and probiotic treatment. *Embo Mol. Med.* **2011**, *3*, 559–572. [CrossRef] [PubMed]

49. Boutagy, N.E.; McMillan, R.P.; Frisard, M.I.; Hulver, M.W. Metabolic endotoxemia with obesity: Is it real and is it relevant? *Biochimie* **2016**, *124*, 11–20. [CrossRef]

50. Cani, P.D.; Bibiloni, R.; Knauf, C.; Waget, A.; Neyrinck, A.M.; Delzenne, N.M.; Burcelin, R. Changes in Gut Microbiota Control Metabolic Endotoxemia-Induced Inflammation in High-Fat Diet–Induced Obesity and Diabetes in Mice. *Diabetes* **2008**, *57*, 1470–1481. [CrossRef] [PubMed]

51. Cani, P.D.; Amar, J.; Iglesias, M.A.; Poggi, M.; Knauf, C.; Bastelica, D.; Neyrinck, A.M.; Fava, F.; Tuohy, K.M.; Chabo, C.; et al. Metabolic Endotoxemia Initiates Obesity and Insulin Resistance. *Diabetes* **2007**, *56*, 1761–1772. [CrossRef] [PubMed]

52. Caesar, R.; Reigstad, C.S.; Bäckhed, H.K.; Reinhardt, C.; Ketonen, M.; Östergren Lundén, G.; Cani, P.D.; Bäckhed, F. Gut-derived lipopolysaccharide augments adipose macrophage accumulation but is not essential for impaired glucose or insulin tolerance in mice. *Gut* **2012**, *61*, 1701–1707. [CrossRef] [PubMed]

53. Caesar, R.; Tremaroli, V.; Kovatcheva-Datchary, P.; Cani, P.D.; Bäckhed, F. Crosstalk between Gut Microbiota and Dietary Lipids Aggravates WAT Inflammation through TLR Signaling. *Cell Metab.* **2015**, *22*, 658–668. [CrossRef] [PubMed]

54. Guerra, C.; Schuhmacher, A.J.; Cañamero, M.; Grippo, P.J.; Verdaguer, L.; Pérez-Gallego, L.; Dubus, P.; Sandgren, E.P.; Barbacid, M. Chronic Pancreatitis Is Essential for Induction of Pancreatic Ductal Adenocarcinoma by K-Ras Oncogenes in Adult Mice. *Cancer Cell* **2007**, *11*, 291–302. [CrossRef] [PubMed]

55. Philip, B.; Roland, C.L.; Daniluk, J.; Liu, Y.; Chatterjee, D.; Gomez, S.B.; Ji, B.; Huang, H.; Wang, H.; Fleming, J.B.; et al. A high-fat diet activates oncogenic Kras and COX2 to induce development of pancreatic ductal adenocarcinoma in mice. *Gastroenterology* **2013**, *145*, 1449–1458. [CrossRef] [PubMed]

56. Dawson, D.W.; Hertzer, K.; Moro, A.; Donald, G.; Chang, H.H.; Go, V.L.; Pandol, S.J.; Lugea, A.; Gukovskaya, A.S.; Li, G.; et al. High-fat, high-calorie diet promotes early pancreatic neoplasia in the conditional KrasG12D mouse model. *Cancer Prev. Res. (Phila)* **2013**, *6*, 1064–1073. [CrossRef] [PubMed]

57. Xu, M.; Chang, H.-H.; Jung, X.; Moro, A.; Chou, C.E.N.; King, J.; Hines, O.J.; Sinnett-Smith, J.; Rozengurt, E.; Eibl, G. Deficiency in hormone-sensitive lipase accelerates the development of pancreatic cancer in conditional KrasG12D mice. *BMC Cancer* **2018**, *18*, 797. [CrossRef]

58. Babic, A.; Bao, Y.; Qian, Z.R.; Yuan, C.; Giovannucci, E.L.; Aschard, H.; Kraft, P.; Amundadottir, L.T.; Stolzenberg-Solomon, R.; Morales-Oyarvide, V.; et al. Pancreatic Cancer Risk Associated with Prediagnostic Plasma Levels of Leptin and Leptin Receptor Genetic Polymorphisms. *Cancer Res.* **2016**, *76*, 7160–7167. [CrossRef]

59. Stolzenberg-Solomon, R.Z.; Newton, C.C.; Silverman, D.T.; Pollak, M.; Nogueira, L.M.; Weinstein, S.J.; Albanes, D.; Männistö, S.; Jacobs, E.J. Circulating Leptin and Risk of Pancreatic Cancer: A Pooled Analysis From 3 Cohorts. *Am. J. Epidemiol.* **2015**, *182*, 187–197. [CrossRef]

60. Bao, Y.; Giovannucci, E.L.; Kraft, P.; Stampfer, M.J.; Ogino, S.; Ma, J.; Buring, J.E.; Sesso, H.D.; Lee, I.M.; Gaziano, J.M.; et al. A prospective study of plasma adiponectin and pancreatic cancer risk in five US cohorts. *J. Natl. Cancer Inst.* **2013**, *105*, 95–103. [CrossRef]

61. Zyromski, N.J.; Mathur, A.; Pitt, H.A.; Wade, T.E.; Wang, S.; Nakshatri, P.; Swartz-Basile, D.A.; Nakshatri, H. Obesity potentiates the growth and dissemination of pancreatic cancer. *Surgery* **2009**, *146*, 258–263. [CrossRef] [PubMed]

62. VanSaun, M.N. Molecular Pathways: Adiponectin and Leptin Signaling in Cancer. *Clin. Cancer Res.* **2013**, *19*, 1926–1932. [CrossRef] [PubMed]

63. Grote, V.A.; Rohrmann, S.; Dossus, L.; Nieters, A.; Halkjær, J.; Tjønneland, A.; Overvad, K.; Stegger, J.; Chabbert-Buffet, N.; Boutron-Ruault, M.-C.; et al. The association of circulating adiponectin levels with pancreatic cancer risk: A study within the prospective EPIC cohort. *Int. J. Cancer* **2012**, *130*, 2428–2437. [CrossRef] [PubMed]

64. Mendonsa, A.M.; Chalfant, M.C.; Gorden, L.D.; VanSaun, M.N. Modulation of the Leptin Receptor Mediates Tumor Growth and Migration of Pancreatic Cancer Cells. *PLoS ONE* **2015**, *10*, e0126686. [CrossRef] [PubMed]

65. Huang, B.; Cheng, X.; Wang, D.; Peng, M.; Xue, Z.; Da, Y.; Zhang, N.; Yao, Z.; Li, M.; Xu, A.; et al. Adiponectin promotes pancreatic cancer progression by inhibiting apoptosis via the activation of AMPK/Sirt1/PGC-1alpha signaling. *Oncotarget* **2014**, *5*, 4732–4745. [CrossRef] [PubMed]

66. Pothuraju, R.; Rachagani, S.; Junker, W.M.; Chaudhary, S.; Saraswathi, V.; Kaur, S.; Batra, S.K. Pancreatic cancer associated with obesity and diabetes: An alternative approach for its targeting. *J. Exp. Clin. Cancer Res.* **2018**, *37*, 319. [CrossRef]

67. Dalamaga, M.; Migdalis, I.; Fargnoli, J.L.; Papadavid, E.; Bloom, E.; Mitsiades, N.; Karmaniolas, K.; Pelecanos, N.; Tseleni-Balafouta, S.; Dionyssiou-Asteriou, A.; et al. Pancreatic Cancer Expresses Adiponectin Receptors and Is Associated with Hypoleptinemia and Hyperadiponectinemia: A Case–Control Study. *Cancer Causes Control* **2009**, *20*, 625–633. [CrossRef] [PubMed]

68. Krechler, T.; Zeman, M.; Vecka, M.; Macasek, J.; Jachymova, M.; Zima, T.; Zak, A. Leptin and adiponectin in pancreatic cancer: Connection with diabetes mellitus. *Neoplasma* **2011**, *58*, 58–64. [CrossRef]

69. Petersen, G.M.; Amundadottir, L.; Fuchs, C.S.; Kraft, P.; Stolzenberg-Solomon, R.Z.; Jacobs, K.B.; Arslan, A.A.; Bueno-de-Mesquita, H.B.; Gallinger, S.; Gross, M.; et al. A genome-wide association study identifies pancreatic cancer susceptibility loci on chromosomes 13q22.1, 1q32.1 and 5p15.33. *Nat. Genet* **2010**, *42*, 224–228. [CrossRef]

70. Mathur, A.; Zyromski, N.J.; Pitt, H.A.; Al-Azzawi, H.; Walker, J.J.; Saxena, R.; Lillemoe, K.D. Pancreatic Steatosis Promotes Dissemination and Lethality of Pancreatic Cancer. *J. Am. Coll. Surg.* **2009**, *208*, 989–994. [CrossRef]

71. White, P.B.; True, E.M.; Ziegler, K.M.; Wang, S.S.; Swartz-Basile, D.A.; Pitt, H.A.; Zyromski, N.J. Insulin, Leptin, and Tumoral Adipocytes Promote Murine Pancreatic Cancer Growth. *J. Gastrointest. Surg.* **2010**, *14*, 1888–1894. [CrossRef] [PubMed]

72. Tilg, H.; Moschen, A.R. Adipocytokines: Mediators linking adipose tissue, inflammation and immunity. *Nat. Rev. Immunol.* **2006**, *6*, 772–783. [CrossRef] [PubMed]

73. Carbone, F.; La Rocca, C.; Matarese, G. Immunological functions of leptin and adiponectin. *Biochimie* **2012**, *94*, 2082–2088. [CrossRef] [PubMed]

74. Auguet, T.; Quintero, Y.; Terra, X.; Martínez, S.; Lucas, A.; Pellitero, S.; Aguilar, C.; Hernández, M.; Del Castillo, D.; Richart, C. Upregulation of Lipocalin 2 in Adipose Tissues of Severely Obese Women: Positive Relationship With Proinflammatory Cytokines. *Obesity* **2011**, *19*, 2295–2300. [CrossRef] [PubMed]

75. Gomez-Chou, S.B.; Swidnicka-Siergiejko, A.K.; Badi, N.; Chavez-Tomar, M.; Lesinski, G.B.; Bekaii-Saab, T.; Farren, M.R.; Mace, T.A.; Schmidt, C.; Liu, Y.; et al. Lipocalin-2 Promotes Pancreatic Ductal Adenocarcinoma by Regulating Inflammation in the Tumor Microenvironment. *Cancer Res.* **2017**, *77*, 2647–2660. [CrossRef] [PubMed]

76. Kim, C.S.; Park, H.S.; Kawada, T.; Kim, J.H.; Lim, D.; Hubbard, N.E.; Kwon, B.S.; Erickson, K.L.; Yu, R. Circulating levels of MCP-1 and IL-8 are elevated in human obese subjects and associated with obesity-related parameters. *Int. J. Obes.* **2006**, *30*, 1347–1355. [CrossRef] [PubMed]

77. Huber, J.; Kiefer, F.W.; Zeyda, M.; Ludvik, B.; Silberhumer, G.R.; Prager, G.; Zlabinger, G.J.; Stulnig, T.M. CC Chemokine and CC Chemokine Receptor Profiles in Visceral and Subcutaneous Adipose Tissue Are Altered in Human Obesity. *J. Clin. Endocrinol. Metab.* **2008**, *93*, 3215–3221. [CrossRef]

78. Hotamisligil, G.S.; Arner, P.; Caro, J.F.; Atkinson, R.L.; Spiegelman, B.M. Increased adipose tissue expression of tumor necrosis factor-alpha in human obesity and insulin resistance. *J. Clin. Investig.* **1995**, *95*, 2409–2415. [CrossRef]

79. Kern, P.A.; Ranganathan, S.; Li, C.; Wood, L.; Ranganathan, G. Adipose tissue tumor necrosis factor and interleukin-6 expression in human obesity and insulin resistance. *Am. J. Physiol. -Endocrinol. Metab.* **2001**, *280*, E745–E751. [CrossRef]

80. Khasawneh, J.; Schulz, M.D.; Walch, A.; Rozman, J.; de Angelis, M.H.; Klingenspor, M.; Buck, A.; Schwaiger, M.; Saur, D.; Schmid, R.M.; et al. Inflammation and mitochondrial fatty acid β-oxidation link obesity to early tumor promotion. *Proc. Natl. Acad. Sci.* **2009**, *106*, 3354–3359. [CrossRef]

81. Wang, W.; Abbruzzese, J.L.; Evans, D.B.; Larry, L.; Cleary, K.R.; Chiao, P.J. The Nuclear Factor-κB RelA Transcription Factor Is Constitutively Activated in Human Pancreatic Adenocarcinoma Cells. *Clin. Cancer Res.* **1999**, *5*, 119–127. [PubMed]

82. Maier, H.J.; Schmidt-Straßburger, U.; Huber, M.A.; Wiedemann, E.M.; Beug, H.; Wirth, T. NF-κB promotes epithelial–mesenchymal transition, migration and invasion of pancreatic carcinoma cells. *Cancer Lett.* **2010**, *295*, 214–228. [CrossRef]

83. Kong, R.; Sun, B.; Jiang, H.; Pan, S.; Chen, H.; Wang, S.; Krissansen, G.W.; Sun, X. Downregulation of nuclear factor-κB p65 subunit by small interfering RNA synergizes with gemcitabine to inhibit the growth of pancreatic cancer. *Cancer Lett.* **2010**, *291*, 90–98. [CrossRef] [PubMed]

84. Maniati, E.; Bossard, M.; Cook, N.; Candido, J.B.; Emami-Shahri, N.; Nedospasov, S.A.; Balkwill, F.R.; Tuveson, D.A.; Hagemann, T. Crosstalk between the canonical NF-κB and Notch signaling pathways inhibits Pparγ expression and promotes pancreatic cancer progression in mice. *J. Clin. Investig.* **2011**, *121*, 4685–4699. [CrossRef] [PubMed]

85. Reilly, S.M.; Saltiel, A.R. Adapting to obesity with adipose tissue inflammation. *Nat. Rev. Endocrinol.* **2017**, *13*, 633. [CrossRef] [PubMed]

86. Xu, H.; Barnes, G.T.; Yang, Q.; Tan, G.; Yang, D.; Chou, C.J.; Sole, J.; Nichols, A.; Ross, J.S.; Tartaglia, L.A.; et al. Chronic inflammation in fat plays a crucial role in the development of obesity-related insulin resistance. *J. Clin. Investig.* **2003**, *112*, 1821–1830. [CrossRef] [PubMed]

87. Taniguchi, K.; Karin, M. IL-6 and related cytokines as the critical lynchpins between inflammation and cancer. *Semin. Immunol.* **2014**, *26*, 54–74. [CrossRef] [PubMed]

88. Lesina, M.; Wörmann, S.M.; Neuhöfer, P.; Song, L.; Algül, H. Interleukin-6 in inflammatory and malignant diseases of the pancreas. *Semin. Immunol.* **2014**, *26*, 80–87. [CrossRef]

89. Lesina, M.; Kurkowski, M.U.; Ludes, K.; Rose-John, S.; Treiber, M.; Klöppel, G.; Yoshimura, A.; Reindl, W.; Sipos, B.; Akira, S.; et al. Stat3/Socs3 Activation by IL-6 Transsignaling Promotes Progression of Pancreatic Intraepithelial Neoplasia and Development of Pancreatic Cancer. *Cancer Cell* **2011**, *19*, 456–469. [CrossRef]

90. Fukuda, A.; Wang, S.C.; Morris, J.P.; Folias, A.E.; Liou, A.; Kim, G.E.; Akira, S.; Boucher, K.M.; Firpo, M.A.; Mulvihill, S.J.; et al. Stat3 and MMP7 Contribute to Pancreatic Ductal Adenocarcinoma Initiation and Progression. *Cancer Cell* **2011**, *19*, 441–455. [CrossRef] [PubMed]

91. Kang, R.; Loux, T.; Tang, D.; Schapiro, N.E.; Vernon, P.; Livesey, K.M.; Krasinskas, A.; Lotze, M.T.; Zeh, H.J. The expression of the receptor for advanced glycation endproducts (RAGE) is permissive for early pancreatic neoplasia. *Proc. Natl. Acad. Sci.* **2012**, *109*, 7031–7036. [CrossRef] [PubMed]

92. Scholz, A.; Heinze, S.; Detjen, K.M.; Peters, M.; Welzel, M.; Hauff, P.; Schirner, M.; Wiedenmann, B.; Rosewicz, S. Activated signal transducer and activator of transcription 3 (STAT3) supports the malignant phenotype of human pancreatic cancer. *Gastroenterology* **2003**, *125*, 891–905. [CrossRef]

93. Huang, C.; Huang, R.; Chang, W.; Jiang, T.; Huang, K.; Cao, J.; Sun, X.; Qiu, Z. The expression and clinical significance of pSTAT3, VEGF and VEGF-C in pancreatic adenocarcinoma. *Neoplasma* **2012**, *59*, 52–61. [CrossRef] [PubMed]

94. Hamacher, R.; Diersch, S.; Scheibel, M.; Eckel, F.; Mayr, M.; Rad, R.; Bajbouj, M.; Schmid, R.M.; Saur, D.; Schneider, G. Interleukin 1 beta gene promoter SNPs are associated with risk of pancreatic cancer. *Cytokine* **2009**, *46*, 182–186. [CrossRef] [PubMed]

95. Hotamisligil, G.; Shargill, N.; Spiegelman, B. Adipose expression of tumor necrosis factor-alpha: Direct role in obesity-linked insulin resistance. *Science* **1993**, *259*, 87–91. [CrossRef] [PubMed]

96. Bruun, J.M.; Lihn, A.S.; Verdich, C.; Pedersen, S.B.; Toubro, S.; Astrup, A.; Richelsen, B. Regulation of adiponectin by adipose tissue-derived cytokines: In vivo and in vitro investigations in humans. *Am. J. Physiol. -Endocrinol. Metab.* **2003**, *285*, E527–E533. [CrossRef] [PubMed]

97. Biffi, G.; Oni, T.E.; Spielman, B.; Hao, Y.; Elyada, E.; Park, Y.; Preall, J.; Tuveson, D.A. IL1-Induced JAK/STAT Signaling Is Antagonized by TGFβ to Shape CAF Heterogeneity in Pancreatic Ductal Adenocarcinoma. *Cancer Discov.* **2019**, *9*, 282–301. [CrossRef] [PubMed]

98. Brunetto, E.; De Monte, L.; Balzano, G.; Camisa, B.; Laino, V.; Riba, M.; Heltai, S.; Bianchi, M.; Bordignon, C.; Falconi, M.; et al. The IL-1/IL-1 receptor axis and tumor cell released inflammasome adaptor ASC are key regulators of TSLP secretion by cancer associated fibroblasts in pancreatic cancer. *J. Immunother. Cancer* **2019**, *7*, 45. [CrossRef] [PubMed]

99. Mitsunaga, S.; Ikeda, M.; Shimizu, S.; Ohno, I.; Furuse, J.; Inagaki, M.; Higashi, S.; Kato, H.; Terao, K.; Ochiai, A. Serum levels of IL-6 and IL-1β can predict the efficacy of gemcitabine in patients with advanced pancreatic cancer. *Br. J. Cancer* **2013**, *108*, 2063. [CrossRef] [PubMed]

100. Denley, S.M.; Jamieson, N.B.; McCall, P.; Oien, K.A.; Morton, J.P.; Carter, C.R.; Edwards, J.; McKay, C.J. Activation of the IL-6R/Jak/Stat Pathway is Associated with a Poor Outcome in Resected Pancreatic Ductal Adenocarcinoma. *J. Gastrointest. Surg.* **2013**, *17*, 887–898. [CrossRef]

101. Zhao, X.; Fan, W.; Xu, Z.; Chen, H.; He, Y.; Yang, G.; Yang, G.; Hu, H.; Tang, S.; Wang, P.; et al. Inhibiting tumor necrosis factor-alpha diminishes desmoplasia and inflammation to overcome chemoresistance in pancreatic ductal adenocarcinoma. *Oncotarget* **2016**, *7*, 81110–81122. [CrossRef] [PubMed]

102. Baumgart, S.; Ellenrieder, V.; Fernandez-Zapico, M.E. Oncogenic transcription factors: Cornerstones of inflammation-linked pancreatic carcinogenesis. *Gut* **2013**, *62*, 310–316. [CrossRef] [PubMed]

103. Schwabe, R.F.; Jobin, C. The microbiome and cancer. *Nat. Rev. Cancer* **2013**, *13*, 800. [CrossRef]

104. Farrell, J.J.; Zhang, L.; Zhou, H.; Chia, D.; Elashoff, D.; Akin, D.; Paster, B.J.; Joshipura, K.; Wong, D.T.W. Variations of oral microbiota are associated with pancreatic diseases including pancreatic cancer. *Gut* **2012**, *61*, 582–588. [CrossRef] [PubMed]

105. Fan, X.; Alekseyenko, A.V.; Wu, J.; Peters, B.A.; Jacobs, E.J.; Gapstur, S.M.; Purdue, M.P.; Abnet, C.C.; Stolzenberg-Solomon, R.; Miller, G.; et al. Human oral microbiome and prospective risk for pancreatic cancer: A population-based nested case-control study. *Gut* **2018**, *67*, 120–127. [CrossRef] [PubMed]

106. Michaud, D.S.; Izard, J.; Wilhelm-Benartzi, C.S.; You, D.-H.; Grote, V.A.; Tjønneland, A.; Dahm, C.C.; Overvad, K.; Jenab, M.; Fedirko, V.; et al. Plasma antibodies to oral bacteria and risk of pancreatic cancer in a large European prospective cohort study. *Gut* **2013**, *62*, 1764–1770. [CrossRef] [PubMed]

107. Ding, S.-P.; Li, J.-C.; Jin, C. A mouse model of severe acute pancreatitis induced with caerulein and lipopolysaccharide. *World J. Gastroenterol* **2003**, *9*, 584–589. [CrossRef]

108. Sharif, R.; Dawra, R.; Wasiluk, K.; Phillips, P.; Dudeja, V.; Kurt-Jones, E.; Finberg, R.; Saluja, A. Impact of toll-like receptor 4 on the severity of acute pancreatitis and pancreatitis-associated lung injury in mice. *Gut* **2009**, *58*, 813–819. [CrossRef] [PubMed]

109. Ochi, A.; Nguyen, A.H.; Bedrosian, A.S.; Mushlin, H.M.; Zarbakhsh, S.; Barilla, R.; Zambirinis, C.P.; Fallon, N.C.; Rehman, A.; Pylayeva-Gupta, Y.; et al. MyD88 inhibition amplifies dendritic cell capacity to promote pancreatic carcinogenesis via Th2 cells. *J. Exp. Med.* **2012**, *209*, 1671–1687. [CrossRef]

110. Ochi, A.; Graffeo, C.S.; Zambirinis, C.P.; Rehman, A.; Hackman, M.; Fallon, N.; Barilla, R.M.; Henning, J.R.; Jamal, M.; Rao, R.; et al. Toll-like receptor 7 regulates pancreatic carcinogenesis in mice and humans. *J. Clin. Investig.* **2012**, *122*, 4118–4129. [CrossRef]

111. Daniluk, J.; Liu, Y.; Deng, D.; Chu, J.; Huang, H.; Gaiser, S.; Cruz-Monserrate, Z.; Wang, H.; Ji, B.; Logsdon, C.D. An NF-κB pathway-mediated positive feedback loop amplifies Ras activity to pathological levels in mice. *J. Clin. Investig.* **2012**, *122*, 1519–1528. [CrossRef] [PubMed]

112. Martinez-Useros, J.; Li, W.; Cabeza-Morales, M.; Garcia-Foncillas, J. Oxidative Stress: A New Target for Pancreatic Cancer Prognosis and Treatment. *J. Clin. Med.* **2017**, *6*, 29. [CrossRef] [PubMed]

113. Zhang, L.; Li, J.; Zong, L.; Chen, X.; Chen, K.; Jiang, Z.; Nan, L.; Li, X.; Li, W.; Shan, T.; et al. Reactive Oxygen Species and Targeted Therapy for Pancreatic Cancer. *Oxidative Med. Cell. Longev.* **2016**, *2016*, 9. [CrossRef] [PubMed]

114. Egberts, J.-H.; Cloosters, V.; Noack, A.; Schniewind, B.; Thon, L.; Klose, S.; Kettler, B.; von Forstner, C.; Kneitz, C.; Tepel, J.; et al. Anti–Tumor Necrosis Factor Therapy Inhibits Pancreatic Tumor Growth and Metastasis. *Cancer Res.* **2008**, *68*, 1443–1450. [CrossRef] [PubMed]

115. Zhuang, Z.; Ju, H.-Q.; Aguilar, M.; Gocho, T.; Li, H.; Iida, T.; Lee, H.; Fan, X.; Zhou, H.; Ling, J.; et al. IL1 Receptor Antagonist Inhibits Pancreatic Cancer Growth by Abrogating NF-κB Activation. *Clin. Cancer Res.* **2016**, *22*, 1432–1444. [CrossRef] [PubMed]

116. Long, K.B.; Tooker, G.; Tooker, E.; Luque, S.L.; Lee, J.W.; Pan, X.; Beatty, G.L. IL6 Receptor Blockade Enhances Chemotherapy Efficacy in Pancreatic Ductal Adenocarcinoma. *Mol. Cancer Ther.* **2017**, *16*, 1898–1908. [CrossRef] [PubMed]

117. Padoan, A.; Plebani, M.; Basso, D. Inflammation and Pancreatic Cancer: Focus on Metabolism, Cytokines, and Immunity. *Int. J. Mol. Sci.* **2019**, *20*, 676. [CrossRef] [PubMed]

118. Sloan-Lancaster, J.; Abu-Raddad, E.; Polzer, J.; Miller, J.W.; Scherer, J.C.; De Gaetano, A.; Berg, J.K.; Landschulz, W.H. Double-Blind, Randomized Study Evaluating the Glycemic and Anti-inflammatory Effects of Subcutaneous LY2189102, a Neutralizing IL-1β Antibody, in Patients With Type 2 Diabetes. *Diabetes Care* **2013**, *36*, 2239–2246. [CrossRef] [PubMed]

119. Moller, D.E. Potential role of TNF-alpha in the pathogenesis of insulin resistance and type 2 diabetes. *Trends Endocrinol. Metab* **2000**, *11*, 212–217. [CrossRef]

120. Herman, J.M.; Wild, A.T.; Wang, H.; Tran, P.T.; Chang, K.J.; Taylor, G.E.; Donehower, R.C.; Pawlik, T.M.; Ziegler, M.A.; Cai, H.; et al. Randomized phase III multi-institutional study of TNFerade biologic with fluorouracil and radiotherapy for locally advanced pancreatic cancer: Final results. *J. Clin. Oncol.* **2013**, *31*, 886–894. [CrossRef]

121. Wu, C.; Fernandez, S.A.; Criswell, T.; Chidiac, T.A.; Guttridge, D.; Villalona-Calero, M.; Bekaii-Saab, T.S. Disrupting cytokine signaling in pancreatic cancer: A phase I/II study of etanercept in combination with gemcitabine in patients with advanced disease. *Pancreas* **2013**, *42*, 813–818. [CrossRef] [PubMed]

122. Taylor, F.; Huffman, M.D.; Macedo, A.F.; Moore, T.H.; Burke, M.; Davey Smith, G.; Ward, K.; Ebrahim, S. Statins for the primary prevention of cardiovascular disease. *Cochrane Database Syst. Rev.* **2013**. [CrossRef]

123. Abdullah, M.I.; de Wolf, E.; Jawad, M.J.; Richardson, A. The poor design of clinical trials of statins in oncology may explain their failure - Lessons for drug repurposing. *Cancer Treat Rev.* **2018**, *69*, 84–89. [CrossRef] [PubMed]

124. Mei, Z.; Liang, M.; Li, L.; Zhang, Y.; Wang, Q.; Yang, W. Effects of statins on cancer mortality and progression: A systematic review and meta-analysis of 95 cohorts including 1,111,407 individuals. *Int. J. Cancer* **2017**, *140*, 1068–1081. [CrossRef] [PubMed]

125. Bonovas, S.; Filioussi, K.; Sitaras, N.M. Statins are not associated with a reduced risk of pancreatic cancer at the population level, when taken at low doses for managing hypercholesterolemia: Evidence from a meta-analysis of 12 studies. *Am. J. Gastroenterol* **2008**, *103*, 2646–2651. [CrossRef] [PubMed]

126. Bradley, M.C.; Hughes, C.M.; Cantwell, M.M.; Murray, L.J. Statins and pancreatic cancer risk: A nested case-control study. *Cancer Causes Control* **2010**, *21*, 2093–2100. [CrossRef] [PubMed]

127. Jacobs, E.J.; Newton, C.C.; Thun, M.J.; Gapstur, S.M. Long-term use of cholesterol-lowering drugs and cancer incidence in a large United States cohort. *Cancer Res.* **2011**, *71*, 1763–1771. [CrossRef]

128. Simon, M.S.; Desai, P.; Wallace, R.; Wu, C.; Howard, B.V.; Martin, L.W.; Schlecht, N.; Liu, S.; Jay, A.; LeBlanc, E.S.; et al. Prospective analysis of association between statins and pancreatic cancer risk in the Women's Health Initiative. *Cancer Causes Control* **2016**, *27*, 415–423. [CrossRef]

129. Zhou, Y.Y.; Zhu, G.Q.; Wang, Y.; Zheng, J.N.; Ruan, L.Y.; Cheng, Z.; Hu, B.; Fu, S.W.; Zheng, M.H. Systematic review with network meta-analysis: Statins and risk of hepatocellular carcinoma. *Oncotarget* **2016**. [CrossRef]

130. Vallianou, N.G.; Kostantinou, A.; Kougias, M.; Kazazis, C. Statins and cancer. *Anticancer Agents Med. Chem.* **2014**, *14*, 706–712. [CrossRef]

131. Gronich, N.; Rennert, G. Beyond aspirin-cancer prevention with statins, metformin and bisphosphonates. *Nat. Rev. Clin. Oncol.* **2013**, *10*, 625–642. [CrossRef] [PubMed]

132. Chen, M.J.; Tsan, Y.T.; Liou, J.M.; Lee, Y.C.; Wu, M.S.; Chiu, H.M.; Wang, H.P.; Chen, P.C. Statins and the risk of pancreatic cancer in Type 2 diabetic patients–A population-based cohort study. *Int. J. Cancer* **2016**, *138*, 594–603. [CrossRef] [PubMed]

133. Walker, E.J.; Ko, A.H.; Holly, E.A.; Bracci, P.M. Statin use and risk of pancreatic cancer: Results from a large, clinic-based case-control study. *Cancer* **2015**, *121*, 1287–1294. [CrossRef] [PubMed]

134. Carey, F.J.; Little, M.W.; Pugh, T.F.; Ndokera, R.; Ing, H.; Clark, A.; Dennison, A.; Metcalfe, M.S.; Robinson, R.J.; Hart, A.R. The differential effects of statins on the risk of developing pancreatic cancer: A case-control study in two centres in the United Kingdom. *Dig. Dis. Sci.* **2013**, *58*, 3308–3312. [CrossRef] [PubMed]

135. Archibugi, L.; Piciucchi, M.; Stigliano, S.; Valente, R.; Zerboni, G.; Barucca, V.; Milella, M.; Maisonneuve, P.; Delle Fave, G.; Capurso, G. Exclusive and Combined Use of Statins and Aspirin and the Risk of Pancreatic Cancer: A Case-Control Study. *Sci. Rep.* **2017**, *7*, 13024. [CrossRef]

136. Zhang, Y.; Liang, M.; Sun, C.; Qu, G.; Shi, T.; Min, M.; Wu, Y.; Sun, Y. Statin Use and Risk of Pancreatic Cancer: An Updated Meta-analysis of 26 Studies. *Pancreas* **2019**, *48*, 142–150. [CrossRef]

137. Bang, U.C.; Watanabe, T.; Bendtsen, F. The relationship between the use of statins and mortality, severity, and pancreatic cancer in Danish patients with chronic pancreatitis. *Eur. J. Gastroenterol Hepatol.* **2018**, *30*, 346–351. [CrossRef]

138. Jeon, C.Y.; Pandol, S.J.; Wu, B.; Cook-Wiens, G.; Gottlieb, R.A.; Merz, C.N.; Goodman, M.T. The association of statin use after cancer diagnosis with survival in pancreatic cancer patients: A SEER-medicare analysis. *PLoS ONE* **2015**, *10*, e0121783. [CrossRef]

139. Wu, B.U.; Chang, J.; Jeon, C.Y.; Pandol, S.J.; Huang, B.; Ngor, E.W.; Difronzo, A.L.; Cooper, R.M. Impact of statin use on survival in patients undergoing resection for early-stage pancreatic cancer. *Am. J. Gastroenterol* **2015**, *110*, 1233–1239. [CrossRef]

140. Lee, H.S.; Lee, S.H.; Lee, H.J.; Chung, M.J.; Park, J.Y.; Park, S.W.; Song, S.Y.; Bang, S. Statin Use and Its Impact on Survival in Pancreatic Cancer Patients. *Med. (Baltim.)* **2016**, *95*, e3607. [CrossRef]

141. Jain, M.K.; Ridker, P.M. Anti-inflammatory effects of statins: Clinical evidence and basic mechanisms. *Nat. Rev. Drug Discov.* **2005**, *4*, 977–987. [CrossRef] [PubMed]

142. Krysiak, R.; Zmuda, W.; Okopien, B. The effect of simvastatin-ezetimibe combination therapy on adipose tissue hormones and systemic inflammation in patients with isolated hypercholesterolemia. *Cardiovasc* **2014**, *32*, 40–46. [CrossRef] [PubMed]

143. Busnelli, M.; Manzini, S.; Froio, A.; Vargiolu, A.; Cerrito, M.G.; Smolenski, R.T.; Giunti, M.; Cinti, A.; Zannoni, A.; Leone, B.E.; et al. Diet induced mild hypercholesterolemia in pigs: Local and systemic inflammation, effects on vascular injury - rescue by high-dose statin treatment. *PLoS ONE* **2013**, *8*, e80588. [CrossRef] [PubMed]

144. Calisto, K.L.; Carvalho Bde, M.; Ropelle, E.R.; Mittestainer, F.C.; Camacho, A.C.; Guadagnini, D.; Carvalheira, J.B.; Saad, M.J. Atorvastatin improves survival in septic rats: Effect on tissue inflammatory pathway and on insulin signaling. *PLoS ONE* **2010**, *5*, e14232. [CrossRef] [PubMed]

145. Khan, T.; Hamilton, M.P.; Mundy, D.I.; Chua, S.C.; Scherer, P.E. Impact of simvastatin on adipose tissue: Pleiotropic effects in vivo. *Endocrinology* **2009**, *150*, 5262–5272. [CrossRef] [PubMed]

146. Blanco-Colio, L.M.; Martin-Ventura, J.L.; de Teresa, E.; Farsang, C.; Gaw, A.; Gensini, G.; Leiter, L.A.; Langer, A.; Martineau, P.; Egido, J.; et al. Elevated ICAM-1 and MCP-1 plasma levels in subjects at high cardiovascular risk are diminished by atorvastatin treatment. Atorvastatin on Inflammatory Markers study: A substudy of Achieve Cholesterol Targets Fast with Atorvastatin Stratified Titration. *Am. Heart J.* **2007**, *153*, 881–888. [CrossRef]

147. Abe, M.; Matsuda, M.; Kobayashi, H.; Miyata, Y.; Nakayama, Y.; Komuro, R.; Fukuhara, A.; Shimomura, I. Effects of statins on adipose tissue inflammation: Their inhibitory effect on MyD88-independent IRF3/IFN-beta pathway in macrophages. *Arter. Thromb Vasc Biol* **2008**, *28*, 871–877. [CrossRef]

148. Hao, F.; Xu, Q.; Wang, J.; Yu, S.; Chang, H.H.; Sinnett-Smith, J.; Eibl, G.; Rozengurt, E. Lipophilic statins inhibit YAP nuclear localization, co-activator activity and colony formation in pancreatic cancer cells and prevent the initial stages of pancreatic ductal adenocarcinoma in KrasG12D mice. *PLoS ONE* **2019**, *14*, e0216603. [CrossRef]

149. McWhorter, F.Y.; Wang, T.; Nguyen, P.; Chung, T.; Liu, W.F. Modulation of macrophage phenotype by cell shape. *Proc. Natl. Acad. Sci. USA* **2013**, *110*, 17253–17258. [CrossRef]

150. Rozengurt, E.; Eibl, G. Central role of Yes-associated protein and WW-domain-containing transcriptional co-activator with PDZ-binding motif in pancreatic cancer development. *World J. Gastroenterol* **2019**, *25*, 1797–1816. [CrossRef]

151. Eibl, G.; Rozengurt, E. KRAS, YAP, and obesity in pancreatic cancer: A signaling network with multiple loops. *Semin Cancer Biol.* **2019**, *54*, 50–62. [CrossRef] [PubMed]

152. Rozengurt, E.; Sinnett-Smith, J.; Eibl, G. Yes-associated protein (YAP) in pancreatic cancer: At the epicenter of a targetable signaling network associated with patient survival. *Signal Transduct Target* **2018**, *3*, 11. [CrossRef] [PubMed]

Review

New Insights into the Liver–Visceral Adipose Axis During Hepatic Resection and Liver Transplantation

María Eugenia Cornide-Petronio [1,†], Mónica B. Jiménez-Castro [1,†], Jordi Gracia-Sancho [2,3] and Carmen Peralta [1,3,*]

1   Institut d'Investigacions Biomèdiques August Pi I Sunyer (IDIBAPS), 08036 Barcelona, Spain;
    cornide@clinic.cat (M.E.C.-P.); monicabjimenez@hotmail.com (M.B.J.-C.)
2   Liver Vascular Biology Research Group, Barcelona Hepatic Hemodynamic Laboratory IDIBAPS,
    08036 Barcelona, Spain; jgracia@clinic.cat
3   Centro de Investigación Biomédica en Red de Enfermedades Hepáticas y Digestivas (CIBERehd),
    08036 Barcelona, Spain
*   Correspondence: cperalta@clinic.cat; Tel.: +34-93-227-5400
†   These authors equally contributed to this work.

Received: 11 July 2019; Accepted: 17 September 2019; Published: 18 September 2019

Abstract: In the last decade, adipose tissue has emerged as an endocrine organ with a key role in energy homeostasis. In addition, there is close crosstalk between the adipose tissue and the liver, since pro- and anti-inflammatory substances produced at the visceral adipose tissue level directly target the liver through the portal vein. During surgical procedures, including hepatic resection and liver transplantation, ischemia–reperfusion injury induces damage and regenerative failure. It has been suggested that adipose tissue is associated with both pathological or, on the contrary, with protective effects on damage and regenerative response after liver surgery. The present review aims to summarize the current knowledge on the crosstalk between the adipose tissue and the liver during liver surgery. Therapeutic strategies as well as the clinical and scientific controversies in this field are discussed. The different experimental models, such as lipectomy, to evaluate the role of adipose tissue in both steatotic and nonsteatotic livers undergoing surgery, are described. Such information may be useful for the establishment of protective strategies aimed at regulating the liver–visceral adipose tissue axis and improving the postoperative outcomes in clinical liver surgery.

Keywords: adipose tissue; liver; inflammation; steatosis; liver resection; liver transplantation; lipectomy

---

1. Introduction

In the last decade, adipose tissue has emerged as an essential and highly active metabolic and endocrine organ [1–3]. The basic function of adipocytes is to take up free fatty acids (FFA) from circulating lipoprotein complexes and esterify them into triacylglycerides [4]. During times of metabolic demand, hydrolysis of triacylglyceride releases FFA to generate adenosine triphosphate (ATP) [5]. These adipocyte processes, termed lipogenesis and lipolysis, respectively, are primarily governed through hormonal pathways [6]. However, one of the most important characteristics of adipose tissue is its function in whole-body energy homeostasis, mediated principally through the endocrine system [4]. Adipose tissue expresses and secretes a variety of bioactive molecules, known as adipokines, which may exert their effects in adipose tissue and in other organs [7]. Adipokines include leptin, interleukin (IL)-6, other cytokines, adiponectin, complement components, adipsin, plasminogen activator inhibitor-1 (PAI-1), and proteins of the renin–angiotensin system, among others [7]. Collectively, adipokines modulate the crosstalk between adipose tissue and other metabolic organs, including the liver [8].

Thus, adipokines directly target the liver through the portal vein [9] and have significant effects on liver diseases [4].

The hypoxia and subsequent oxygen delivery restoration to the liver, namely, hepatic ischemia–reperfusion (I/R), is one of the major pathophysiological events and causes of morbidity and mortality in liver resections and transplantation, being more evident in the presence of hepatic steatosis [10–14]. Despite the attempts to solve this issue, hepatic I/R is an unresolved problem in clinical practice [15]. The cellular mechanisms involved in liver I/R injury are numerous and complicated [14], which led to discrepancies in our understanding of this pathology [16]. For instance, the mechanisms underlying I/R injury in conditions of cold ischemia associated with liver transplantation (LT) are different from those that occur in conditions of warm ischemia associated with liver resections. In addition, hepatic steatosis is associated with an increased postoperative complication index and mortality after liver resection and transplantation, and the mechanisms responsible for hepatic damage and regenerative failure are different in steatotic versus nonsteatotic livers [15]. The investigations focused on the role of adipose tissue are of clinical and scientific relevance since the prevalence of obesity ranges from 24–45% of the population and consequently is expected to increase the number of steatotic livers submitted to surgery, which poorly tolerate I/R damage, resulting in liver dysfunction and regenerative failure [17–23]. In addition, it has been reported that adipose tissue exerts both pathological or, on the contrary, protective effects on damage and regenerative response [24]. It should be noted that functional differences between lean and obese adipose tissue have been extensively described [25–27] and summarized, as seen in Figure 1. Briefly, adipose tissue from lean individuals is a connective tissue of low density with small insulin-sensitive adipocytes that secrete adipokines involved in energy homeostasis, angiogenesis, and antioxidant processes. However, the rigidity of adipose tissue from obese individuals is caused by the increment of connective fiber content. Hypertrophic insulin-resistant adipocytes secrete different inflammatory mediators, resulting in adipose tissue dysfunction, impaired angiogenesis, and cell death [25–27]. Moreover, obesity induces changes in the secretion of adipokines from adipose tissue to the circulation [28–30] and increases the inflammatory response and oxidative stress in adipose tissue [31–35]. Therefore, investigations focused on evaluating the liver–adipose tissue axis in steatotic and nonsteatotic livers subjected to hepatic resections or transplants are highly useful in the establishment of specific therapies to prevent both hepatic I/R injury and regenerative failure in liver surgery.

In the first part of this review, we highlight the actual knowledge of the crosstalk between the adipose tissue and the liver during liver surgery. In addition, the different experimental models and pharmacological strategies aimed at regulating potential dysfunctions in the adipose tissue–liver axis in liver surgery are presented, focusing on the strengths and limitations. Clinical results on the role of adipose tissue in the postoperative outcomes after liver surgery are also discussed.

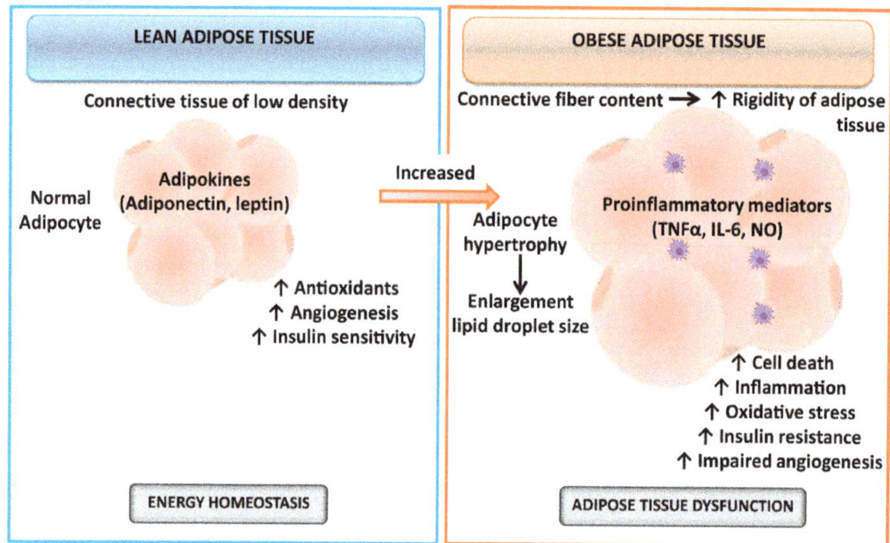

Figure 1. Schematic illustration of functional differences between lean and obese adipose tissue. Abbreviations: IL, interleukin; NO, nitric oxide; TNFα, tumor necrosis factor α.

2. Relevance of Adipose Tissue in Experimental Models of Liver Resection

Crosstalk between the adipose tissue and the liver is an important event both in the physiological function of the liver and in the development of liver diseases [4]. Indeed, adipose tissue inflammation is a well-recognized sign of obesity [36,37], one of its major consequences being the alteration of the secretion of adipokines that drain to the liver via the portal vein, a notion known as the "portal theory" [37–39]. This dysfunctional adipose tissue–liver axis is supported by the specific disruption in adipocytes of inflammatory mediators (apoptosis antigen 1 or cluster of differentiation 95; Fas/CD95) and/or inflammatory signals (c-Jun N-terminal kinase-1, JNK1) in different mouse models, resulting in protection against hepatic steatosis [40,41].

It has been suggested that lipids are the preferred energy substrate for nonsteatotic livers in conditions of partial hepatectomy (PH) without I/R [42–44]. Briefly, hypoglycemia that follows PH induces catabolism of peripheral adipose stores followed by hepatic accumulation of systemically derived fat and subsequent liver regeneration [45–47]. This is supported by the fact that glucose administration could block the mobilization of fatty acids from adipose tissue by the liver to obtain energy [48]. Parameters of lipid metabolism have been reported during hepatic regeneration: esterification rate of fatty acids from adipose tissue is higher and lipogenesis is raised [49,50]. In fact, the remaining liver after PH expresses the lipoprotein lipase, which could take up fatty acids from circulating triacylglycerides [51]. Moreover, different studies have described that mice lacking lipid metabolism-associated genes have reduced hepatic adipogenesis and regeneration liver failure [52,53]. In line with this, the surgical relevance of the lipid lowering effect of omega-3 fatty acids has been studied in steatotic livers in the setting of hepatic I/R injury without PH [54] or in PH without I/R [55] (Figure 2). It should be noted that the process of liver regeneration requires careful regulation of lipid accumulation. In fact, Yang et al. suggested that Smad interacting protein-1 (SIP1) (a key factor linked to the transforming growth factor-β (TGF-β), bone morphogenetic protein (BMP), and Wnt signaling pathways) is one of the mechanisms involved in the hepatic lipid accumulation and, consequently, in the process of liver regeneration under PH without I/R [56] (Figure 2). It is well known that in order for the liver to regenerate, the provision of fatty acids, phospholipids, and cholesterol to the liver is essential for the maintenance of the rate of formation of the membranes of dividing liver cells [50,51]. On the other hand, important questions for steatotic livers arise from these data since further studies will be required to elucidate whether steatosis may be reduced to avoid the vulnerability of steatotic livers to I/R and the regenerative failure, or instead if drugs aimed at increasing the levels of hepatic triglycerides should be used during surgery and thus conserve the energy required for liver regeneration.

The studies mentioned above in liver surgery have been reported in settings of I/R without PH or in PH without I/R. However, it should be noted that in the clinical setting, PH is usually performed under vascular occlusion. Thus, if our aim is the establishment of new protective strategies in the clinical setting of hepatic resection, the experimental conditions used at the bench-side should simulate as close as possible the clinical reality. Fortunately, some studies have evaluated the contribution of adipose tissue to liver injury and regeneration in PH with I/R conditions. Firstly, Mendes-Braz et al. demonstrated that the relevance of adipose tissue in hepatic damage and regeneration depends on the type of liver [57]. In this sense, adipose tissue is not required for the regeneration of nonsteatotic livers subjected to PH with I/R. In contrast, it is necessary to promote regeneration and reduce injury in steatotic livers. Taking these data into account, glucose or lipid emulsion was administered in obese and lean animals undergoing PH + I/R. Glucose or lipid treatment in nonsteatotic livers protected against hepatic damage and regenerative failure. In obese animals, glucose treatment did not protect steatotic livers against damage but improved their regeneration. However, lipid treatment conferred protection against damage and regenerative failure [57]. Mendes-Braz et al. suggest that in addition to the function of adipose tissue as a lipid precursor for new membrane synthesis, the requirement of systemic adipose stores during regeneration of steatotic livers might be based on the endocrine role of adipose tissue as a source of different adipokines, which are essential signals for liver regeneration [57].

In addition to the studies related to the role of adipose tissue as a source of energy substrates or inductors of hepatic lipid accumulation, other studies have investigated the potential contribution of adipose tissue as a source of bioactive molecules such as visfatin, cortisol, and soluble forms of the VEGF receptor 1 (sFlt1).

Elias-Miró et al. found the injurious effects of visfatin in PH with I/R and that steatotic livers were more vulnerable to upregulated visfatin than nonsteatotic livers. The administration of visfatin exacerbated damage and regenerative failure in steatotic livers following PH with I/R. Treatment with resistin maintained low levels of visfatin in steatotic livers by blocking its hepatic reuptake from adipose tissue and consequently prevented the injurious effects of visfatin on hepatic damage and regenerative failure [58] (Figure 2).

In pathologic states, adipose tissue may also secret a range of hormones including cortisol, which may be taken up from the circulation by the liver [1,59]. Cornide-Petronio et al. reported that in instances of PH with I/R, the contributory potential of adipose tissue (as a cortisol source) is dependent on baseline liver status (steatotic versus nonsteatotic livers). In such surgical conditions, the authors found that cortisol levels in adipose tissue and liver were elevated only in obese animals [59]. In addition, cortisol administration under PH with I/R conditions exacerbated hepatic damage and regenerative failure only in obese animals. Indeed, in obese animals, alterations in enzymatic regulation of cortisol metabolism caused cortisol accumulation in steatotic livers, whereas in lean animals, compensatory mechanisms mainly based on the clearance of hepatic cortisol were shown to prevent intrahepatic cortisol and its deleterious effects [59].

Interestingly, Bujaldon et al. recently examined the effects of vascular endothelial growth factor type A (VEGFA) on damage and regeneration in steatotic and nonsteatotic livers submitted to PH with I/R. The authors reported that VEGFA levels were decreased in both steatotic and nonsteatotic livers after surgery, but the exogenous VEGFA administrated was only able to reach nonsteatotic livers, reducing the incidence of postoperative complications following surgery. Unexpectedly, the authors found that circulating VEGFA was sequestered by the high circulating levels of the sFlt1 released from adipose tissue, so VEGFA could not reach the steatotic liver to exert its effects, ultimately exacerbating damage and regenerative failure [60]. Thus, the concomitant administration of VEGFA and an antibody against sFlt1 was required to avoid binding of sFlt1 to VEGFA. This was associated with high VEGFA levels in steatotic livers and protection against damage and regenerative failure [60] (Figure 2).

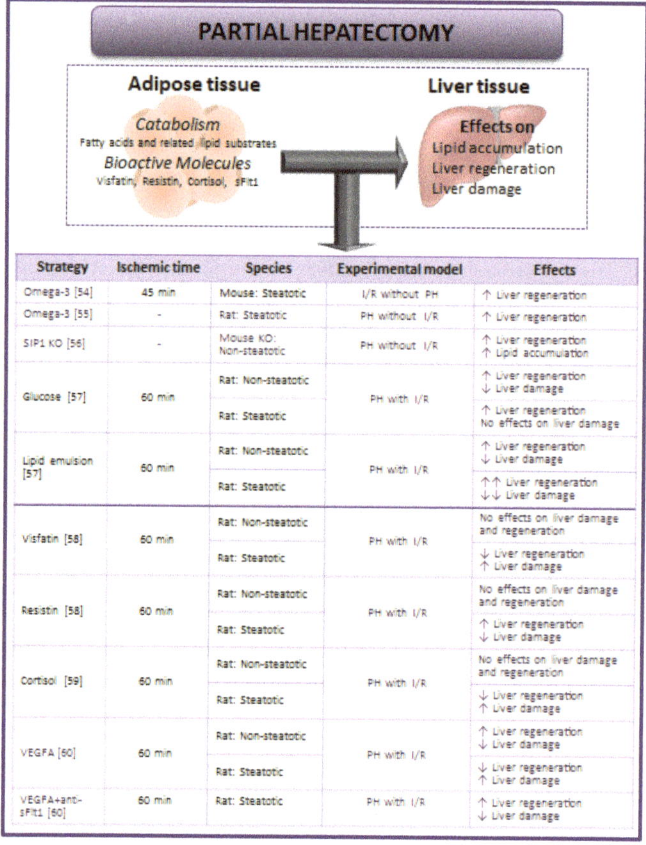

Strategy	Ischemic time	Species	Experimental model	Effects
Omega-3 [54]	45 min	Mouse: Steatotic	I/R without PH	↑ Liver regeneration
Omega-3 [55]	-	Rat: Steatotic	PH without I/R	↑ Liver regeneration
SIP1 KO [56]	-	Mouse KO: Non-steatotic	PH without I/R	↑ Liver regeneration ↑ Lipid accumulation
Glucose [57]	60 min	Rat: Non-steatotic	PH with I/R	↑ Liver regeneration ↓ Liver damage
		Rat: Steatotic		↑ Liver regeneration No effects on liver damage
Lipid emulsion [57]	60 min	Rat: Non-steatotic	PH with I/R	↑ Liver regeneration ↓ Liver damage
		Rat: Steatotic		↑↑ Liver regeneration ↓↓ Liver damage
Visfatin [58]	60 min	Rat: Non-steatotic	PH with I/R	No effects on liver damage and regeneration
		Rat: Steatotic		↓ Liver regeneration ↑ Liver damage
Resistin [58]	60 min	Rat: Non-steatotic	PH with I/R	No effects on liver damage and regeneration
		Rat: Steatotic		↑ Liver regeneration ↓ Liver damage
Cortisol [59]	60 min	Rat: Non-steatotic	PH with I/R	No effects on liver damage and regeneration
		Rat: Steatotic		↓ Liver regeneration ↑ Liver damage
VEGFA [60]	60 min	Rat: Non-steatotic	PH with I/R	↑ Liver regeneration ↓ Liver damage
		Rat: Steatotic		↓ Liver regeneration ↑ Liver damage
VEGFA+anti-sFlt1 [60]	60 min	Rat: Steatotic	PH with I/R	↑ Liver regeneration ↓ Liver damage

Figure 2. Strategies aimed at regulating hepatic damage and regenerative failure considering the adipose tissue–liver crosstalk during partial hepatectomy (PH) with or without I/R [54–60]. Abbreviations: I/R, ischemia–reperfusion; PH, partial hepatectomy; sFlt1, soluble form of the VEGF receptor 1; SIP1, smad interacting protein 1; VEGFA, vascular endothelial growth factor type A.

3. Relevance of Adipose Tissue in Experimental Models of Liver Transplantation

To our knowledge, the relevance of adipose tissue (as a source of fatty acids and related lipid substrates as well as bioactive molecules) on lipid metabolism, hepatic damage, and regeneration associated with transplantation remains to be elucidated (Figure 3). In addition, the few experimental studies on LT [61–63] have described the levels of adipokines in the liver but not in adipose tissue. In this vein, it has been demonstrated in experimental studies that adiponectin, resistin, and visfatin levels were not modified in recipients when nonsteatotic livers were subjected to transplantation, whereas in recipients of steatotic liver grafts, the presence of hepatic steatosis down-regulated both adiponectin and resistin levels under such surgical conditions, whereas no changes in visfatin levels were observed [61,62]. The role of adipose tissue as a potential source of adiponectin, resistin, or visfatin was unexplored. Nevertheless, the effects of such bioactive molecules on hepatic damage and regenerative failure were investigated in steatotic and nonsteatotic livers under PH with I/R conditions. As expected, hepatic damage in recipients of steatotic liver grafts was unaltered under pharmacological regulation of visfatin. However, the treatment with either exogenous adiponectin or resistin in steatotic liver grafts improved the postoperative outcomes after transplantation. In addition, the activation of adenosine monophosphate-activated protein kinase (AMPK) by pharmacological drugs such as AICAR (cell-permeable adenosine analog that is a selective activator of AMPK) or ischemic preconditioning (PC)—which increased both adiponectin and resistin in steatotic liver grafts of recipients submitted to transplantation—resulted in protection against hepatic damage [62]. In experimental models of ex vivo LT, it has been reported that the addition of leptin to preservation solutions was able to increase the signal transducer and activator of transcription-3 levels and to reduce damage in nonsteatotic grafts submitted to transplantation [63].

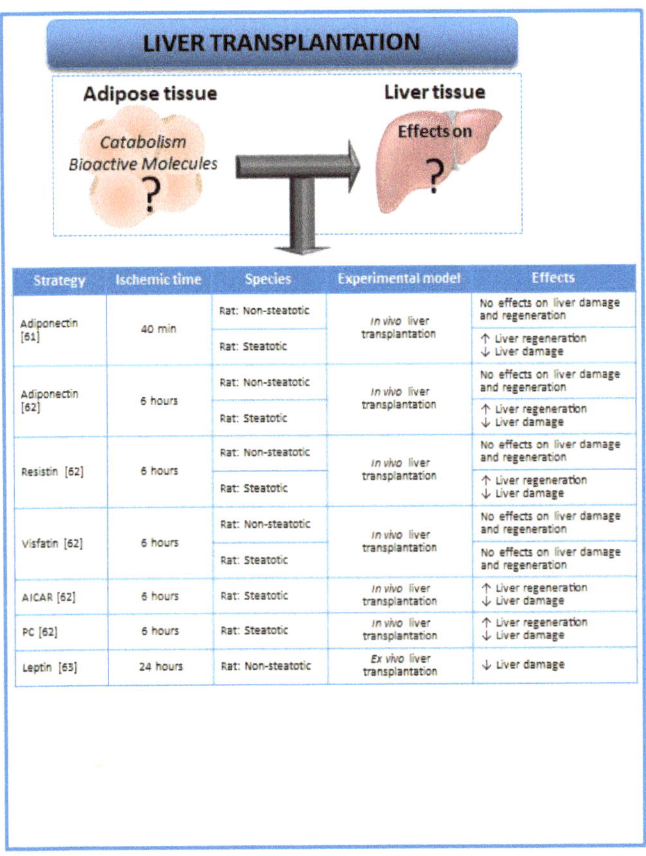

Strategy	Ischemic time	Species	Experimental model	Effects
Adiponectin [61]	40 min	Rat: Non-steatotic	*In vivo* liver transplantation	No effects on liver damage and regeneration
		Rat: Steatotic		↑ Liver regeneration ↓ Liver damage
Adiponectin [62]	6 hours	Rat: Non-steatotic	*In vivo* liver transplantation	No effects on liver damage and regeneration
		Rat: Steatotic		↑ Liver regeneration ↓ Liver damage
Resistin [62]	6 hours	Rat: Non-steatotic	*In vivo* liver transplantation	No effects on liver damage and regeneration
		Rat: Steatotic		↑ Liver regeneration ↓ Liver damage
Visfatin [62]	6 hours	Rat: Non-steatotic	*In vivo* liver transplantation	No effects on liver damage and regeneration
		Rat: Steatotic		No effects on liver damage and regeneration
AICAR [62]	6 hours	Rat: Steatotic	*In vivo* liver transplantation	↑ Liver regeneration ↓ Liver damage
PC [62]	6 hours	Rat: Steatotic	*In vivo* liver transplantation	↑ Liver regeneration ↓ Liver damage
Leptin [63]	24 hours	Rat: Non-steatotic	*Ex vivo* liver transplantation	↓ Liver damage

Figure 3. Strategies aimed at evaluating the role of adipokines on hepatic damage and regenerative failure in LT [61–63]. The adipose tissue–liver crosstalk during LT is still unknown. Abbreviations: AICAR, Cell-permeable adenosine analog that is a selective activator of AMPK; LT, liver transplantation; PC, preconditioning.

All the studies mentioned above [61–63] reported the levels as well as the role of adipokines in liver grafts in experimental models of LT. However, the levels of adipokines in adipose tissue as well as the potential involvement of adipose tissue in the hepatic levels of adipokines following transplantation were not evaluated in such studies. In our view, further investigations to address this issue are of clinical and scientific relevance. In fact, in clinical practice, most of the liver grafts are obtained from brain-dead donors [64]. It is well known that brain death is associated with cerebral trauma and usually caused by hypoxia [65]. In addition, an experimental study aimed at evaluating the role of adipose tissue in the metabolism of rats with brain injury showed that resistin levels were increased in subcutaneous fat of rats with traumatic brain injury [66]. Moreover, it should be taken into account that adipose tissue is considered to be an important source of adipokines, such as leptin, adiponectin, and resistin, and that adipokines directly access the liver from adipose tissue through the portal vein [9,67]. Therefore, the relevance of adipose tissue in the hepatic damage associated with transplantation should be considered in further experimental research of LT to elucidate whether the regulation of adipose tissue functions could improve the quality of donor organs and postoperative outcomes after transplantation.

Altogether, the current knowledge emphasizes the relevance of further characterizing the role of mediators released from peripheral adipose tissue on damage and regenerative failure in both steatotic and nonsteatotic livers undergoing surgery. All of this is required to provide novel therapeutic approaches that can be transferred to clinical liver surgery and consequently increase the number of available donors for transplantation and improve recovery for patients subjected to liver resections.

4. Relevance of Adipose Tissue in Patients Undergoing Liver Surgery

Interestingly, the extent of visceral adipose tissue as well as serum levels of adipokines have been evaluated in patients undergoing general surgery. Nevertheless, from our knowledge, pharmacological modulation of adipokine actions has not been reported in the clinical practice of liver surgery [68–71]. Indeed, the significance of visceral adipose tissue remains controversial in the surgical setting [71].

In liver resections associated with hepatocellular carcinoma (HCC), preoperative visceral adiposity, as well as low muscularity (since obesity might be associated with a decrease in muscle mass), was closely related to postoperative death and HCC recurrence [72–74]. In addition, it has been reported that greater fat accumulation in skeletal muscle has been associated with a worse prognosis and survival after PH in patients with HCC, even with adjustment for other known predictors [75]. Moreover, prospective studies and meta-analyses have suggested that obese patients have increased risk and a poorer prognosis for many types of cancer [72–74]. All of these results in PH are in line with those observed in living donor liver transplantation (LDLT), since patients with a high degree of muscle steatosis and visceral adiposity show worse survival rates after transplantation compared with patients without obesity or with normal musculature [76]. Nevertheless, these findings are challenged by opposite observations. Indeed, preoperative abdominal computed tomographic (CT) scans in patients undergoing major hepatic resection associated with cancer suggest that obesity does not correlate with poor outcomes after major surgery [77]. Interestingly, neither preoperative visceral adiposity nor low muscularity were poor prognostic factors in patients undergoing liver resection for colorectal liver metastases [78]. In addition, some studies focused on liver resections of different cancer types showed that patients with a higher body mass index (BMI) survive longer than normal-weight patients after surgery [79–82]. It should be noted that CT measurement enables specific quantification of visceral adipose tissue, which is not reflected by BMI.

The contradictory results in clinical practice, the so-called obesity paradox, might occur due to the different methodologies used to evaluate and measure adipose tissue [71], but different types of surgery (resection vs. transplantation) as well as liver pathologies should also be noted.

Changes in adipokine levels in patients subjected to PH have been reported, suggesting that early-phase elevation of serum levels of hepatocyte growth factor (HGF), leptin, and macrophage colony-stimulating factor (M-CSF) could be associated with the acceleration of liver regeneration [83]. In line with this, plasmatic adipokines after LDLT have been mainly reported as biochemical markers to evaluate the risk of fibrosis progression in patients transplanted due to hepatitis C [84]. However, in these studies, the role of adipose tissue as a source of adipokines was not evaluated.

5. Experimental Strategies to Evaluate Adipose Tissue in Liver Surgery

5.1. Lipectomy

The literature describes the surgical excision of adipose tissue (lipectomy) to evaluate the function of adipose tissue in physiological conditions and different pathologies [37,40,41]. However, few studies have attempted to discern the role of adipose tissue on adipokine levels and hepatic damage and regenerative failure in liver surgery of PH with I/R [56–60], and no studies have evaluated the effects of a lipectomy in livers submitted for transplantation (Figure 4).

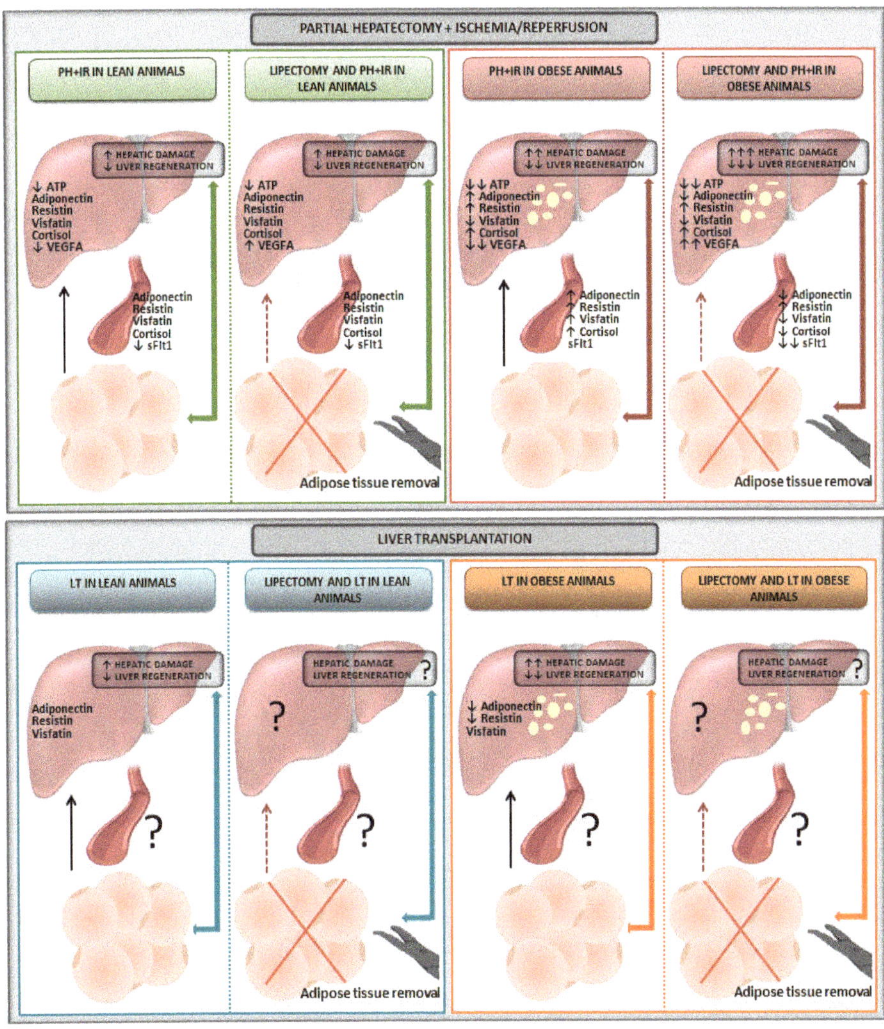

Figure 4. Schematic illustration of lipectomy effects in liver surgery. Abbreviations: ATP, adenosine triphosphate; I/R, ischemia–reperfusion; LT, liver transplantation; PH, partial hepatectomy; sFlt1, soluble form of the VEGF receptor 1; VEGFA, vascular endothelial growth factor type A.

The results obtained using lipectomy in PH with I/R indicate that, in contrast with nonsteatotic livers, adipose tissue is required for liver regeneration and to reduce damage in the presence of steatosis [57] (Figure 4). Interestingly, adipose tissue does not seem to be an energy source for the nonsteatotic liver since ATP levels were unchanged after lipectomy [57]. Regarding adipokines, removal of adipose tissue using lipectomy in lean animals undergoing PH with I/R resulted in plasmatic and hepatic levels of adiponectin, resistin, visfatin, and cortisol, similar to those observed under PH with I/R conditions [57–59]. The levels of sFlt1 were reduced in plasma in lean animals lipectomized and undergoing PH + I/R, indicating that adipose tissue might be a potential source of sFlt1 [60]. As it has been suggested that circulating VEGFA is sequestered by the sFlt1 released by adipose tissue in lean animals in conditions of PH + I/R [60], this was associated with increases in hepatic VEGFA (Figure 4). In obese animals, the reduced hepatic ATP levels were more evident than in lean animals. However, similar to that occurring in lean animals, adipose tissue does not seem to be an energy source for the steatotic liver, since ATP levels were unchanged after lipectomy. Adipose tissue seems to be an adiponectin source for the steatotic liver, since the induction of lipectomy in obese animals reduced both plasmatic and hepatic levels of adiponectin compared with the results obtained under PH with I/R conditions [57]. The contribution of adipose tissue as a source of resistin was irrelevant since obese animals undergoing PH with I/R showed high levels of plasmatic and hepatic resistin levels, whereas these resistin levels were unaltered under lipectomy conditions [58] (Figure 4). In obese animals undergoing PH with I/R, the visfatin levels were increased and reduced in liver and plasma, respectively. When adipose tissue was removed in obese animals undergoing PH with I/R, circulating visfatin levels were reduced, whereas hepatic visfatin accumulation was unaltered [58]. Thus, PH with I/R induced the release of visfatin from adipose tissue to circulation and reduced the generation of visfatin by the liver [58]. In obese animals undergoing PH with I/R, plasmatic and hepatic cortisol levels were increased [61]. In contrast, lipectomy in obese animals reduced cortisol levels in plasma but not in the steatotic liver [59], although the potential contribution of adipose tissue in the hepatic levels of cortisol of obese animals cannot be discounted. Indeed, changes in the enzymes engaged in cortisol generation and clearance were detected in adipose tissue of obese animals undergoing PH with I/R [59]. The reduced plasmatic levels of sFlt1 in obese animals lipectomized and undergoing PH with I/R were more evident than in lean animals [60]. This increased the circulating VEGFA bioavailability and, consequently, increased the opportunity of VEGFA to be taken up by the steatotic liver [60] (Figure 4).

Reduced adiponectin and resistin levels were observed only in steatotic livers when obese animals were subjected to LT [62] (Figure 4). The role of adipose tissue on hepatic adipokine levels, hepatic damage, and regeneration following LT remains to be elucidated. In our view, given the key role of adipose tissue in steatotic and nonsteatotic livers undergoing PH with I/R, strategies based in the adipose tissue removal should been used to study the crosstalk liver–adipose tissue in LT, mainly in the presence of steatosis.

5.2. Transgenic Animal Models

The use of transgenic animal models has improved our understanding of the pathophysiology of adipose tissue. The main focus of transgenic animal models has been the expression or knockout of selected genes, specifically in adipose tissue, identifying and characterizing promoter regions that confer adipose–tissue specific expression [7]. For instance, to target both white and brown adipose tissue, the promoters for adipocyte lipid binding protein aP215 and for phosphoenolpyruvate carboxykinase are usually used, whereas to target only brown adipose tissue, the mitochondrial uncoupling protein-1 (UCP-1) promoter is used [85–87]. In our view, the potential applications of transgenic animal models with overexpression or knockout of adipose tissue-selected genes might be of scientific and clinical interest to evaluate the adipose tissue–liver axis in hepatic resections and transplantation, since in different surgical conditions, hepatic diseases might be improved by directly targeting adipose tissue, rather than liver tissue per se.

6. Conclusions

The role of adipose tissue on damage and regenerative failure in experimental liver surgery depends on the type of surgical procedure (PH with or without I/R) as well as the type of liver (steatotic versus nonsteatotic) submitted to liver surgery. This should be taken into account for the establishment of protective strategies modulating the liver–adipose tissue axis, which would be specific for each surgical procedure and type of liver, as it has been reported in the present review. Further clinical studies and appropriate methods for adipose tissue measurement will be required to elucidate the significance of visceral adipose tissue in the clinical scenario of surgical hepatic resections. The use of experimental models of lipectomy as well as transgenic animal models with expression or knockout of adipose tissue-selected genes might be of scientific and clinical relevance to elucidate the contribution of the adipose tissue–liver axis, as well as the role of adipose tissue as an energy substrate and/or a source of different adipokines and hormones in livers subjected to transplantation. This would provide novel therapeutic approaches to be transferred to clinical conditions to improve the post-transplantation outcomes and consequently increase the number of available donors for transplantation.

Funding: This research was supported by the Ministerio de Ciencia, Innovación y Universidades (RTI2018-095114-B-I00) Madrid, Spain; the European Union (Fondos Feder, 'una manera dehacer Europa'); CERCA Program/Generalitat de Catalunya; and Secretaria d'Universitats I Recerca del Departament d'Economia i Coneixement (2017 SGR-551) Barcelona, Spain.

Conflicts of Interest: The authors declare no conflict of interest.

Abbreviations

AICAR	Cell-permeable adenosine analog that is a selective activator of AMPK
AMPK	Adenosine monophosphate-activated protein kinase
ATP	Adenosine triphosphate
BMI	Body mass index
BMP	Bone morphogenetic protein
CT	Computed tomography
Fas/CD95	Apoptosis antigen 1 or cluster of differentiation 95
FFA	Free fatty acids
HCC	Hepatocellular carcinoma
HGF	Hepatocyte growth factor
I/R	Ischemia–reperfusion
IL	Interleukin
JNK1	c-Jun N-terminal kinase-1
LDLT	Living donor liver transplantation
LT	Liver transplantation
M-CSF	Macrophage colony-stimulating factor
NO	Nitric oxide
PAI-1	Plasminogen activator inhibitor-1
PC	Preconditioning
PH	Partial hepatectomy
sFlt1	Soluble form of the VEGF receptor 1
SIP1	Smad interacting protein 1
TGF-β	Transforming growth factor-β
TNFα	Tumor necrosis factor α
UCP-1	Mitochondrial uncoupling protein-1
VEGFA	Vascular endothelial growth factor type A

References

1. Kershaw, E.E.; Flier, J.S. Adipose tissue as an endocrine organ. *J. Clin. Endocrinol. Metab.* **2004**, *89*, 2548–2556. [CrossRef] [PubMed]
2. Frühbeck, G.; Gómez-Ambrosi, J.; Muruzábal, F.J.; Burrell, M.A. The adipocyte: A model for integration of endocrine and metabolic signaling in energy metabolism regulation. *Am. J. Physiol. Endocrinol. Metab.* **2001**, *280*, E827–E847. [CrossRef] [PubMed]
3. Wang, Z.G.; Dou, X.B.; Zhou, Z.X.; Song, Z.Y. Adipose tissue-liver axis in alcoholic liver disease. *World J. Gastrointest. Pathophysiol.* **2016**, *7*, 17–26. [CrossRef] [PubMed]
4. Parker, R. The role of adipose tissue in fatty liver diseases. *Liver Res.* **2018**, *2*, 35–42. [CrossRef]
5. Large, V.; Peroni, O.; Letexier, D.; Ray, H.; Beylot, M. Metabolism of lipids in human white adipocyte. *Diabetes Metab.* **2004**, *30*, 294–309. [CrossRef]
6. Nielsen, T.S.; Jessen, N.; Jørgensen, J.O.; Møller, N.; Lund, S. Dissecting adipose tissue lipolysis: Molecular regulation and implications for metabolic disease. *J. Mol. Endocrinol.* **2014**, *52*, R199–R222. [CrossRef] [PubMed]
7. Blüher, M. Transgenic animal models for the study of adipose tissue biology. *Best Pract. Res. Clin. Endocrinol. Metab.* **2005**, *19*, 605–623. [CrossRef] [PubMed]
8. Ye, D.W.; Rong, X.L.; Xu, A.M.; Guo, J. Liver-adipose tissue crosstalk: A key player in the pathogenesis of glucolipid metabolic disease. *Chin. J. Integr. Med.* **2017**, *23*, 410–414. [CrossRef] [PubMed]
9. Schäffler, A.; Schölmerich, J.; Büchler, C. Mechanisms of disease: Adipocytokines and visceral adipose tissue-emerging role in nonalcoholic fatty liver disease. *Nat. Clin. Pract. Gastroenterol. Hepatol.* **2005**, *2*, 273–280. [CrossRef]
10. Casillas-Ramírez, A.; Mosbah, I.B.; Ramalho, F.; Rosselló-Catafau, J.; Peralta, C. Past and future approaches to ischemia-reperfusion lesion associated with liver transplantation. *Life Sci.* **2006**, *79*, 1881–1894. [CrossRef]
11. Serafin, A.; Rosello-Catafau, J.; Prats, N.; Xaus, C.; Gelpi, E.; Peralta, C. Ischemic preconditioning increases the tolerance of fatty liver to hepatic ischemia-reperfusion injury in the rat. *Am. J. Pathol.* **2002**, *161*, 587–601. [CrossRef]
12. Clavien, P.; Harvey, P.; Strasberg, S. Preservation and reperfusion injuries in liver allografts. An overview and synthesis of current studies. *Transplantation* **1992**, *53*, 957–978. [CrossRef] [PubMed]
13. Huguet, C.; Gavelli, A.; Chieco, P.; Bona, S.; Harb, J.; Joseph, J.M.; Jobard, J.; Gramaglia, M.; Lasserre, M. Liver ischemia for hepatic resection: Where is the limit? *Surgery* **1992**, *111*, 251–259. [PubMed]
14. Fu, P.; Li, W. Nitric oxide in liver ischemia-reperfusion injury. In *Liver Pathophysiology*; Muriel, P., Ed.; Elsevier Inc.: London, UK, 2017; Volume 8, pp. 125–127.
15. Cornide-Petronio, M.E.; Jiménez-Castro, M.B.; Gracia-Sancho, J.; Peralta, C. Ischemic preconditioning directly or remotely applied on the liver to reduce ischemia-reperfusion injury in resections and transplantation. In *Liver Disease and Surgery*; Tsoulfas, G., Ed.; IntechOpen: London, UK, 2019; in press.
16. Gracia-Sancho, J.; Casillas-Ramírez, A.; Peralta, C. Molecular pathways in protecting the liver from ischaemia/reperfusion injury: A 2015 update. *Clin. Sci. (Lond.)* **2015**, *129*, 345–362. [CrossRef] [PubMed]
17. Pratap, A.; Panakanti, R.; Yang, N.; Lakshmi, R.; Modanlou, K.A.; Eason, J.D.; Mahato, R.I. Cyclopamine attenuates acute warm ischemia reperfusion injury in cholestatic rat liver: Hope for marginal livers. *Mol. Pharm.* **2011**, *8*, 958–968. [CrossRef] [PubMed]
18. Nadig, S.N.; Periyasamy, B.; Shafizadeh, S.F.; Polito, C.; Fiorini, R.N.; Rodwell, D.; Evans, Z.; Cheng, G.; Dunkelberger, D.; Schmidt, M.; et al. Hepatocellular ultrastructure after ischemia/reperfusion injury in human orthotopic liver transplantation. *J. Gastrointest. Surg.* **2004**, *8*, 695–700. [CrossRef] [PubMed]
19. Verran, D.; Kusyk, T.; Painter, D.; Fisher, J.; Koorey, D.; Strasser, S.; Stewart, G.; McCaughan, G. Clinical experience gained from the use of 120 steatotic donor livers for orthotopic liver transplantation. *Liver Transplant.* **2003**, *9*, 500–505. [CrossRef] [PubMed]
20. Schaubel, D.E.; Sima, C.S.; Goodrich, N.P.; Feng, S.; Merion, R.M. The survival benefit of deceased donor liver transplantation as a function of candidate disease severity and donor quality. *Am. J. Transplant.* **2008**, *8*, 419–425. [CrossRef]
21. Ploeg, R.J.; D'Alessandro, A.M.; Knechtle, S.J.; Stegall, M.D.; Pirsch, J.D.; Hoffmann, R.M.; Sasaki, T.; Sollinger, H.W.; Belzer, F.O.; Kalayoglu, M. Risk factors for primary dysfunction after liver transplantation-a multivariate analysis. *Transplantation* **1993**, *55*, 807–813. [CrossRef]

22. Behrns, K.E.; Tsiotos, G.G.; DeSouza, N.F.; Krishna, M.K.; Ludwig, J.; Nagorney, D.M. Hepatic steatosis as a potential risk factor for major hepatic resection. *J. Gastrointest. Surg.* **1998**, *2*, 292–298. [CrossRef]

23. Safwan, M.; Collins, K.M.; Abouljoud, M.S.; Salgia, R. Outcome of liver transplantation in patients with prior bariatric surgery. *Liver Transplant.* **2017**, *23*, 1415–1421. [CrossRef] [PubMed]

24. Shook, B.; Rivera Gonzalez, G.; Ebmeier, S.; Grisotti, G.; Zwick, R.; Horsley, V. The role of adipocytes in tissue regeneration and stem cell niches. *Annu. Rev. Cell Dev. Biol.* **2016**, *32*, 609–631. [CrossRef] [PubMed]

25. Odegaard, J.I.; Chawla, A. Alternative macrophage activation and metabolism. *Annu. Rev. Pathol.* **2011**, *6*, 275–297. [CrossRef] [PubMed]

26. Khan, T.; Muise, E.S.; Iyengar, P.; Wang, Z.V.; Chandalia, M.; Abate, N.; Zhang, B.B.; Bonaldo, P.; Chua, S.; Scherer, P.E. Metabolic dys-regulation and adipose tissue fibrosis: Role of collagen VI. *Mol. Cell. Biol.* **2009**, *29*, 1575–1591. [CrossRef] [PubMed]

27. Hosogai, N.; Fukuhara, A.; Oshima, K.; Miyata, Y.; Tanaka, S.; Segawa, K.; Furukawa, S.; Tochino, Y.; Komuro, R.; Matsuda, M.; et al. Adipose tissue hypoxia in obesity and its impact on adipocytokine dysregulation. *Diabetes* **2007**, *56*, 901–911. [CrossRef] [PubMed]

28. Elias-Miro, M.; Massip-Salcedo, M.; Jimenez-Castro, M.; Peralta, C. Does adiponectin benefit steatotic liver transplantation? *Liver Transplant.* **2011**, *17*, 993–1004. [CrossRef]

29. Tilg, H.; Hotamisligil, G.S. Nonalcoholic fatty liver disease: Cytokine-adipokine interplay and regulation of insulin resistance. *Gastroenterology* **2006**, *131*, 934–945. [CrossRef]

30. Farrell, G.C. Probing Prometheus: Fat fueling the fire? *Hepatology* **2004**, *40*, 1252–1255. [CrossRef]

31. Gordon, S. Macrophage heterogeneity and tissue lipids. *J. Clin. Investig.* **2007**, *117*, 89–93. [CrossRef]

32. Wellen, K.E.; Hotamisligil, G.S. Inflammation, stress, and diabetes. *J. Clin. Investig.* **2005**, *115*, 1111–1119. [CrossRef]

33. Wellen, K.E.; Hotamisligil, G.S. Obesity-induced inflammatory changes in adipose tissue. *J. Clin. Investig.* **2003**, *112*, 1785–1788. [CrossRef] [PubMed]

34. Weisberg, S.P.; Mccann, D.; Desai, M.; Rosenbaum, M.; Leibel, R.L.; Ferrante, A.W., Jr. Obesity is associated with macrophage accumulation in adipose tissue. *J. Clin. Investig.* **2003**, *112*, 1796–1808. [CrossRef] [PubMed]

35. Duwaerts, C.C.; Maher, J.J. Macronutrients and the Adipose-Liver Axis in Obesity and Fatty Liver. *Cell. Mol. Gastroenterol. Hepatol.* **2019**, *7*, 749–761. [CrossRef] [PubMed]

36. Harman-Boehm, I.; Bluher, M.; Redel, H.; Sion-Vardy, N.; Ovadia, S.; Avinoach, E.; Shai, I.; Klöting, N.; Stumvoll, M.; Bashan, N.; et al. Macrophage infiltration into omental versus subcutaneous fat across different populations: Effect of regional adiposity and the comorbidities of obesity. *J. Clin. Endocrinol. Metab.* **2007**, *92*, 2240–2247. [CrossRef] [PubMed]

37. Nov, O.; Shapiro, H.; Ovadia, H.; Tarnovscki, T.; Dvir, I.; Shemesh, E.; Kovsan, J.; Shelef, I.; Carmi, Y.; Voronov, E.; et al. Interleukin-1β regulates fat-liver crosstalk in obesity by auto-paracrine modulation of adipose tissue inflammation and expandability. *PLoS ONE* **2013**, *8*, e53626. [CrossRef] [PubMed]

38. Bergman, R.N.; Kim, S.P.; Hsu, I.R.; Catalano, K.J.; Chiu, J.D.; Kabir, M.; Richey, J.M.; Ader, M. Abdominal obesity: Role in the pathophysiology of metabolic disease and cardiovascular risk. *Am. J. Med.* **2007**, *120*, S3–S8; discussion S29–S32. [CrossRef] [PubMed]

39. Kabir, M.; Catalano, K.J.; Ananthnarayan, S.; Kim, S.P.; Van Citters, G.W.; Dea, M.K.; Bergman, R.N. Molecular evidence supporting the portal theory: A causative link between visceral adiposity and hepatic insulin resistance. *Am. J. Physiol. Endocrinol. Metab.* **2005**, *288*, E454–E461. [CrossRef]

40. Sabio, G.; Das, M.; Mora, A.; Zhang, Z.; Jun, J.Y.; Ko, H.J.; Barrett, T.; Kim, J.K.; Davis, R.J. A stress signalling pathway in adipose tissue regulates hepatic insulin resistance. *Science* **2008**, *322*, 1539–1543. [CrossRef] [PubMed]

41. Wueest, S.; Rapold, R.A.; Schumann, D.M.; Rytka, J.M.; Schildknecht, A.; Nov, O.; Chervonsky, A.V.; Rudich, A.; Schoenle, E.J.; Donath, M.Y.; et al. Deletion of Fas in adipocytes relieves adipose tissue inflammation and hepatic manifestations of obesity in mice. *J. Clin. Investig.* **2010**, *120*, 191–202. [CrossRef] [PubMed]

42. Caruana, J.A.; Whalen, D.A., Jr.; Anthony, W.P.; Sunby, C.R.; Ciechoski, M.P. Paradoxical effects of glucose feeding on liver regeneration and survival after partial hepatectomy. *Endocr. Res.* **1986**, *12*, 147–156. [CrossRef] [PubMed]

43. Nakatani, T.; Ozawa, K.; Asano, M.; Ukikusa, M.; Kamiyama, Y.; Tobe, T. Differences in predominant energy substrate in relation to the resected hepatic mass in the phase immediately after hepatectomy. *J. Lab. Clin. Med.* **1981**, *97*, 887–898. [PubMed]

44. Nakatani, T.; Yasuda, K.; Ozawa, K.; Kawashima, S.; Tobe, T. Effects of (+)-octanoylcarnitine on deoxyribonucleic acid synthesis in regenerating rabbit liver. *Clin. Sci. (Lond.)* **1982**, *62*, 295–297. [CrossRef] [PubMed]

45. Gazit, V.; Weymann, A.; Hartman, E.; Finck, B.N.; Hruz, P.W.; Tzekov, A.; Rudnick, D.A. Liver regeneration is impaired in lipodystrophic fatty liver dystrophy mice. *Hepatology* **2010**, *52*, 2109–2117. [CrossRef] [PubMed]

46. Tijburg, L.B.; Nyathi, C.B.; Meijer, G.W.; Geelen, M.J. Biosynthesis and secretion of triacylglycerol in rat liver after partial hepatectomy. *Biochem. J.* **1991**, *277*, 723–728. [CrossRef] [PubMed]

47. Brasaemle, D.L. Cell biology. A metabolic push to proliferate. *Science* **2006**, *313*, 1581–1582. [CrossRef] [PubMed]

48. Baker, N.; Garfinkel, A.S.; Schotz, M.C. Hepatic triglyceride secretion in relation to lipogenesis and free fatty acid mobilization in fasted and glucose-refed rats. *J. Lipid Res.* **1968**, *9*, 1–7. [PubMed]

49. Michalopoulos, G.K.; DeFrances, M.C. Liver regeneration. *Science* **1997**, *276*, 60–66. [CrossRef]

50. Palmero, E.; Ricart, D.; Llobera, M.; Peinado-Onsurbe, J. Partial hepatectomy and/or surgical stress provoke changes in the expression of lipoprotein lipase and actin in liver and extrahepatic tissues. *Biochim. Biophys. Acta* **1999**, *1441*, 61–68. [CrossRef]

51. Sabugal, R.; Robert, M.Q.; Julve, J.; Auwerx, J.; Llobera, M.; Peinado-Onsurbe, J. Hepatic regeneration induces changes in lipoprotein lipase activity in several tissues and its re-expression in the liver. *Biochem. J.* **1996**, *318*, 597–602. [CrossRef]

52. Shteyer, E.; Liao, Y.; Muglia, L.J.; Hruz, P.W.; Rudnick, D.A. Disruption of hepaticadipogenesis is associated with impaired liver regeneration in mice. *Hepatology* **2004**, *40*, 1322–1332. [CrossRef]

53. Kohjima, M.; Tsai, T.H.; Tackett, B.C.; Thevananther, S.; Li, L.; Chang, B.H.; Chan, L. Delayed liver regeneration after partial hepatectomy in adipose differentiation related protein-null mice. *J. Hepatol.* **2013**, *59*, 1246–1254. [CrossRef]

54. El-Badry, A.M.; Moritz, W.; Contaldo, C.; Tian, Y.; Graf, R.; Clavien, P.A. Prevention of reperfusion injury and microcirculatory failure in macrosteatotic mouse liver by omega-3 fatty acids. *Hepatology* **2007**, *45*, 855–863. [CrossRef] [PubMed]

55. Marsman, H.A.; de Graaf, W.; Heger, M.; van Golen, R.F.; Ten Kate, F.J.; Bennink, R.; van Gulik, T.M. Hepatic regeneration and functional recovery following partial liver resection in an experimental model of hepatic steatosis treated with omega-3 fatty acids. *Br. J. Surg.* **2013**, *100*, 674–683. [CrossRef] [PubMed]

56. Yang, L.; Wang, Y.; Shi, Y.; Bu, H.; Ye, F. Deletion of SIP1 promotes liver regeneration and lipid accumulation. *Pathol. Res. Pract.* **2016**, *212*, 421–425. [CrossRef] [PubMed]

57. Mendes-Braz, M.; Elias-Miró, M.; Kleuser, B.; Fayyaz, S.; Jiménez-Castro, M.B.; Massip-Salcedo, M.; Gracia-Sancho, J.; Ramalho, F.S.; Rodes, J.; Peralta, C. The effects of glucose and lipids in steatotic and nonsteatotic livers in conditions of partial hepatectomy under ischaemia-reperfusion. *Liver Int.* **2014**, *34*, e271–e289. [CrossRef] [PubMed]

58. Elias-Miró, M.; Mendes-Braz, M.; Cereijo, R.; Villarroya, F.; Jiménez-Castro, M.B.; Gracia-Sancho, J.; Guixé-Muntet, S.; Massip-Salcedo, M.; Domingo, J.C.; Bermudo, R.; et al. Resistin and visfatin in steatotic and nonsteatotic livers in the setting of partial hepatectomy under ischemia-reperfusion. *J. Hepatol.* **2014**, *60*, 87–95. [CrossRef] [PubMed]

59. Cornide-Petronio, M.E.; Bujaldon, E.; Mendes-Braz, M.; Avalos de León, C.G.; Jiménez-Castro, M.B.; Álvarez-Mercado, A.I.; Gracia-Sancho, J.; Rodés, J.; Peralta, C. The impact of cortisol in steatotic and nonsteatotic liver surgery. *J. Cell. Mol. Med.* **2017**, *21*, 2344–2358. [CrossRef] [PubMed]

60. Bujaldon, E.; Cornide-Petronio, M.E.; Gulfo, J.; Rotondo, F.; Ávalos de León, C.; Negrete-Sánchez, E.; Gracia-Sancho, J.; Novials, A.; Jiménez-Castro, M.B.; Peralta Uroz, C. Relevance of VEGFA in rat livers subjected to partial hepatectomy under ischemia-reperfusion. *J. Mol. Med. (Berl.)* **2019**, *97*, 1299–1314. [CrossRef]

61. Man, K.; Zhao, Y.; Xu, A.; Lo, C.M.; Lam, K.S.; Ng, K.T.; Ho, J.W.; Sun, C.K.; Lee, T.K.; Li, X.L.; et al. Fat-derived hormone adiponectin combined with FTY720 significantly improves small-for-size fatty liver graft survival. *Am. J. Transplant.* **2006**, *6*, 467–476. [CrossRef]

62. Jiménez-Castro, M.B.; Casillas-Ramírez, A.; Mendes-Braz, M.; Massip-Salcedo, M.; Gracia-Sancho, J.; Elias-Miró, M.; Rodés, J.; Peralta, C. Adiponectin and resistin protect steatotic livers undergoing transplantation. *J. Hepatol.* **2013**, *59*, 1208–1214. [CrossRef]

63. Carbone, M.; Campagnolo, L.; Angelico, M.; Tisone, G.; Almerighi, C.; Telesca, C.; Lenci, I.; Moscatelli, I.; Massoud, R.; Baiocchi, L. Leptin attenuates ischemia-reperfusion injury in the rat liver. *Transpl. Int.* **2012**, *25*, 1282–1288. [CrossRef] [PubMed]

64. Jiménez-Castro, M.B.; Meroño, N.; Mendes-Braz, M.; Gracia-Sancho, J.; Martínez-Carreres, L.; Cornide-Petronio, M.E.; Casillas-Ramirez, A.; Rodés, J.; Peralta, C. The effect of brain death in rat steatotic and nonsteatotic liver transplantation with previous ischemic preconditioning. *J. Hepatol.* **2015**, *62*, 83–91. [CrossRef] [PubMed]

65. Jiménez-Castro, M.B.; Gracia-Sancho, J.; Peralta, C. Brain death and marginal grafts in liver transplantation. *Cell Death Dis.* **2015**, *6*, e1777. [CrossRef] [PubMed]

66. Zhang, R.; Wang, Z.Y.; Zhu, L.L.; Wu, F.; Chen, D.Q.; Sun, L.F.; Lu, Z.Q. Resistin expression in adipose tissues and its effect on glucose metabolism in rats with brain injury. *Genet. Mol. Res.* **2016**, *15*. [CrossRef] [PubMed]

67. Marra, F.; Bertolani, C. Adipokines in liver diseases. *Hepatology* **2009**, *50*, 957–969. [CrossRef] [PubMed]

68. Mullen, J.T.; Moorman, D.W.; Davenport, D.L. The obesity paradox: Body mass index and outcomes in patients undergoing nonbariatric general surgery. *Ann. Surg.* **2009**, *250*, 166–172. [CrossRef] [PubMed]

69. Dindo, D.; Muller, M.K.; Weber, M.; Clavien, P.A. Obesity in general elective surgery. *Lancet* **2003**, *361*, 2032–2035. [CrossRef]

70. Barone, M.; Viggiani, M.T.; Avolio, A.W.; Iannone, A.; Rendina, M.; Di Leo, A. Obesity as predictor of postoperative outcomes in liver transplant candidates: Review of the literature and future perspectives. *Dig. Liver Dis.* **2017**, *49*, 957–966. [CrossRef]

71. Hamaguchi, Y.; Kaido, T.; Okumura, S.; Kobayashi, A.; Shirai, H.; Yao, S.; Yagi, S.; Kamo, N.; Seo, S.; Taura, K.; et al. Preoperative Visceral Adiposity and Muscularity Predict Poor Outcomes after Hepatectomy for Hepatocellular Carcinoma. *Liver Cancer* **2019**, *8*, 92–109. [CrossRef]

72. Calle, E.E.; Rodriguez, C.; Walker-Thurmond, K.; Thun, M.J. Overweight, obesity, and mortality from cancer in a prospectively studied cohort of US adults. *N. Engl. J. Med.* **2003**, *348*, 1625–1638. [CrossRef]

73. Renehan, A.G.; Tyson, M.; Egger, M.; Heller, R.F.; Zwahlen, M. Body-mass index and incidence of cancer: A systematic review and meta-analysis of prospective observational studies. *Lancet* **2008**, *371*, 569–578. [CrossRef]

74. Arnold, M.; Pandeya, N.; Byrnes, G.; Renehan, A.G.; Stevens, G.A.; Ezzati, M.; Ferlay, J.; Miranda, J.J.; Romieu, I.; Dikshit, R.; et al. Global burden of cancer attributable to high body-mass index in 2012: A population-based study. *Lancet Oncol.* **2015**, *16*, 36–46. [CrossRef]

75. Kaibori, M.; Ishizaki, M.; Iida, H.; Matsui, K.; Sakaguchi, T.; Inoue, K.; Mizuta, T.; Ide, Y.; Iwasaka, J.; Kimura, Y.; et al. Effect of Intramuscular Adipose Tissue Content on Prognosis in Patients Undergoing Hepatocellular Carcinoma Resection. *J. Gastrointest. Surg.* **2015**, *19*, 1315–1323. [CrossRef] [PubMed]

76. Kamo, N.; Kaido, T.; Hamaguchi, Y.; Okumura, S.; Kobayashi, A.; Shirai, H.; Yao, S.; Yagi, S.; Uemoto, S. Impact of sarcopenic obesity on outcomes in patients undergoing living donor liver transplantation. *Clin. Nutr.* **2018**. [CrossRef] [PubMed]

77. Morris, K.; Tuorto, S.; Gönen, M.; Schwartz, L.; DeMatteo, R.; D'Angelica, M.; Jarnagin, W.R.; Fong, Y. Simple measurement of intra-abdominal fat for abdominal surgery outcome prediction. *Arch. Surg.* **2010**, *145*, 1069–1073. [CrossRef] [PubMed]

78. Kobayashi, A.; Kaido, T.; Hamaguchi, Y.; Okumura, S.; Shirai, H.; Kamo, N.; Yagi, S.; Taura, K.; Okajima, H.; Uemoto, S. Impact of Visceral Adiposity as Well as Sarcopenic Factors on Outcomes in Patients Undergoing Liver Resection for Colorectal Liver Metastases. *World J. Surg.* **2018**, *42*, 1180–1191. [CrossRef] [PubMed]

79. Hakimi, A.A.; Furberg, H.; Zabor, E.C.; Jacobsen, A.; Schultz, N.; Ciriello, G.; Mikklineni, N.; Fiegoli, B.; Kim, P.H.; Voss, M.H.; et al. An epidemiologic and genomic investigation into the obesity paradox in renal cell carcinoma. *J. Natl. Cancer Inst.* **2013**, *105*, 1862–1870. [CrossRef] [PubMed]

80. Martin, L.; Birdsell, L.; Macdonald, N.; Reiman, T.; Clandinin, M.T.; McCargar, L.J.; Murphy, R.; Ghosh, S.; Sawyer, M.B.; Baracos, V.E. Cancer cachexia in the age of obesity: Skeletal muscle depletion is a powerful prognostic factor, independent of body mass index. *J. Clin. Oncol.* **2013**, *31*, 1539–1547. [CrossRef] [PubMed]

81. Schlesinger, S.; Siegert, S.; Koch, M.; Walter, J.; Heits, N.; Hinz, S.; Jacobs, G.; Hampe, J.; Schafmayer, C.; Nöthlings, U. Postdiagnosis body mass index and risk of mortality in colorectal cancer survivors: A prospective study and metaanalysis. *Cancer Causes Control* **2014**, *25*, 1407–1418. [CrossRef] [PubMed]

82. Amptoulach, S.; Gross, G.; Kalaitzakis, E. Differential impact of obesity and diabetes mellitus on survival after liver resection for colorectal cancer metastases. *J. Surg. Res.* **2015**, *199*, 378–385. [CrossRef]

83. Matsumoto, K.; Miyake, Y.; Umeda, Y.; Matsushita, H.; Matsuda, H.; Takaki, A.; Sadamori, H.; Nouso, K.; Yagi, T.; Fujiwara, T.; et al. Serial changes of serum growth factor levels and liver regeneration after partial hepatectomy in healthy humans. *Int. J. Mol. Sci.* **2013**, *14*, 20877–20889. [CrossRef] [PubMed]

84. Veldt, B.J.; Poterucha, J.J.; Watt, K.D.; Wiesner, R.H.; Hay, J.E.; Rosen, C.B.; Heimbach, J.K.; Janssen, H.L.; Charlton, M.R. Insulin resistance, serum adipokines and risk of fibrosis progression in patients transplanted for hepatitis C. *Am. J. Transplant.* **2009**, *9*, 1406–1413. [CrossRef] [PubMed]

85. Short, M.K.; Clouthier, D.E.; Schaefer, I.M.; Hammer, R.E.; Magnuson, M.A.; Beale, E.G. Tissue-specific, developmental, hormonal, and dietary regulation of rat phosphoenolpyruvate carboxykinase-human growth hormone fusion genes in transgenic mice. *Mol. Cell. Biol.* **1992**, *12*, 1007–1020. [CrossRef] [PubMed]

86. Cassard-Doulcier, A.M.; Gelly, C.; Fox, N.; Schrementi, J.; Raimbault, S.; Klaus, S.; Forest, C.; Bouillaud, F.; Ricquier, D. Tissue-specific and beta-adrenergic regulation of the mitochondrial uncoupling protein gene: Control by cis-acting elements in the 50-flanking region. *Mol. Endocrinol.* **1993**, *7*, 497–506. [PubMed]

87. Valet, P.; Tavernier, G.; Castan-Laurell, I.; Saulnier-Blache, J.S.; Langin, D. Understanding adipose tissue development from transgenic animal models. *J. Lipid Res.* **2002**, *43*, 835–860. [PubMed]

Review

Obesity as a Risk Factor for Severe COVID-19 and Complications: A Review

Fien Demeulemeester *, Karin de Punder , Marloes van Heijningen and Femke van Doesburg

Natura Foundation, 3281 NC Numansdorp, The Netherlands; k.depunder@naturafoundation.com (K.d.P.); m.vanheijningen@naturafoundation.com (M.v.H.); f.vandoesburg@naturafoundation.com (F.v.D.)
* Correspondence: f.demeulemeester@naturafoundation.com

Abstract: Emerging data suggest that obesity is a major risk factor for the progression of major complications such as acute respiratory distress syndrome (ARDS), cytokine storm and coagulopathy in COVID-19. Understanding the mechanisms underlying the link between obesity and disease severity as a result of SARS-CoV-2 infection is crucial for the development of new therapeutic interventions and preventive measures in this high-risk group. We propose that multiple features of obesity contribute to the prevalence of severe COVID-19 and complications. First, viral entry can be facilitated by the upregulation of viral entry receptors, like angiotensin-converting enzyme 2 (ACE2), among others. Second, obesity-induced chronic inflammation and disruptions of insulin and leptin signaling can result in impaired viral clearance and a disproportionate or hyper-inflammatory response, which together with elevated ferritin levels can be a direct cause for ARDS and cytokine storm. Third, the negative consequences of obesity on blood coagulation can contribute to the progression of thrombus formation and hemorrhage. In this review we first summarize clinical findings on the relationship between obesity and COVID-19 disease severity and then further discuss potential mechanisms that could explain the risk for major complications in patients suffering from obesity.

Keywords: coagulopathy; COVID-19; cytokine storm; inflammation; leptin; obesity; SARS-CoV-2

Citation: Demeulemeester, F.; de Punder, K.; van Heijningen, M.; van Doesburg, F. Obesity as a Risk Factor for Severe COVID-19 and Complications: A Review. *Cells* **2021**, *10*, 933. https://doi.org/10.3390/cells10040933

Academic Editor: Javier Gómez-Ambrosi

Received: 17 March 2021
Accepted: 14 April 2021
Published: 17 April 2021

Publisher's Note: MDPI stays neutral with regard to jurisdictional claims in published maps and institutional affiliations.

1. Introduction

The novel coronavirus SARS-CoV-2 that first appeared in the Chinese city of Wuhan is still causing a global pandemic with a death toll that already exceeds 2,430,000 people and continues to grow every day (number from John Hopkins University). SARS-CoV-2 is the virus that causes COVID-19, a clinical picture characterized by fever, coughing, muscle pain and fatigue and can evolve into hyperinflammation, cytokine storm, ARDS and COVID-associated-coagulopathy (CAC) [1,2]. A large number of patients severely ill with COVID-19 arriving at the ICU are overweight or suffer from obesity [3]. These conditions, as well as smoking, age, type II diabetes and cardiovascular diseases, appear to be major risk factors for serious complications and increased mortality in COVID-19 patients [4–6].

Understanding the contributing factors to COVID-19 disease severity and complications in the context of obesity is of key importance for the development of therapeutic interventions, as well as for advancing preventative strategies in this high-risk group. Therefore, in this review we first summarize clinical findings on obesity and COVID-19 disease outcomes. Next, we discuss possible underlying mechanisms linking obesity to major disease complications as a result of SARS-CoV-2 infection Here we focus on the metabolic- and immune-related consequences of obesity on COVID-19 disease course.

2. Obesity

2.1. Obesity Is a Common Disease Associated with Chronic Inflammation and Insulin and Leptin Resistance

The prevalence of obesity has increased worldwide over the last 50 years. In 2015 the mean prevalence of obesity in adults of selected countries was 19.5% and ranged from 3.7% in Japan to 38.2% in the United States [7]. Obesity (BMI $\geq$ 30 kg/m^2) is a major risk factor for the development of non-communicable diseases and is defined by the WHO as abnormal or excessive fat accumulation that might impair health [8].

Fat or adipose tissue, originally regarded as a simple organ for storing energy, is currently viewed as one of the most important endocrine organs [9–11]. Fat cells, or adipocytes, produce cytokine-like hormones, called adipokines, which play a major role in metabolism and inflammation [11]. Adipocytes in different regions fulfill different functions. Distinctions should be made between brown and white adipocytes, between ectopic and non-ectopic and between visceral and subcutaneous adipose tissue to accurately estimate the impact of the adipose tissue on the patient's health. For example, visceral fat, but not so much subcutaneous fat accumulation, promotes systemic inflammation and is associated with impairments of glucose and lipid metabolism accompanied by insulin resistance [12–14].

Obesity is associated with chronic inflammation, resulting from immune cell activity in dysfunctional (visceral) adipose tissue. In obesity, the excessive presence and hypertrophy of adipocytes result in hypoxia, cell stress and apoptosis. The hypoxic environment induces the infiltration of immune cells into the adipose tissue as a result of the expression of chemo attractive molecules [15]. Also, hypertrophic adipocytes produce multiple pro-inflammatory adipokines, such as such as tumor necrosis factor-α (TNF-α), interleukin (IL)-6 and leptin [16].

Leptin informs the brain about the amount of energy stored in the adipose tissue [17,18]. The percentage of total body fat is the most important determinant of leptin levels [19]. Given their high percentage of total body fat, individuals suffering from obesity show elevated leptin concentrations [20], a condition also referred to as hyperleptinemia. Persistent hyperleptinemia is often accompanied with central leptin resistance, due to which appetite and satiety regulation is disrupted [21]. In addition to the regulation of hunger and satiety, leptin also has pro-inflammatory properties [22], further contributing to a chronic inflammatory state in individuals suffering from obesity [23].

2.2. Obesity and COVID-19 Disease Severity

In the following section, we summarize clinical findings supporting an association between obesity and COVID-19 disease severity (see also Table 1). We searched the Pubmed database up to March 15, 2021. In our search strategy, the combination of the following keywords was used: obesity OR BMI OR overweight OR adiposity OR adipose tissue AND COVID-19 OR SARS-CoV-2. Our search was limited by published full-text article in English language. Most studies investigating this association used BMI categories as the predictor variable [24–37]. Four cross-sectional studies did not specify the used classification of obesity [38–41] presumably using the WHO guidelines defining obesity as BMI of 30 or higher. Outcome variables included: hospitalization [36,38,42], ICU admission [31,35,37,38,43–45], intubation [24,25,29,37,38], invasive mechanical ventilation [26,31,34], disease severity [27,28,30,33] and death [24–29,32,38–41]. Of the eleven studies investigating the association between BMI and mortality in hospitalized COVID-19 patients, ten studies observed an increased mortality rate in patients that were overweight (BMI $\geq$ 25 to <30) [24,32], or suffering from obesity (BMI $\geq$ 30) [25,29,38–41], or severe obesity (BMI $\geq$ 35) [26,27]. One study observed no difference in in-hospital deaths between normal and overweight (BMI > 28) patients, but overweight patients did show more severe disease symptoms [28]. Because BMI does not discriminate between fat and lean body mass and poorly reflects fat distribution, four studies used measures of visceral adipose tissue (VAT) obtained by high-resolution computed tomography [42–45]. In COVID-19 patients reporting at the emergency department, subsequent admission to the ICU was associated with 30% higher VAT and 30% lower subcutaneous adipose tissue (SAT) [43] with VAT scores being the best ICU admission predictor. Another study observed no differences in BMI but did see higher VAT and a higher VAT/total adipose tissue (TAT) ratio in hospitalized patients compared to outpatients that did not require hospitalization [42]. Finally, in a different study it was observed that both VAT and TAT were associated with ICU admission [45]. To summarize, numerous studies indicate that obesity is associated with increased disease severity and mortality in COVID-19.

Table 1. Clinical Studies Investigating the Association Between Obesity and Disease Severity in COVID-19 Patients.

Reference	Study Design	Predictor	Outcome	Effect
Anderson et al., 2020 [29]	Retrospective cohort study ($n = 2466$)	BMI categories: underweight (BMI < 18.5), normal weight (BMI $\geq$ 18.5 to <25), overweight (BMI $\geq$ 25 to <30), class 1 obesity (BMI $\geq$ 30 to <35), class 2 obesity (BMI $\geq$ 35 to <40), and class 3 obesity ($\geq$ 40)	Intubation death	Patients younger than 65 with obesity were at higher risk for intubation or death, with the highest risk among those with class 3 obesity (BMI $\geq$ 40).
Battisti et al., 2020 [43]	Cohort study ($n = 441$)	VAT/SAT ratio	ICU admission	VAT/SAT was associated with increased risk of ICU admission.
Chandarana et al., 2020 [42]	Retrospective study ($n = 51$)	VAT, SAT, TAT, VAT/TAT and BMI	Hospi-talization	Higher VAT and VAT/TAT in hospitalized patients.
Deng et al., 2020 [30]	Retrospective cohort study ($n = 65$)	BMI, subcutaneous fat thickness, epicardial fat and visceral fat	Disease severity	High BMI was a risk factor for severe COVID-19.
Frank et al., 2020 [25]	Retrospective cohort study ($n = 305$)	BMI categories: BMI < 25, BMI $\geq$ 25 to < 30, BMI $\geq$ 30 to < 35, and BMI $\geq$ 35	Intubation, death	BMI $\geq$ 30 was associated with an increased risk of intubation or death.
Hernàndez-Galdamez et al., 2020 [38]	Cross-sectional study ($n = 212{,}802$)	Obesity (not specified)	Hospi-talization ICU admission Intubation death	Obesity was associated with an increased risk of hospitalization, ICU admission, intubation and death.
Kalligeros et al., 2020 [31]	Retrospective cohort study ($n = 103$)	BMI categories: BMI < 25, BMI $\geq$ 25 to <30, BMI $\geq$ 30 to <35, BMI $\geq$ 35	ICU admission, IMV	Severe obesity (BMI $\geq$ 35) was positively associated with ICU admission. Obesity (BMI $\geq$ 30 to <35) and severe obesity (BMI $\geq$ 35) were positively associated with the use of IMV.
Kim et al., 2020 [26]	Retrospective cohort study ($n = 10{,}861$)	BMI categories: underweight (BMI < 18.5), normal weight (BMI $\geq$ 18.5 to <25), overweight (BMI $\geq$ 25 to < 30), obesity class I (BMI $\geq$ 30 to < 35), obesity class II (BMI $\geq$ 35 to <40), and obesity class III (BMI $\geq$ 40)	IMV, death	Categories: overweight, obesity class I, II and III were associated with increased risk of requiring IMV. Underweight and obesity classes II and III were associated with increased risk of death.
Mash et al., 2021 [32]	Descriptive observational cross-sectional study ($n = 1376$)	BMI categories: normal (BMI $\geq$ 18.5 to <25), overweight/obese (BMI $\geq$ 25)	Death	Overweight/obesity (BMI $\geq$ 25) was significantly linked with mortality.

Table 1. *Cont.*

Reference	Study Design	Predictor	Outcome	Effect
Nakeshbandi et al., 2020 [24]	Retrospective cohort study (*n* = 504)	BMI categories: normal (BMI $\geq$ 18.5 to <25), overweight (BMI $\geq$ 25 to <30), and obese (BMI $\geq$ 30)	Mortality, intubation	Patients with overweight and obesity were at increased risk for mortality and intubation.
Palaiodimos et al., 2020 [27]	Retrospective cohort study (*n* = 200)	BMI categories: BMI < 25, BMI $\geq$ 25 to <35, BMI $\geq$ 35	In-hospital mortality, Worse in-hospital outcomes	Severe obesity (BMI $\geq$ 35) was associated with higher in-hospital mortality and worse in-hospital outcomes.
Parra-Bracamonte et al., 2020 [39]	Cross-sectional study (*n* = 331,298)	Obesity (not specified)	Death	Obesity was associated with higher risk of mortality.
Pediconi et al., 2020 [44]	Retrospective cohort study (*n* = 62)	VAT score (overweight: VAT area 100–129 cm^2 or VAT score 1. Obesity: VAT area $\geq$ 130 cm^2 or VAT score 2)	ICU admission	VAT score was the best ICU admission predictor.
Peña et al., 2020 [40]	Cross-sectional study (*n* = 323,671)	Obesity (not specified)	Death	Obesity was a major risk factor for mortality.
Randhawa et al., 2021 [33]	Retrospective cohort study (*n* = 302)	BMI categories: normal weight (BMI < 30), obesity BMI $\geq$ 30)	Compli-cations	Patients with obesity were more likely to suffer severe complications.
Rao et al., 2020 [28]	Retrospective cohort study (*n* = 240)	BMI (overweight, BMI > 28)	In-hospital death, Disease severity	Being overweight was related to COVID-19 severity but not to in-hospital death.
Salinas Aguirre et al., 2021 [41]	Cross-sectional study (*n* = 17,479)	Obesity (not specified)	Death	Obesity was associated with mortality.
Simonnet et al., 2020 [34]	Retrospective cohort study (*n* = 124)	BMI categories: lean (BMI $\geq$ 18.5 to <25), overweight (BMI $\geq$ 25 to < 30), moderate obesity (BMI $\geq$ 30 to < 35) and severe obesity (BMI $\geq$ 35)	Need for IMV	Need for IMV was associated with BMI.
Suleyman et al., 2020 [35]	Case series (*n* = 463)	BMI categories: severe obesity (BMI $\geq$ 40)	ICU admission	Severe obesity was independently associated with ICU admission.
van Zelst et al., 2020 [37]	Prospective observational cohort study (*n* = 166)	BMI Abdominal adiposity (waist-to-hip-ratio)	Unfavorable outcome	Abdominal adiposity and BMI were associated with an increased risk for unfavorable outcome (respiratory support of 3 L/min, intubation, ICU admission).
Watanabe et al., 2020 [45]	Retrospective cohort study (*n* = 150)	TAT VAT	ICU admission	TAT and VAT had a univariate association with ICU admission.
Zhu et al., 2020 [36]	Retrospective cohort study (*n* = 489,769)	BMI, categories: normal weight (BMI $\geq$ 18.5 to <25), overweight (BMI $\geq$ 25.0 to <30), and obese (BMI $\geq$ 30); waist circumference and waist-to-hip ratio	Hospi-talization with 'severe COVID-19'	BMI, waist circumference and waist-to-hip ratio were positively associated with the risk of severe COVID-19.

Abbreviations: VAT = visceral adipose tissue; SAT = subcutaneous adipose tissue; TAT = total adipose tissue; BMI = body mass index (kg/m^2); IMV = invasive mechanical ventilation; ICU = intensive care unit.

3. Underlying Mechanisms Linking Obesity to Major Complications as a Result of SARS-CoV-2 Infection

To gain a better understanding of the pathophysiology of COVID-19 in patients suffering from obesity, in the next section and in Figure 1 we provide an overview of the mechanistic pathways linking obesity with COVID-19 disease severity, with a focus on the metabolic and immunological consequences of obesity on COVID-19 disease course. Other features of obesity, like impaired respiratory mechanics and pulmonary function and the co-existence of metabolic disorders like diabetes and cardiovascular disease within a single individual also increase the risk for severe COVID-19 and complications [46], but are beyond the scope of this review. In this overview we discuss, *in vitro*, animal and human *in vivo* studies, including clinical trials. Studies were excluded when not indexed and when methodology did not reach minimal criteria.

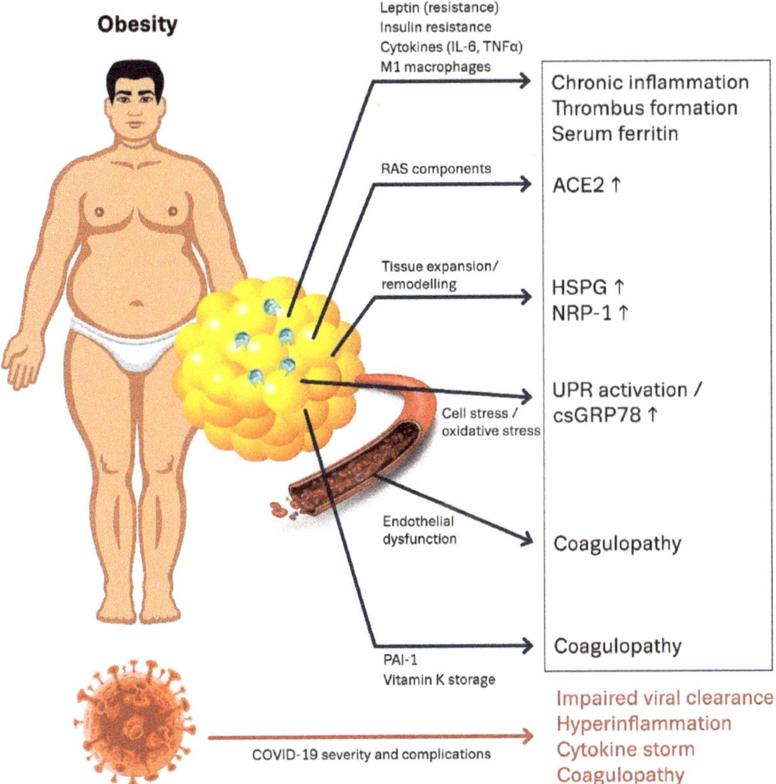

Figure 1. Schematic representation of different mechanisms through which obesity can promote COVID-19 disease severity and risk for complications. Obesity is often accompanied by insulin and leptin resistance which impairs viral clearance. Next, obesity is characterized by large hypoxic adipocytes infiltrated with immune cells and M1 macrophages leading to a chronic inflammatory state, hypercoagulability and hyperferritinemia. ACE2 produced by adipocytes could provide viral entry into the adipose tissue. In this way the adipose tissue could possibly function as a reservoir for the virus. The constant tissue expansion and tissue remodeling accompanying obesity in concert with high cell stress can upregulate the expression of other potential SARS-CoV-2 receptors, such as csGRP78, HSPG and NRP-1 in adipose tissue and other organs. Obesity-associated endothelial dysfunction, enhanced production of PAI-1 and vitamin K deficiency all increase the risk of developing COVID-19 associated coagulopathy. Abbreviations: IL-6 = Interleukin-6; TNFα = Tumor necrosis factor α; RAS = Renin angiotensin system; ACE = Angiotensin converting enzyme; HSPG = Heparan sulfate proteoglycan; NRP-1 = Neuropilin-1; UPR = Unfolded protein response; csGRP78 = Cell surface glucose related protein 78; PAI-1 = Plasminogen activator inhibitor 1.

3.1. Obesity Facilitates Viral Entry of SARS-CoV-2

Cell entry of SARS-CoV-2 virus depends on binding of the viral spike (S) proteins to cellular receptors and on S protein priming by host cell proteases. Obesity can promote viral entry by the upregulation of some of these host cell receptors.

3.1.1. Obesity Increases ACE2 Expression

Viral entry into target cells is facilitated by binding of the SARS-CoV-2 spike (S) protein to the membrane protein angiotensin-converting enzyme 2 (ACE2) and the priming of the S protein by serine protease TMPRSS2 and the endosomal cysteine proteases cathepsin B and L expressed by host cells [47,48]. ACE2 is part of the renin-angiotensin system (RAS) and plays an important role in regulating systemic blood pressure. Renin induces proteolytic cleavage of angiotensinogen into angiotensin (Ang)-I, which is then converted into Ang-II by ACE1. The action of ACE1 induces vasoconstriction and inflammation while ACE2 acts as a counter-regulator of the RAS. ACE2 induces vasodilatation and has an anti-inflammatory effect, by converting the pro-inflammatory Ang-II into Ang-1-7 which opposes the actions of Ang-II. ACE2 is expressed in a wide variety of human tissues such as the small intestine, testis, kidneys, heart, thyroid, lungs and brain [49]. Adipose tissue also expresses ACE2, together with all other components of the RAS. The RAS has been shown to play an important role in lipid metabolism and vice versa [50–53]. For example, the expression of ACE2 in adipocytes seems to be promoted by a high-fat diet [54]. Also, analysis of available transcriptome data showed that diet-induced obese mice displayed increased expression of ACE2 in the lung epithelium [55]. The expression of ACE2 was correlated with the expression of genes that code for sterol response element binding proteins 1 and 2, transcription factors controlling lipid synthesis and adipogenesis [55].

In addition, a number of studies indicate that ACE2 expression can be upregulated by a wide range of pro-inflammatory cytokines [56,57]. The chronic inflammatory state that characterizes obesity could therefore potentially facilitate viral entry [56]. Moreover, it has been suggested that viral pools can lodge in adipose tissue and promote shedding, immune activation, and chronic excessive cytokine release [58]. Earlier studies demonstrated that other viruses such as influenza A, HIV and cytomegalovirus use the adipose tissue as a reservoir [59].

Binding of SARS-CoV-2 to ACE2, which prevents ACE2 from exerting its enzymatic activities [60], has further consequences for immune system functioning. Decreased enzymatic ACE2 activity results in a reduction of anti-inflammatory Ang-1-7 and an accumulation of pro-inflammatory Ang-II. This process further increases the risk for COVID-19-related immune system complications such as ARDS.

In conclusion, obesity and its related pro-inflammatory state can promote viral entry, viral shedding and excessive immune activation through the upregulation of ACE2.

3.1.2. Activation of the Unfolded Protein Response by GRP78

Another receptor that provides viral entry for SARS-CoV-2 is glucose-regulated protein 78 (GRP78) [61,62]. GRP78, also called immunoglobulin heavy chain binding protein (BiP), is part of the heat shock protein 70 family and is a molecular chaperone found in the lumen of the endoplasmic reticulum (ER). GRP78 functions as an anti-apoptotic regulator that protects cells against cell death induced by ER stress [62]. GRP78 protects against cell death by ensuring the correct folding and assembly of proteins in the ER and by initiating the degradation of misfolded proteins. This evolutionarily preserved cellular homeostatic response to ER stress, also named the ER-stress or unfolded protein response (UPR) [63] can be disrupted by viruses in the establishment of acute, chronic and latent infections [64].

Normally, new foreign viral proteins produced inside host cells would be immediately degraded by the UPR. Therefore, viruses that acquired the ability to disrupt the process designed to degrade unrecognized protein have a clear advantage. Binding GRP78 is a mechanism commonly used by viruses like the Ebola virus, Zika virus, Dengue virus, Japanese encephalitis virus, Coxsackie A9, MERS-CoV and the Borna disease virus to ensure safe entry into host cells and to facilitate viral replication [65–71]. Also, the S proteins of the SARS-CoV-2 virus binds GRP78 with high affinity [61,72].

Obesity and various factors associated with obesity induce ER stress [73] and consequently UPR activation. These factors, including hypoxia, reactive oxygen species, insulin resistance, nutritional imbalance and excessive fat storage [74–76], stimulate GRP78 expression as a mechanism of the UPR to restore normal cell functions [77,78]. For example, evidence from *in vitro* and *in vivo* studies shows that dyslipidemia is associated with GRP78 overexpression in adipocytes (especially in white adipose tissue) [79], pneumocytes [80], neurons of the hypothalamus [81] and hepatocytes [82].

GRP78 overexpression induces its translocation to the cell membrane or cell surface. When GRP78 is located on the cell surface, it is referred to as cell surface (cs)-GRP78. This acts as a multi-functional receptor that can bind various ligands and plays a crucial role in apoptosis and inflammation, among other things [62]. By binding to cs-GRP78, SARS-CoV-2 could ensure safe entry into host cells [61,72].

3.1.3. Heparan Sulfate Proteoglycans and Neuropilin-1

Heparan sulfate proteoglycans (HSPGs) are located in the extracellular matrix and at the surface of the cell, where they act as co-receptors for various ligands. HSPGs are widely expressed and mediate many biological activities, including angiogenesis, blood coagulation, developmental processes, and cell homeostasis. The binding of cytokines, chemokines and growth factors to HSPGs at the cell surface prevents their degradation [83]. A significant amount of research has established that many different types of viruses can interact with HSPGs. For some of these viruses, these interactions are essential for internalization into host cells [84]. This is also the case for the internalization of SARS-CoV-2 [85].

Syndecans are the major family of transmembrane HSPGs and are present on virtually all nucleated human cells. Syndecans have been shown to facilitate the cellular entry of SARS-CoV-2 *in vitro*. Among syndecans, syndecan-4 was most efficient in mediating SARS-CoV-2 uptake, yet overexpression of other isoforms, including syndecan-1 and the neuronal syndecan-3, also increased SARS-CoV-2 internalization [85,86]. Syndecan 4 is widely expressed in most adult tissues [87] and, among other functions, is involved in lipid metabolism, by clearing pro-atherogenic remnant lipoproteins from the circulation [88]. In addition, syndecan-4 is crucial for adipocyte differentiation and proliferation [89]. Syndecan 4 is upregulated upon activation of the pro-inflammatory transcription factor nuclear factor-κB (NF-κB) [90,91] and its expression is therefore markedly increased under inflammatory conditions [92].

Neuropilin-1 (NRP-1), a membrane-bound co-receptor for growth factors such as vascular endothelial growth factor (VEGF), also facilitates viral entry of SARS-CoV-2 *in vitro* [93]. NRP-1 is strongly involved in adipogenesis and is highly upregulated during adipose-derived stem cell differentiation [94].

In obesity, chronic inflammation, higher circulating pro-atherogenic lipoprotein levels [95] and the constant expansion of adipose tissue accompanied with the differentiation and proliferation of adipose-derived stem cells, could increase syndecan (especially syndecan-4) and NRP-1 expression in various tissues. The up-regulation of syndecans and NRP-1 could facilitate cellular entry of SARS-CoV-2. However, further studies need to confirm this theory.

3.2. Obesity Related Insulin Resistance Contributes to an Impaired Immune Response to SARS-CoV-2 Infection

3.2.1. Obesity Induces Insulin Resistance

Insulin resistance is a consequence of the impairment of insulin signaling in insulin-responsive cells, like hepatocytes, myocytes and adipocytes. In the adipose tissue, insulin resistance is promoted by excess lipid accumulation, causing hypoxia and inflammation. The subsequent invasion of macrophages in the adipose tissue that release pro-inflammatory cytokines further impair insulin signaling (see also Section 3.2.2). As a consequence of adipocyte insulin resistance higher levels of free fatty acids (FFA) leave the fat tissue and enter into the circulation. The increase in circulating FFA and pro-inflammatory mediators further impairs insulin action in other metabolically active organs and tissues, including skeletal muscle and the liver, leading to systemic insulin resistance, which is associated with impaired glucose transport [96,97]. It has been shown that the size of the visceral adipose tissue and adipocyte size in humans is directly associated with systemic insulin resistance [14].

3.2.2. Insulin Resistance Is Induced by Adipocytes and Related Immune Cells

Adipose tissue macrophages can be divided into the classical M1 and alternatively activated M2 macrophages. Adipose tissue from obese individuals contains elevated numbers of M1-like macrophages, which produce pro-inflammatory cytokines, such as TNF-α and IL-6 [98,99]. These pro-inflammatory cytokines inhibit insulin signaling pathways in multiple tissues [100].

The production of TNF-α by M1 macrophages is positively related to the size of the adipose tissue mass. In the adipose tissue TNF-α induces phosphorylation and inactivation of insulin receptors and activates lipolysis, which increases the FFA load. The production of IL-6 by adipocytes and related immune cells is also associated with the amount of body fat. IL-6 induces the production of the pro-inflammatory acute-phase protein C-reactive protein (CRP) and increases fibrinogen levels, resulting in a prothrombotic state. It also promotes adhesion molecule expression by endothelial cells and activates local RAS pathways [101].

Chronic elevations of pro-inflammatory cytokines, such as TNF-α and IL-6, also directly influence COVID-19 disease course. Because TNF-α plays an important role in promoting ARDS, obese individuals with chronically elevated serum TNF-α are at greater risk of developing this life-threatening complication [102]. In patients with COVID-19, IL-6 levels are significantly elevated and associated with adverse clinical outcomes. Meta-analysis of mean IL-6 concentrations demonstrated 2.9-fold higher levels of IL-6 in hospitalized COVID-19 patients with complications compared to patients without complications [103].

3.2.3. Systemic Insulin Resistance Impairs the Immune Response

Insulin acts upon immune cells and therefore, systemic insulin resistance can have a substantial impact on the functioning of the adaptive and innate immune system [100]. Animal studies have shown that insulin signaling is essential for optimal T cell effector function [104]. In humans, insulin-resistant individuals displayed delayed innate immune-related pathway activation after respiratory viral infection compared to insulin-sensitive individuals, suggesting an impairment of the early acting innate immune response [105].

It is likely to assume that chronic inflammation and impairments of the immune response as a consequence of obesity-induced insulin resistance may reduce efficient viral clearance and drive organ injury in the development of severe COVID-19 [106].

3.3. Disrupted Leptin Signaling in Obesity can Induce Hyperinflammation during SARS-CoV-2 Infection

3.3.1. Leptin Is an Important Regulator of Energy Metabolism

Leptin is a pleiotropic protein secreted primarily from white adipose tissue into the bloodstream and can be transported across the blood-brain barrier. Through its effects on the central nervous system (CNS) and peripheral tissues, leptin is an important regulator of energy homeostasis, metabolism, neuroendocrine and immune system function [107].

Obesity proceeds from a chronic energy imbalance and is characterized by persistent hyperleptinemia and central leptin resistance. Under physiological conditions, leptin informs the brain about the energy status in the periphery, but in obesity, signaling to regulatory centers in the brain that normally inhibit food intake and regulate body weight and energy homeostasis is disrupted. Mechanisms underlying leptin resistance, include disruption of leptin signaling in hypothalamic and other CNS neurons, impaired leptin transport across blood-brain barrier, hypothalamic inflammation, ER stress, and autophagy [107,108].

3.3.2. Leptin Modulates the Immune System

Leptin has pro-inflammatory properties and upregulates the secretion of pro-inflammatory cytokines [18,22]. Leptin signaling results in the modulation of both the innate and adaptive immune system on multiple levels. Leptin signaling occurs primarily through the binding of leptin to the long isoform of the leptin receptor, followed by activation of the JAK/STAT pathway [2]. In the innate immune system, increased leptin production stimulates chemotaxis and neutrophil survival, induces pro-inflammatory cytokine production [109,110], and higher expression of adhesion molecules by eosinophils and basophils [111,112]. Monocyte activation and proliferation as well as the production of pro-inflammatory cytokines and chemotaxis are also stimulated by leptin [22]. Certain immune cells, more specifically those that contain the long isoform of the leptin receptor, may become unresponsive to leptin, or leptin resistant, when exposed to high leptin levels for an extended period of time [113,114]. Therefore, chronic hyperleptinemia, as seen in obesity, can have detrimental effects on the immune response due to both chronic pro-inflammatory effects and immune cell dysfunction [21].

Additionally, the adaptive immune system responds to leptin. Leptin induces a shift towards the more pro-inflammatory Th1 response [21], activates T and B lymphocytes, and inhibits regulatory T cells [18,115,116]. Regulatory T cells are involved in suppressing an excessive immune response [117,118]. It was demonstrated in patients suffering from obesity that leptin levels and BMI were inversely correlated with the number of regulatory T cells [119].

In obesity, leptin levels can be further increased due to infection or sepsis [120]. Elevated leptin levels, in combination with the obesity-induced pro-inflammatory state, further increases the risk for the development of a disproportionate or hyper-inflammatory response upon SARS-CoV-2 infection. Furthermore, SARS-CoV-2 infection has been shown to increase the expression of the gene that encodes for suppressor of cytokine signaling 3 (SOCS3) in lung epithelium [55]. This gene is a key regulator of inflammation and an inhibitor of leptin signaling. Increased SOCS3 expression as a result of SARS-CoV-2 infection could therefore further impair leptin signaling and negatively influence the immune response in patients suffering from obesity [121].

3.4. Hyperferritinemia as a Result of Hyperinflammation can Induce a Cytokine Storm

Ferritin is an iron-binding molecule that stores iron in a biologically available form for vital cellular processes and protects proteins, lipids and DNA from the potential toxicity of this metal element. Ferritin is a marker of the acute-phase response, and its secretion is regulated by pro-inflammatory cytokines. However, the origin of circulating serum ferritin during inflammatory conditions is still debated [122]. While some describe serum ferritin as a leakage product of damaged cells [123], increasing evidence shows that circulating serum ferritin levels may play a critical role in the inflammatory process [91]. Serum ferritin may protect the host during active infection by limiting the availability of iron to pathogens, a phenomenon called 'nutritional immunity' [124–126].

COVID-19 is also accompanied by a rise in circulating ferritin levels. Serum ferritin levels can be used as a diagnostic marker and even a predictor of COVID-19 severity [127–131]. The elevated ferritin levels observed in COVID-19 patients are probably a consequence of the inflammatory process induced by SARS-CoV-2 infection and actively act as enhancer of the inflammatory process in more severe COVID-19 [132]. Alternatively, it has been suggested that ferritin levels increase due to the break-down of red blood cells by the SARS-CoV-2 virus. By breaking down red blood cells and then attacking the hemoglobin 1-beta chain, the virus could separate iron from the porphyrin ring to eventually hijack the porphyrin ring, producing a rise in free iron and subsequently increasing ferritin levels [133]. Although hemoglobin-related biomarkers such as serum ferritin, progressively increase as the severity of COVID-19 increases [134–136], there is very limited evidence supporting this hypothesis [133]. This hypothesis has subsequently been refuted by DeMartino and colleagues, confirming that disease markers such as hemoglobin, iron and ferritin did not differ between critically ill COVID-19 patients and non-COVID ARDS patients [137].

Several studies have shown a relationship between obesity and elevated ferritin serum levels [138]. Chronic inflammation, as a result of the increased release of leptin and pro-inflammatory cytokines by adipocytes and related immune cells, [11,21,23,120] likely contributes to the manifestation of elevated ferritin levels in individuals with obesity.

Infection with SARS-CoV-2 in individuals with obesity as pre-existing condition can progress into a hyperferritinaemic state as both SARS-CoV-2 and obesity initiate the release of ferritin. Ferritin further stimulates macrophages to produce pro-inflammatory cytokines, mainly IL-1, IL-6 and IL-17. At overly elevated serum ferritin levels, macrophages produce so many cytokines that the situation can evolve into a cytokine storm [122]. The cytokine storm occurring with COVID-19 could be considered a hyperferritinemia syndrome [139] and is a significant adverse development in the course of the disease, translating into a marked increased risk of death [140].

3.5. Obesity-Related Risk Factors for Developing Coagulopathy in COVID-19 Patients

About a third of ICU patients with COVID-19 develop thrombotic complications [141,142]. The characteristics of CAC seem to be more complex than the development of thromboin-flammation triggered by systemic inflammation in response to infection [143]. For example, the presence of systemic microthrombi and hemorrhage in SARS-CoV-2 affected organs indicates a malfunction in the coordination of coagulation and fibrinolysis [144]. Moreover, CAC is commonly associated with increased D-dimer and fibrinogen levels, indicating there is initially no suppression of fibrinolysis [145]. Abnormalities in other coagulation biomarker such as prothrombin time and platelet count are less frequent and seem less affected by SARS-CoV-2 infection [146].

In the following section we provide an overview of parameters that are dysregulated in obesity and could contribute to increase the risk of developing CAC (See also Figure 1).

3.5.1. Hyperleptinemia

Hyperleptinemia is a risk factor for developing thrombi. Leptin affects blood clotting by enhancing prothrombotic and antifibrinolytic protein expression in vascular and inflammatory cells and thereby producing a more hypofibrinolytic state [147]. A large population-based cohort study from The Netherlands, examining associations between serum leptin concentrations and coagulation factor concentrations and parameters of platelet activation, showed that serum leptin concentrations were positively associated with concentrations of coagulation factor VIII and IX [148]. Hyperleptinemia and leptin resistance have also been described as risk factors for the development of cardiovascular diseases [149,150]. As elevated leptin levels are a key feature of obesity [149,151,152], the prothrombotic state produced by hyperleptinemia, increases the risk for developing thrombotic complications in COVID-19.

3.5.2. PAI-1 Production by Adipocytes

Plasminogen activator inhibitor 1 (PAI-1) is the primary physiological inhibitor of plasminogen activation. Elevations in plasma PAI-1 compromise normal fibrin clearance mechanisms and promote thrombosis. PAI-1 is also produced by adipocytes and its production is dramatically upregulated in obesity [153,154]. PAI-1 mRNA expression was demonstrated in the visceral and subcutaneous fat of obese rats [155] and in adipose tissue from human subjects [156]. In both rats and humans, VAT produced significantly more PAI-1 compared to SAT. These results are consistent with the observation that cardiovascular risk is most closely correlated with central obesity [157]. PAI-1 produced by adipocytes and vascular endothelium is involved in tissue expansion and angiogenesis necessary during adipose tissue development [158].

Several molecular mechanisms are involved in the upregulation of PAI-1 mRNA expression in obesity. For example, different cytokines including TNF-α and transforming growth factor (TGF)-ß, triglycerides, free fatty acids and insulin all stimulate PAI-1 expression in adipose tissue [153]. Leptin, additionally, contributes to a prothrombotic state by increasing the expression of PAI-1 in vascular endothelium [159]. Thus, obesity-induced upregulation of PAI-1 further increases the risk for coagulopathy, following SARS-CoV-2 infection.

3.5.3. Endothelial Dysfunction

The endothelium serves as a dynamic barrier that separates blood from interstitia. Endothelial cells respond rapidly to changes in the circulation and become activated according to environmental needs [160]. The multitude of physiological functions of the endothelium are still a topic of extended research. Endothelium is recently described as an active regulator of lipid and glucose homeostasis [161].

Endothelial cells play a major role in vascular homeostasis and blood coagulation. Endothelial dysfunction is considered a hallmark of metabolic diseases and is characterized by a loss of molecular cell functions and inevitably causing coagulopathy. Endothelial dysfunction contributes to various pathological states such as atherothrombosis, arterial thrombosis (stroke, visceral and peripheral artery occlusive diseases), venous thrombosis, intravascular coagulation and thrombotic microangiopathies [160].

Oxidative stress is considered the major cause of endothelial dysfunction. Obesity generates oxidative stress through different pathways, such as hyperglycemia, known to trigger vascular damage by inducing the accumulation of reactive oxygen species (ROS) [162]. Hyperglycemia also activates NF-κB, a transcription factor that mediates vascular inflammation [162]. The exacerbated production of pro-inflammatory cytokines by adipose tissue further increases oxidative stress levels and promotes the upregulation of procoagulant factors and adhesion molecules in the endothelium, the downregulation of anticoagulant regulatory proteins, increases thrombin generation, and enhances platelet activation [163]. Not coincidentally, endothelial dysfunction is a common feature of comorbidities that increase the risk for severe COVID-19, including hypertension, obesity, diabetes mellitus, coronary artery disease and heart failure.

Furthermore, because ACE2 is expressed abundantly on vascular endothelial cells of both small and large arteries and veins [164], SARS-CoV-2 infection of endothelial cells can further aggravate endothelial dysfunction. Endothelial damage and dysfunction may thus be the result of cellular infection by SARS-CoV-2, as well as a consequence of obesity-associated excessive systemic inflammation [165]. Evidently, patients with preexisting endothelial dysfunction are more vulnerable to develop severe complications, including coagulopathy, following SARS-CoV-2 infection (See also Figure 1).

3.5.4. Vitamin K

Vitamin K is a fat-soluble vitamin, required for the carboxylation of vitamin K-dependent proteins (VKDP). Intrahepatic VKDP include coagulation factors II, VII, IX, X, and different anticoagulant proteins. Extra-hepatic VKDP include diverse gamma-carboxyglutamate (Gla) proteins involved in maintaining bone homeostasis, as well as inhibiting ectopic calcification [166]. Vitamin K is not a single entity but a family of structurally related molecules such as phylloquinone, also referred to as vitamin K1, and menaquinone or vitamin K2.

Obesity is linked with vitamin K deficiency [167–169]. Vitamin K deficiency results in decreased vitamin K-dependent carboxylation and phosphorylation of Gla-proteins and, as a consequence, into elevated levels of circulating desphosphorylated-uncarboxylated matrix Gla protein (dp-ucMGP). Recent research studied the relationship between obesity and serum dp-ucMGP in a cohort of 278 Chinese Han people. The results demonstrated that serum dp-ucMGP level was positively associated with visceral fat index, waist height ratio, but not BMI [167]. In addition, lower vitamin K2 levels were observed in obese hemodialysis patients compared to non-obese patients [168]. Also, a lower vitamin K1 status was observed in patients with obesity compared to healthy individuals [169]. Yet, differences in dietary K1 intake could not fully explain this observation. It has been suggested that vitamin K accumulates in adipose tissue, thereby reducing the bioavailability of this vitamin in individuals with obesity [169]. This theory could be plausible since a similar mechanism has been established for other fat-soluble vitamins [170,171]. However, understanding the mechanisms underlying the association between obesity and vitamin K deficiency remains a topic for further investigation.

Vitamin K status was also assessed in COVID-19 patients by measuring dp-ucMGP. Levels of dp-ucMGP were significantly elevated in COVID-19 patients compared to controls and were higher in COVID-19 patients with unfavorable disease outcome. Carboxylated matrix Gla-protein protect against degradation and calcification of vasculature and elastic fibers in the extracellular matrix of the lungs. In COVID-19, elastic fiber degradation combined with calcification of these fibers due to low vitamin K status could aggravate lung injury and lung fibrosis [172].

Thus, to summarize, we can argue that several factors associated with obesity increase the risk for CAC, including elevated leptin levels, increased PAI-1 levels, endothelial dysfunction and low vitamin K levels. The latter is possibly also involved in the development of more severe lung injury in COVID-19.

4. Conclusions and Implications for Further Research

Obesity cannot simply be defined as an excess of fat cells. Adipose tissue releases many active substances, such as adipokines and components of the RAS, all influencing the brain and metabolic- and immune system. Being obese increases the risk of SARS-CoV-2 infection and complications via several mechanisms. First, viral entry is enhanced due to increased ACE2, csGRP78 and presumably HSPG and NRP-1 expression levels in various cell types, like pneumocytes and adipocytes. Second, the immune system is unable to provide an adequate immune response leading to impaired viral clearance. Eventually, the immune system can overreact as a result of pro-inflammatory 'priming' due to excessive cytokine production by adipose tissue and its related immune cells and high ferritin levels, eventually triggering a cytokine storm [122,139]. Finally, hyperleptinemia, PAI-1 production by adipocytes and endothelial cells, endothelial dysfunction and low levels, or low bioavailability, of vitamin K, all increase the risk for the development of thrombus formation and hemorrhage.

An important lesson learned from the coronavirus pandemic is the importance of a healthy lifestyle to positively influence the course of COVID-19 disease. A non-processed nutrient-rich diet [173–175], limited excessive or overly energy-rich food, sufficient and intensive exercise [176], sufficient sleep [177] and avoiding chronic psycho-emotional stress [178] are all efficient health-promoting measures in the prevention of obesity [179]. We also advocate an integrated multidisciplinary approach in the fight against COVID-19. Future research must identify causes of severity and complications to develop efficient preventive measures and curative interventions.

Author Contributions: F.D. developed the concepts, reviewed the literature and drafted the manuscript. M.v.H. provided editorial assistance. K.d.P. and F.v.D. developed the concepts and provided final editorial oversight. All authors have read and agreed to the published version of the manuscript.

Funding: This research received no external funding.

Data Availability Statement: Not applicable.

Acknowledgments: We are grateful for the input of all the members of the Natura Foundation research team including Michaëla Van Langeveld, Margo Peinemann, Aldert Hoogland, Lotte Bos, Iris Brunsmann and André Frankhuizen.

Conflicts of Interest: Authors F.D., K.d.P., M.v.H. and F.v.D. were sponsored by Bonusan B.V. The funders had no role in the design of the study; in the collection, analyses, or interpretation of data; in the writing of the manuscript, or in the decision to publish the results.

References

1. Connors, J.M.; Levy, J.H. COVID-19 and its implications for thrombosis and anticoagulation. *Blood* **2020**, *135*, 2033–2040. [CrossRef]
2. Harapan, H.; Itoh, N.; Yufika, A.; Winardi, W.; Keam, S.; Te, H. Coronavirus disease 2019 (COVID-19): A literature review. *J. Infect. Public Health* **2020**, *13*, 667–673. [CrossRef]
3. Garg, S.; Kim, L.; Whitaker, M.; O'Halloran, A.; Cummings, C.; Holstein, R. Hospitalization Rates and Characteristics of Patients Hospitalized with Laboratory-Confirmed Coronavirus Disease 2019—COVID-NET, 14 States, 1–30 March 2020. *MMWR Morb Mortal Wkly Rep.* **2020**, *69*, 458–464. [CrossRef]
4. Engin, A.B.; Engin, E.D.; Engin, A. Two important controversial risk factors in SARS-CoV-2 infection: Obesity and smoking. *Environ. Toxicol. Pharmacol.* **2020**, *78*, 103411. [CrossRef] [PubMed]
5. Vardavas, C.; Nikitara, K. COVID-19 and Smoking: A Systematic Review of the Evidence. *Tob Induc Dis.* **2020**. Available online: http://www.journalssystem.com/tid/COVID-19-and-smoking-A-systematic-review-of-the-evidence,119324,0,2.html (accessed on 20 March 2020). [CrossRef] [PubMed]
6. Yang, J.; Zheng, Y.; Gou, X.; Pu, K.; Chen, Z.; Guo, Q. Prevalence of comorbidities and its effects in patients infected with SARS-CoV-2: A systematic review and meta-analysis. *Int. J. Infect. Dis.* **2020**, *94*, 91–95. [CrossRef]
7. Blüher, M. Obesity: Global epidemiology and pathogenesis. *Nat. Rev. Endocrinol.* **2019**, *15*, 288–298. [CrossRef] [PubMed]
8. Obesity. Available online: https://www.who.int/westernpacific/health-topics/obesity (accessed on 26 November 2020).
9. Chaldakov, G.; Stankulov, I.; Hristova, M.; Ghenev, P. Adipobiology of Disease: Adipokines and Adipokine-Targeted Pharmacology. *Curr. Pharm. Des.* **2003**, *9*, 1023–1031. [CrossRef]
10. Kershaw, E.E.; Flier, J.S. Adipose Tissue as an Endocrine Organ. *J. Clin. Endocrinol. Metab.* **2004**, *89*, 2548–2556. [CrossRef]
11. Scherer, P.E. Adipose Tissue: From Lipid Storage Compartment to Endocrine Organ. *Diabetes* **2006**, *55*, 1537–1545. [CrossRef]
12. Fontana, L.; Eagon, J.C.; Trujillo, M.E.; Scherer, P.E.; Klein, S. Visceral Fat Adipokine Secretion Is Associated With Systemic Inflammation in Obese Humans. *Diabetes* **2007**, *56*, 1010–1013. [CrossRef] [PubMed]
13. Matsuzawa, Y.; Shimomura, I.; Nakamura, T.; Keno, Y.; Kotani, K.; Tokunaga, K. Pathophysiology and Pathogenesis of Visceral Fat Obesity. *Obes. Res.* **1995**, *3*, 187s–194s. [CrossRef]
14. Hardy, O.T.; Czech, M.P.; Corvera, S. What causes the insulin resistance underlying obesity? *Curr. Opin. Endocrinol. Diabetes Obes.* **2012**, *19*, 81–87. [CrossRef]
15. Vieira-Potter, V.J. Inflammation and macrophage modulation in adipose tissues: Adipose tissue macrophage modulation. *Cell Microbiol.* **2014**, *16*, 1484–1492. [CrossRef]
16. Tilg, H.; Moschen, A.R. Adipocytokines: Mediators linking adipose tissue, inflammation and immunity. *Nat. Rev. Immunol.* **2006**, *6*, 772–783. [CrossRef] [PubMed]
17. Pandit, R.; Beerens, S.; Adan, R.H. Role of leptin in energy expenditure: The hypothalamic perspective. *Am. J. Physiol. Regul. Integr. Comp. Physiol.* **2017**, *312*, R938–R947. [CrossRef]
18. Iikuni, N.; Kwan Lam, Q.; Lu, L.; Matarese, G.; Cava, A. Leptin and Inflammation. *Curr. Immunol. Rev.* **2008**, *4*, 70–79. [CrossRef]
19. Wauters, M.; Mertens, I.; Considine, R.; De Leeuw, I.; Van Gaal, L. Are leptin levels dependent on body fat distribution in obese men and women? *Eat Weight Disord.* **1998**, *3*, 124–130. [CrossRef]
20. Considine, R.V.; Sinha, M.K.; Heiman, M.L.; Kriauciunas, A.; Stephens, T.W.; Nyce, M.R. Serum Immunoreactive-Leptin Concentrations in Normal-Weight and Obese Humans. *N. Engl. J. Med.* **1996**, *334*, 292–295. [CrossRef]
21. Abella, V.; Scotece, M.; Conde, J.; Pino, J.; Gonzalez-Gay, M.A.; Gómez-Reino, J.J. Leptin in the interplay of inflammation, metabolism and immune system disorders. *Nat. Rev. Rheumatol.* **2017**, *13*, 100–109. [CrossRef] [PubMed]
22. Conde, J.; Scotece, M.; Gómez, R.; Gómez-Reino, J.J.; Lago, F.; Gualillo, O. At the crossroad between immunity and metabolism: Focus on leptin. *Expert Rev. Clin. Immunol.* **2010**, *6*, 801–808. [CrossRef] [PubMed]
23. Divella, R.; De Luca, R.; Abbate, I.; Naglieri, E.; Daniele, A. Obesity and cancer: The role of adipose tissue and adipo-cytokines-induced chronic inflammation. *J. Cancer* **2016**, *7*, 2346–2359. [CrossRef] [PubMed]
24. Nakeshbandi, M.; Maini, R.; Daniel, P.; Rosengarten, S.; Parmar, P.; Wilson, C. The impact of obesity on COVID-19 complications: A retrospective cohort study. *Int. J. Obes.* **2020**, *44*, 1832–1837. [CrossRef]
25. Frank, R.C.; Mendez, S.R.; Stevenson, E.K.; Guseh, J.S.; Chung, M.; Silverman, M.G. Obesity and the Risk of Intubation or Death in Patients With Coronavirus Disease 2019. *Crit. Care Med.* **2020**, *48*, e1097–e1101. [CrossRef]
26. Kim, T.; Roslin, M.; Wang, J.J.; Kane, J.; Hirsch, J.S.; Ji Kim, E. Body Mass Index as a Risk Factor for Clinical Outcomes in Patients Hospitalized with COVID-19 in New York. *Obesity* **2021**, *29*, 279–284. [CrossRef]
27. Palaiodimos, L.; Kokkinidis, D.G.; Li, W.; Karamanis, D.; Ognibene, J.; Arora, S. Severe obesity, increasing age and male sex are independently associated with worse in-hospital outcomes, and higher in-hospital mortality, in a cohort of patients with COVID-19 in the Bronx, New York. *Metabolism* **2020**, *108*, 154262. [CrossRef]
28. Rao, X.; Wu, C.; Wang, S.; Tong, S.; Wang, G.; Wu, G. The importance of overweight in COVID-19: A retrospective analysis in a single center of Wuhan, China. *Medicine* **2020**, *99*, e22766. [CrossRef]

29. Anderson, M.R.; Geleris, J.; Anderson, D.R.; Zucker, J.; Nobel, Y.R.; Freedberg, D. Body Mass Index and Risk for Intubation or Death in SARS-CoV-2 Infection: A Retrospective Cohort Study. *Ann. Intern. Med.* **2020**, *173*, 782–790. [CrossRef]
30. Deng, M.; Qi, Y.; Deng, L.; Wang, H.; Xu, Y.; Li, Z. Obesity as a Potential Predictor of Disease Severity in Young COVID-19 Patients: A Retrospective Study. *Obesity* **2020**, *28*, 1815–1825. [CrossRef]
31. Kalligeros, M.; Shehadeh, F.; Mylona, E.K.; Benitez, G.; Beckwith, C.G.; Chan, P.A. Association of Obesity with Disease Severity among Patients with COVID-19. *Obesity* **2020**. Available online: https://www.ncbi.nlm.nih.gov/pmc/articles/PMC7267224/ (accessed on 29 October 2020).
32. Mash, R.J.; Presence-Vollenhoven, M.; Adeniji, A.; Christoffels, R.; Doubell, K.; Eksteen, L. Evaluation of patient characteristics, management and outcomes for COVID-19 at district hospitals in the Western Cape, South Africa: Descriptive observational study. *BMJ Open* **2021**, *11*, e047016. [CrossRef] [PubMed]
33. Randhawa, G.; Syed, K.A.; Singh, K.; Kundal, S.V.; Oli, S.; Silver, M. The relationship between obesity, hemoglobin A1c and the severity of COVID-19 at an urban tertiary care center in New York City: A retrospective cohort study. *BMJ Open* **2021**, *11*, e044526. [CrossRef] [PubMed]
34. Simonnet, A.; Chetboun, M.; Poissy, J.; Raverdy, V.; Noulette, J.; Duhamel, A. High Prevalence of Obesity in Severe Acute Respiratory Syndrome Coronavirus-2 (SARS-CoV-2) Requiring Invasive Mechanical Ventilation. *Obesity* **2020**, *28*, 1195–1199. [CrossRef]
35. Suleyman, G.; Fadel, R.A.; Malette, K.M.; Hammond, C.; Abdulla, H.; Entz, A. Clinical Characteristics and Morbidity Associated With Coronavirus Disease 2019 in a Series of Patients in Metropolitan Detroit. *JAMA Netw. Open* **2020**, *3*, e2012270. [CrossRef]
36. Zhu, Z.; Hasegawa, K.; Ma, B.; Fujiogi, M.; Camargo, C.A.; Liang, L. Association of obesity and its genetic predisposition with the risk of severe COVID-19: Analysis of population-based cohort data. *Metabolism* **2020**, *112*, 154345. [CrossRef] [PubMed]
37. van Zelst, C.M.; Janssen, M.L.; Pouw, N.; Birnie, E.; Castro Cabezas, M.; Braunstahl, G.-J. Analyses of abdominal adiposity and metabolic syndrome as risk factors for respiratory distress in COVID-19. *BMJ Open Resp. Res.* **2020**, *7*, e000792. [CrossRef] [PubMed]
38. Hernández-Galdamez, D.R.; González-Block, M.Á.; Romo-Dueñas, D.K.; Lima-Morales, R.; Hernández-Vicente, I.A.; Lumbreras-Guzmán, M. Increased Risk of Hospitalization and Death in Patients with COVID-19 and Pre-existing Noncommunicable Diseases and Modifiable Risk Factors in Mexico. *Arch. Med. Res.* **2020**, *51*, 683–689. [CrossRef]
39. Parra-Bracamonte, G.M.; Lopez-Villalobos, N.; Parra-Bracamonte, F.E. Clinical characteristics and risk factors for mortality of patients with COVID-19 in a large data set from Mexico. *Ann. Epidemiol.* **2020**, *52*, 93–98.e2. [CrossRef] [PubMed]
40. Peña, J.E.; la Rascón-Pacheco, R.A.; Ascencio-Montiel, I.; González-Figueroa, E.; Fernández-Gárate, J.E.; Medina-Gómez, O.S. Hypertension, Diabetes and Obesity, Major Risk Factors for Death in Patients With COVID-19 in Mexico. *Arch. Med. Res.* **2020**. [CrossRef]
41. Salinas Aguirre, J.E.; Sánchez García, C.; Rodríguez Sanchez, R.; Rodríguez Muñoz, L.; Díaz Castaño, A.; Bernal Gómez, R. Clinical characteristics and comorbidities associated with mortality in patients with COVID-19 in Coahuila (Mexico). *Rev. Clin. Esp.* **2021**. [CrossRef]
42. Chandarana, H.; Dane, B.; Mikheev, A.; Taffel, M.T.; Feng, Y.; Rusinek, H. Visceral adipose tissue in patients with COVID-19: Risk stratification for severity. *Abdom. Radiol.* **2020**, *46*, 816–825. [CrossRef] [PubMed]
43. Battisti, S.; Pedone, C.; Napoli, N.; Russo, E.; Agnoletti, V.; Nigra, S.G. Computed Tomography Highlights Increased Visceral Adiposity Associated With Critical Illness in COVID-19. *Diabetic Care* **2020**, *43*, e129–e130. [CrossRef]
44. Pediconi, F.; Rizzo, V.; Schiaffino, S.; Cozzi, A.; Della Pepa, G.; Galati, F. Visceral adipose tissue area predicts intensive care unit admission in COVID-19 patients. *Obes. Res. Clin. Pract.* **2020**, *15*, 82–96.
45. Watanabe, M.; Caruso, D.; Tuccinardi, D.; Risi, R.; Zerunian, M.; Polici, M. Visceral fat shows the strongest association with the need of intensive care in patients with COVID-19. *Metabolism* **2020**, *111*, 154319. [CrossRef]
46. Dietz, W.; Santos-Burgoa, C. Obesity and its Implications for COVID-19 Mortality. *Obesity* **2020**, *28*, 1005. [CrossRef]
47. Hoffmann, M.; Kleine-Weber, H.; Schroeder, S.; Krüger, N.; Herrler, T.; Erichsen, S. SARS-CoV-2 Cell Entry Depends on ACE2 and TMPRSS2 and Is Blocked by a Clinically Proven Protease Inhibitor. *Cell* **2020**, *181*, 271–280.e8. [CrossRef] [PubMed]
48. Yan, R.; Zhang, Y.; Li, Y.; Xia, L.; Guo, Y.; Zhou, Q. Structural basis for the recognition of SARS-CoV-2 by full-length human ACE2. *Science* **2020**, *367*, 1444–1448. [CrossRef] [PubMed]
49. Li, M.-Y.; Li, L.; Zhang, Y.; Wang, X.-S. Expression of the SARS-CoV-2 cell receptor gene ACE2 in a wide variety of human tissues. *Infect. Dis. Poverty* **2020**, *9*, 45. [CrossRef] [PubMed]
50. Cassis, L.A. Fat cell metabolism: Insulin, fatty acids, and renin. *Curr. Hypertens Rep.* **2000**, *2*, 132–138. [CrossRef]
51. Cassis, L.A.; Police, S.B.; Yiannikouris, F.; Thatcher, S.E. Local adipose tissue renin-angiotensin system. *Curr. Sci.* **2008**, *10*, 93–98. [CrossRef]
52. Massiéra, F.; Bloch-Faure, M.; Ceiler, D.; Murakami, K.; Fukamizu, A.; Gasc, J.-M. Adipose angiotensinogen is involved in adipose tissue growth and blood pressure regulation. *FASEB J.* **2001**, *15*, 1–25. [CrossRef] [PubMed]
53. Yvan-Charvet, L.; Quignard-Boulangé, A. Role of adipose tissue renin-angiotensin system in metabolic and inflammatory diseases associated with obesity. *Kidney Int.* **2011**, *79*, 162–168. [CrossRef]
54. Gupte, M.; Boustany-Kari, C.M.; Bharadwaj, K.; Police, S.; Thatcher, S.; Gong, M.C. ACE2 is expressed in mouse adipocytes and regulated by a high-fat diet. *Am. J. Physiol. Regul. Integr. Comp. Physiol.* **2008**, *295*, R781–R788. [CrossRef] [PubMed]
55. Heialy, S.A.; Hachim, M.; Senok, A.; Tayoun, A.A.; Hamoudi, R.; Alsheikh-Ali, A. Regulation of angiotensin converting enzyme 2 (ACE2) in obesity: Implications for COVID-19. 2020. Available online: http://biorxiv.org/lookup/doi/10.1101/2020.04.17.046938 (accessed on 4 May 2020).

56. Hennighausen, L.; Lee, H.K. Activation of the SARS-CoV-2 Receptor *Ace2* by Cytokines through pan JAK-STAT Enhancers. *Genomics* **2020**. Available online: http://biorxiv.org/lookup/doi/10.1101/2020.05.11.089045 (accessed on 4 November 2020). [CrossRef] [PubMed]

57. Ziegler, C.G.K.; Allon, S.J.; Nyquist, S.K.; Mbano, I.M.; Miao, V.N.; Tzouanas, C.N. SARS-CoV-2 receptor ACE2 is an interferon-stimulated gene in human airway epithelial cells and is detected in specific cell subsets across tissues. *Cell* **2020**, *181*, 1016–1035.e19. [CrossRef]

58. Ryan, P.M.; Caplice, N.M. Is Adipose Tissue a Reservoir for Viral Spread, Immune Activation, and Cytokine Amplification in Coronavirus Disease 2019? *Obesity* **2020**, *28*, 1191–1194. [CrossRef] [PubMed]

59. Kassir, R. Risk of COVID-19 for patients with obesity. *Obes. Rev.* **2020**. Available online: https://onlinelibrary.wiley.com/doi/abs/10.1111/obr.13034 (accessed on 26 May 2020). [CrossRef]

60. Costa, L.B.; Perez, L.G.; Palmeira, V.A.; Macedo e Cordeiro, T.; Ribeiro, V.T.; Lanza, K.; Simões e Silva, A.C. Insights on SARS-CoV-2 Molecular Interactions With the Renin-Angiotensin System. *Front Cell Dev. Biol.* **2020**, *8*. [CrossRef]

61. Ha, D.P.; Van Krieken, R.; Carlos, A.J.; Lee, A.S. The stress-inducible molecular chaperone GRP78 as potential therapeutic target for coronavirus infection. *J. Infect.* **2020**, *81*, 452–482. [CrossRef] [PubMed]

62. Ibrahim, I.M.; Abdelmalek, D.H.; Elfiky, A.A. GRP78: A cell's response to stress. *Life Sci.* **2019**, *226*, 156–163. [CrossRef] [PubMed]

63. Hetz, C.; Zhang, K.; Kaufman, R.J. Mechanisms, regulation and functions of the unfolded protein response. *Nat. Rev. Mol. Cell Biol.* **2020**. Available online: http://www.nature.com/articles/s41580-020-0250-z (accessed on 16 June 2020). [CrossRef]

64. Chan, S.-W. The unfolded protein response in virus infections. *Front. Microbiol.* **2014**. Available online: http://journal.frontiersin.org/article/10.3389/fmicb.2014.00518/abstract (accessed on 23 November 2020). [CrossRef]

65. Chu, H.; Chan, C.-M.; Zhang, X.; Wang, Y.; Yuan, S.; Zhou, J. Middle East respiratory syndrome coronavirus and bat coronavirus HKU9 both can utilize GRP78 for attachment onto host cells. *J. Biol. Chem.* **2018**, *293*, 11709–11726. [CrossRef] [PubMed]

66. Honda, T.; Horie, M.; Daito, T.; Ikuta, K.; Tomonaga, K. Molecular Chaperone BiP Interacts with Borna Disease Virus Glycoprotein at the Cell Surface. *J. Virol.* **2009**, *83*, 12622–12625. [CrossRef] [PubMed]

67. Mairiang, D.; Zhang, H.; Sodja, A.; Murali, T.; Suriyaphol, P.; Malasit, P. Identification of New Protein Interactions between Dengue Fever Virus and Its Hosts, Human and Mosquito. *PLoS ONE* **2013**, *8*, e53535. [CrossRef]

68. Nain, M.; Mukherjee, S.; Karmakar, S.P.; Paton, A.W.; Paton, J.C.; Abdin, M.Z. GRP78 Is an Important Host Factor for Japanese Encephalitis Virus Entry and Replication in Mammalian Cells. *J. Virol.* **2017**, *91*, e02274-e16. [CrossRef]

69. Sager, G.; Gabaglio, S.; Sztul, E.; Belov, G. Role of Host Cell Secretory Machinery in Zika Virus Life Cycle. *Viruses* **2018**, *10*, 559. [CrossRef] [PubMed]

70. Triantafilou, K.; Fradelizi, D.; Wilson, K.; Triantafilou, M. GRP78, a Coreceptor for Coxsackievirus A9, Interacts with Major Histocompatibility Complex Class I Molecules Which Mediate Virus Internalization. *J. Virol.* **2002**, *76*, 633–643. [CrossRef] [PubMed]

71. Yu, D.-S.; Weng, T.-H.; Hu, C.-Y.; Wu, Z.-G.; Li, Y.-H.; Cheng, L.-F. Chaperones, Membrane Trafficking and Signal Transduction Proteins Regulate Zaire Ebola Virus trVLPs and Interact With trVLP Elements. *Front. Microbiol.* **2018**, *9*, 2724. [CrossRef]

72. Ibrahim, I.M.; Abdelmalek, D.H.; Elshahat, M.E.; Elfiky, A.A. COVID-19 spike-host cell receptor GRP78 binding site prediction. *J. Infect.* **2020**, *80*, 554–562. [CrossRef]

73. Kawasaki, N.; Asada, R.; Saito, A.; Kanemoto, S.; Imaizumi, K. Obesity-induced endoplasmic reticulum stress causes chronic inflammation in adipose tissue. *Sci. Rep.* **2012**, *2*, 799. [CrossRef] [PubMed]

74. Amen, O.M.; Sarker, S.D.; Ghildyal, R.; Arya, A. Endoplasmic Reticulum Stress Activates Unfolded Protein Response Signaling and Mediates Inflammation, Obesity, and Cardiac Dysfunction: Therapeutic and Molecular Approach. *Front. Pharmacol.* **2019**, *10*, 977. [CrossRef] [PubMed]

75. de Ferranti, S.; Mozaffarian, D. The Perfect Storm: Obesity, Adipocyte Dysfunction, and Metabolic Consequences. *Clin. Chem.* **2008**, *54*, 945–955. [CrossRef] [PubMed]

76. Gregor, M.F.; Hotamisligil, G.S. Thematic review series: Adipocyte Biology. Adipocyte stress: The endoplasmic reticulum and metabolic disease. *J. Lipid Res.* **2007**, *48*, 1905–1914. [CrossRef] [PubMed]

77. Fang, L.; Kojima, K.; Zhou, L.; Crossman, D.K.; Mobley, J.A.; Grams, J. Analysis of the Human Proteome in Subcutaneous and Visceral Fat Depots in Diabetic and Non-diabetic Patients with Morbid Obesity. *J. Proteom. Bioinform.* **2015**. Available online: https://www.omicsonline.org/open-access/analysis-of-the-human-proteome-in-subcutaneous-and-visceral-fat-depotsin-diabetic-and-nondiabetic-patients-with-morbid-obesity-jpb-1000361.php?aid=53283 (accessed on 11 June 2020).

78. Girona, J.; Rodríguez-Borjabad, C.; Ibarretxe, D.; Vallvé, J.-C.; Ferré, R.; Heras, M. The Curr. Immunol. Rev.culating GRP78/BiP Is a Marker of Metabolic Diseases and Atherosclerosis: Bringing Endoplasmic Reticulum Stress into the Clinical Scenario. *JCM* **2019**, *8*, 1793. [CrossRef]

79. Zhu, W.; Niu, X.; Wang, M.; Li, Z.; Jiang, H.-K.; Li, C. Endoplasmic reticulum stress may be involved in insulin resistance and lipid metabolism disorders of the white adipose tissues induced by high-fat diet containing industrial trans-fatty acids. *Diabetes Metab. Syndr. Obes.* **2019**, *12*, 1625–1638. [CrossRef] [PubMed]

80. Aslani, M.R.; Ghobadi, H.; Panahpour, H.; Ahmadi, M.; Khaksar, M.; Heidarzadeh, M. Modification of lung endoplasmic reticulum genes expression and NF-kB protein levels in obese ovalbumin-sensitized male and female rats. *Life Sci.* **2020**, *247*, 117446. [CrossRef]

81. Zhang, X.; Zhang, G.; Zhang, H.; Karin, M.; Bai, H.; Cai, D. Hypothalamic IKKβ/NF-κB and ER Stress Link Overnutrition to Energy Imbalance and Obesity. *Cell* **2008**, *135*, 61–73. [CrossRef]

82. Boden, G.; Song, W.; Duan, X.; Cheung, P.; Kresge, K.; Barrero, C. Infusion of Glucose and Lipids at Physiological Rates Causes Acute Endoplasmic Reticulum Stress in Rat Liver. *Obesity* **2011**, *19*, 1366–1373. [CrossRef]

83. Sarrazin, S.; Lamanna, W.C.; Esko, J.D. Heparan Sulfate Proteoglycans. *Cold Spring Harbor Perspect. Biol.* **2011**, *3*, a004952. [CrossRef]

84. Cagno, V.; Tseligka, E.D.; Jones, S.T.; Tapparel, C. Heparan Sulfate Proteoglycans and Viral Attachment: True Receptors or Adaptation Bias? *Viruses* **2019**, *11*, 596. [CrossRef] [PubMed]

85. Hudák, A.; Szilák, L.; Letoha, T. Contribution of Syndecans to the Cellular Entry of SARS-CoV-2. In *Review*; 2020; Available online: https://www.researchsquare.com/article/rs-70340/v1 (accessed on 10 November 2020).

86. Bermejo-Jambrina, M.; Eder, J.; Kaptein, T.M.; Helgers, L.C.; Brouwer, P.J.M.; Hamme, J.L. SARS-CoV-2 Infection and Transmission Depends on Heparan Sulfates and Is Blocked by Low Molecular Weight Heparins. *bioRxiv* **2020**. bioRxiv:2020.08.18.255810.

87. Bellin, R.; Capila, I.; Lincecum, J.; Park, P.W.; Reizes, O.; Bernfield, M.R. Unlocking the secrets of syndecans: Transgenic organisms as a potential key. *Glycoconj. J.* **2002**, *19*, 295–304. [CrossRef] [PubMed]

88. Gordts, P.L.S.M.; Esko, J.D. The heparan sulfate proteoglycan grip on hyperlipidemia and atherosclerosis. *Matrix Biol.* **2018**, *71–72*, 262–282. [CrossRef] [PubMed]

89. Landry, R.; Rioux, V.; Bensadoun, A. Characterization of syndecan-4 expression in 3T3-F442A mouse adipocytes: Link between syndecan-4 induction and cell proliferation. *Cell Growth Differ. Mol. Biol. J. Am. Assoc. Cancer Res.* **2001**, *12*, 497–504.

90. Okuyama, E.; Suzuki, A.; Murata, M.; Ando, Y.; Kato, I.; Takagi, Y. Molecular mechanisms of syndecan-4 upregulation by TNF- in the endothelium-like EAhy926 cells. *J. Biochem.* **2013**, *154*, 41–50. [CrossRef]

91. Zhang, Y.; Pasparakis, M.; Kollias, G.; Simons, M. Myocyte-dependent Regulation of Endothelial Cell Syndecan-4 Expression. *J. Biol. Chem.* **1999**, *274*, 14786–14790. [CrossRef]

92. Vuong, T.T.; Reine, T.M.; Sudworth, A.; Jenssen, T.G.; Kolset, S.O. Syndecan-4 Is a Major Syndecan in Primary Human Endothelial Cells In Vitro, Modulated by Inflammatory Stimuli and Involved in Wound Healing. *J. Histochem. Cytochem.* **2015**, *63*, 280–292. [CrossRef]

93. Cantuti-Castelvetri, L.; Ojha, R.; Pedro, L.D.; Djannatian, M.; Franz, J.; Kuivanen, S. Neuropilin-1 facilitates SARS-CoV-2 cell entry and infectivity. *Science* **2020**, *357*, 856–860. [CrossRef]

94. Ceccarelli, S.; Nodale, C.; Vescarelli, E.; Pontecorvi, P.; Manganelli, V.; Casella, G. Neuropilin 1 Mediates Keratinocyte Growth Factor Signaling in Adipose-Derived Stem Cells: Potential Involvement in Adipogenesis. *Stem Cells Int.* **2018**, *2018*, e1075156. Available online: https://www.hindawi.com/journals/sci/2018/1075156/ (accessed on 6 November 2020). [CrossRef]

95. Magkos, F.; Mohammed, B.S.; Mittendorfer, B. Effect of obesity on the plasma lipoprotein subclass profile in normoglycemic and normolipidemic men and women. *Int. J. Obes.* **2008**, *32*, 1655–1664. [CrossRef]

96. Sears, B.; Perry, M. The role of fatty acids in insulin resistance. *Lipids Health Dis.* **2015**, *14*, 121. [CrossRef] [PubMed]

97. de Luca, C.; Olefsky, J.M. Inflammation and insulin resistance. *FEBS Lett.* **2008**, *582*, 97–105. [CrossRef] [PubMed]

98. Castoldi, A.; Naffah de Souza, C.; Câmara, N.O.S.; Moraes-Vieira, P.M. The Macrophage Switch in Obesity Development. *Front. Immunol.* **2016**. Available online: http://journal.frontiersin.org/Article/10.3389/fimmu.2015.00637/abstract (accessed on 13 April 2021). [CrossRef]

99. Lumeng, C.N.; Bodzin, J.L.; Saltiel, A.R. Obesity induces a phenotypic switch in adipose tissue macrophage polarization. *J. Clin. Investig.* **2007**, *117*, 175–184. [CrossRef] [PubMed]

100. Sun, Q.; Li, J.; Gao, F. New insights into insulin: The anti-inflammatory effect and its clinical relevance. *World J. Diabetes* **2014**, *5*, 89–96. [CrossRef] [PubMed]

101. Rochlani, Y.; Pothineni, N.V.; Kovelamudi, S.; Mehta, J.L. Metabolic syndrome: Pathophysiology, management, and modulation by natural compounds. *Ther. Adv. Cardiovasc. Dis.* **2017**, *11*, 215–225. [CrossRef]

102. Leija-Martínez, J.J.; Huang, F.; Del-Río-Navarro, B.E.; Sanchéz-Muñoz, F.; Muñoz-Hernández, O.; Giacoman-Martínez, A. IL-17A and TNF-α as potential biomarkers for acute respiratory distress syndrome and mortality in patients with obesity and COVID-19. *Med. Hypotheses* **2020**, *144*, 109935. [CrossRef]

103. Coomes, E.A.; Haghbayan, H. Interleukin-6 in Covid-19: A systematic review and META-ANALYSIS. *Rev. Med. Virol.* **2020**, *30*, 1–9. [CrossRef]

104. Tsai, S.; Clemente-Casares, X.; Zhou, A.C.; Lei, H.; Ahn, J.J.; Chan, Y.T. Insulin Receptor-Mediated Stimulation Boosts T Cell Immunity during Inflammation and Infection. *Cell Metab.* **2018**, *28*, 922–934.e4. [CrossRef] [PubMed]

105. Zhou, W.; Sailani, M.R.; Contrepois, K.; Zhou, Y.; Ahadi, S.; Leopold, S.R. Longitudinal multi-omics of host–microbe dynamics in prediabetes. *Nature* **2019**, *569*, 663–671. [CrossRef] [PubMed]

106. Huizinga, G.P.; Singer, B.H.; Singer, K. The Collision of Meta-Inflammation and SARS-CoV-2 Pandemic Infection. *Endocrinology* **2020**, *161*, bqaa154. [CrossRef] [PubMed]

107. Park, H.-K.; Ahima, R.S. Physiology of leptin: Energy homeostasis, neuroendocrine function and metabolism. *Metabolism* **2015**, *64*, 24–34. [CrossRef] [PubMed]

108. Jung, C.H.; Kim, M.-S. Molecular mechanisms of central leptin resistance in obesity. *Arch. Pharm. Res.* **2013**, *36*, 201–207. [CrossRef] [PubMed]

109. Bruno, A.; Conus, S.; Schmid, I.; Simon, H.-U. Apoptotic Pathways Are Inhibited by Leptin Receptor Activation in Neutrophils. *J. Immunol.* **2005**, *174*, 8090–8096. [CrossRef]

110. Caldefie-Chezet, F.; Poulin, A.; Vasson, M.-P. Leptin Regulates Functional Capacities of Polymorphonuclear Neutrophils. *Free Radic. Res.* **2003**, *37*, 809–814. [CrossRef]

111. Suzukawa, M.; Nagase, H.; Ogahara, I.; Han, K.; Tashimo, H.; Shibui, A. Leptin Enhances Survival and Induces Migration, Degranulation, and Cytokine Synthesis of Human Basophils. *J. Immunol.* **2011**, *186*, 5254–5260. [CrossRef]

112. Wong, C.K.; Cheung, P.F.-Y.; Lam, C.W.K. Leptin-mediated cytokine release and migration of eosinophils: Implications for immunopathophysiology of allergic inflammation. *Eur. J. Immunol.* **2007**, *37*, 2337–2348. [CrossRef]

113. Wrann, C.D.; Laue, T.; Hübner, L.; Kuhlmann, S.; Jacobs, R.; Goudeva, L. Short-term and long-term leptin exposure differentially affect human natural killer cell immune functions. *Am. J. Physiol. Endocrinol. Metab.* **2012**, *302*, E108–E116. [CrossRef] [PubMed]

114. Lamas, B.; Goncalves-Mendes, N.; Nachat-Kappes, R.; Rossary, A.; Caldefie-Chezet, F.; Vasson, M.-P. Leptin modulates dose-dependently the metabolic and cytolytic activities of NK-92 cells. *J. Cell Physiol.* **2013**, *228*, 1202–1209. [CrossRef]

115. Agrawal, S.; Gollapudi, S.; Su, H.; Gupta, S. Leptin Activates Human B Cells to Secrete TNF-α, IL-6, and IL-10 via JAK2/STAT3 and p38MAPK/ERK1/2 Signaling Pathway. *J. Clin. Immunol.* **2011**, *31*, 472–478. [CrossRef] [PubMed]

116. De Rosa, V.; Procaccini, C.; Calì, G.; Pirozzi, G.; Fontana, S.; Zappacosta, S. A Key Role of Leptin in the Control of Regulatory T Cell Proliferation. *Immunity* **2007**, *26*, 241–255. [CrossRef]

117. Smigiel, K.S.; Srivastava, S.; Stolley, J.M.; Campbell, D.J. Regulatory T cell homeostasis: Steady-state maintenance and modulation during inflammation. *Immunol. Rev.* **2014**, *259*, 40–59. [CrossRef] [PubMed]

118. Gogishvili, T.; Langenhorst, D.; Lühder, F.; Elias, F.; Elflein, K.; Dennehy, K.M. Rapid regulatory T-cell response prevents cytokine storm in CD28 superagonist treated mice. *PLoS ONE* **2009**, *4*, e4643. [CrossRef]

119. Matarese, G.; Procaccini, C.; De Rosa, V.; Horvath, T.L.; La Cava, A. Regulatory T cells in obesity: The leptin connection. *Trends Mol. Med.* **2010**, *16*, 247–256. [CrossRef] [PubMed]

120. Conde, J.; Scotece, M.; Abella, V.; López, V.; Pino, J.; Gómez-Reino, J.J. An update on leptin as immunomodulator. *Expert Rev. Clin. Immunol.* **2014**, *10*, 1165–1170. [CrossRef]

121. Howard, J.K.; Flier, J.S. Attenuation of leptin and insulin signaling by SOCS proteins. *Trends Endocrinol. Metab.* **2006**, *17*, 365–371. [CrossRef]

122. Rosário, C.; Zandman-Goddard, G.; Meyron-Holtz, E.G.; D'Cruz, D.P.; Shoenfeld, Y. The Hyperferritinemic Syndrome: Macrophage activation syndrome, Still's disease, septic shock and catastrophic antiphospholipid syndrome. *BMC Med.* **2013**, *11*, 185. [CrossRef]

123. Kell, D.B.; Pretorius, E. Serum ferritin is an important inflammatory disease marker, as it is mainly a leakage product from damaged cells. *Metallomics* **2014**, *6*, 748–773. [CrossRef]

124. Kernan, K.F.; Carcillo, J.A. Hyperferritinemia and inflammation. *Int. Immunol.* **2017**, *29*, 401–409. [CrossRef]

125. Pieracci, F.M.; Barie, P.S. Iron and the risk of infection. *Surg. Infect.* **2005**, *6* (Suppl. 1), S41–S46. [CrossRef]

126. Núñez, G.; Sakamoto, K.; Soares, M.P. Innate Nutritional Immunity. *J. Immunol.* **2018**, *201*, 11–18. [CrossRef]

127. Cavezzi, A.; Troiani, E.; Corrao, S. COVID-19: Hemoglobin, Iron, and Hypoxia Beyond Inflammation. A Narrative Review. *Clin. Pract.* **2020**. Available online: https://www.clinicsandpractice.org/index.php/cp/article/view/1271 (accessed on 11 June 2020). [CrossRef]

128. Ruan, Q.; Yang, K.; Wang, W.; Jiang, L.; Song, J. Clinical predictors of mortality due to COVID-19 based on an analysis of data of 150 patients from Wuhan, China. *Intensive Care Med.* **2020**, *46*, 846–848. [CrossRef] [PubMed]

129. Hanif, W.; Ali, O.; Shahzad, H.; Younas, M.; Iqbal, H.; Afzal, K. Biochemical Markers in COVID-19 in Multan. *J. Coll. Physicians Surg. Pak.* **2020**, *30*, 1026–1029. [PubMed]

130. Lin, Z.; Long, F.; Yang, Y.; Chen, X.; Xu, L.; Yang, M. Serum ferritin as an independent risk factor for severity in COVID-19 patients. *J. Infection.* **2020**, *81*, 647–679. [CrossRef] [PubMed]

131. Taneri, P.E.; Gómez-Ochoa, S.A.; Llanaj, E.; Raguindin, P.F.; Rojas, L.Z.; Roa-Díaz, Z.M. Anemia and iron metabolism in COVID-19: A systematic review and meta-analysis. *Eur. J. Epidemiol.* **2020**, *35*, 763–773. [CrossRef] [PubMed]

132. Ruscitti, P.; Ursini, F.; Cipriani, P.; Greco, M.; Alvaro, S.; Vasiliki, L. IL-1 inhibition improves insulin resistance and adipokines in rheumatoid arthritis patients with comorbid type 2 diabetes: An observational study. *Medicine* **2019**, *98*, e14587. [CrossRef]

133. Wenzhong, L.; Hualan, L. COVID-19: Attacks the 1-Beta Chain of Hemoglobin and Captures the Porphyrin to Inhibit Human Heme Metabolism. 2020. Available online: https://chemrxiv.org/articles/COVID-19_Disease_ORF8_and_Surface_Glycoprotein_Inhibit_Heme_Metabolism_by_Binding_to_Porphyrin/11938173 (accessed on 8 April 2021).

134. Lippi, G.; Mattiuzzi, C. Hemoglobin value may be decreased in patients with severe coronavirus disease 2019. *Hematol. Transfus. Cell Ther.* **2020**, *42*, 116–117. [CrossRef]

135. Wang, D.; Hu, B.; Hu, C.; Zhu, F.; Liu, X.; Zhang, J. Clinical Characteristics of 138 Hospitalized Patients With 2019 Novel Coronavirus–Infected Pneumonia in Wuhan, China. *JAMA* **2020**, *323*, 1061. [CrossRef]

136. Zhou, C.; Chen, Y.; Ji, Y.; He, X.; Xue, D. Increased Serum Levels of Hepcidin and Ferritin Are Associated with Severity of COVID-19. *Med. Sci. Monit.* **2020**, *26*, e926178. [CrossRef] [PubMed]

137. DeMartino, A.W.; Rose, J.J.; Amdahl, M.B.; Dent, M.R.; Shah, F.A.; Bain, W. No evidence of hemoglobin damage by SARS-CoV-2 infection. *Haematologica* **2020**, *105*, 2769–2773. [CrossRef] [PubMed]

138. Lecube, A.; Hernández, C.; Pelegrí, D.; Simó, R. Factors accounting for high ferritin levels in obesity. *Int. J. Obes.* **2008**, *32*, 1665–1669. [CrossRef]

139. Shoenfeld, Y. Corona (COVID-19) time musings: Our involvement in COVID-19 pathogenesis, diagnosis, treatment and vaccine planning. *Autoimmun. Rev.* **2020**, *19*, 102538. [CrossRef] [PubMed]
140. Tang, D.; Comish, P.; Kang, R. The hallmarks of COVID-19 disease. *PLoS Pathog.* **2020**, *16*, e1008536. [CrossRef] [PubMed]
141. Klok, F.A.; Kruip, M.J.H.A.; van der Meer, N.J.M.; Arbous, M.S.; Gommers, D.A.M.P.J.; Kant, K.M. Incidence of thrombotic complications in critically ill ICU patients with COVID-19. *Thrombosis Res.* **2020**, *191*, 145–147. [CrossRef]
142. Piazza, G.; Campia, U.; Hurwitz, S.; Snyder, J.E.; Rizzo, S.M.; Pfeferman, M.B. Registry of Arterial and Venous Thromboembolic Complications in Patients With COVID-19. *J. Am. Coll. Cardiol.* **2020**, *76*, 2060–2072. [CrossRef]
143. Wang, X.; Sahu, K.K.; Cerny, J. Coagulopathy, endothelial dysfunction, thrombotic microangiopathy and complement activation: Potential role of complement system inhibition in COVID-19. *J. Thromb. Thrombolysis* **2020**. Available online: https://doi.org/10.1007/s11239-020-02297-z (accessed on 11 November 2020). [CrossRef]
144. Ji, H.-L.; Zhao, R.; Matalon, S.; Matthay, M.A. Elevated Plasmin(ogen) as a Common Risk Factor for COVID-19 Susceptibility. *Physiol Rev.* **2020**, *100*, 1065–1075. [CrossRef]
145. Madoiwa, S. Recent advances in disseminated intravascular coagulation: Endothelial cells and fibrinolysis in sepsis-induced DIC. *J. Intensive Care* **2015**. Available online: https://www.ncbi.nlm.nih.gov/pmc/articles/PMC4940964/ (accessed on 11 November 2020). [CrossRef]
146. Iba, T.; Levy, J.H.; Connors, J.M.; Warkentin, T.E.; Thachil, J.; Levi, M. The unique characteristics of COVID-19 coagulopathy. *Crit. Care.* **2020**, *24*, 360. [CrossRef] [PubMed]
147. Schafer, K.; Konstantinides, S. Mechanisms linking leptin to arterial and venous thrombosis: Potential pharmacological targets. *Curr. Pharm. Des.* **2014**, *20*, 635–640. [CrossRef]
148. Buis, D.T.P.; Christen, T.; Smit, R.J.; de Mutsert, R.; Jukema, J.W.; Cannegieter, S.C. The association between leptin concentration and blood coagulation: Results from the NEO study. *Thromb. Res.* **2020**, *188*, 44–48. [CrossRef]
149. Martin, S.S.; Qasim, A.; Reilly, M.P. Leptin resistance: A possible interface of inflammation and metabolism in obesity-related cardiovascular disease. *J. Am. Coll. Cardiol.* **2008**, *52*, 1201–1210. [CrossRef]
150. Wang, H.; Luo, W.; Eitzman, D. Leptin in Thrombosis and Atherosclerosis. *Curr. Pharm. Des.* **2014**, *20*, 641–645. [CrossRef]
151. Scarpace, P.J. Elevated leptin: Consequence or cause of obesity? *Front. Biosci.* **2007**, *12*, 3531. [CrossRef]
152. Francisco, V.; Pino, J.; Campos-Cabaleiro, V.; Ruiz-Fernández, C.; Mera, A.; Gonzalez-Gay, M.A. Obesity, Fat Mass and Immune System: Role for Leptin. *Front. Physiol.* **2018**, *9*, 640. [CrossRef] [PubMed]
153. Loskutoff David, J.; Samad, F. The Adipocyte and Hemostatic Balance in Obesity. *Arterioscler. Thromb. Vasc. Biol.* **1998**, *18*, 1–6. [CrossRef]
154. Skurk, T.; Hauner, H. Obesity and impaired fibrinolysis: Role of adipose production of plasminogen activator inhibitor-1. *Int. J. Obes.* **2004**, *28*, 1357–1364. [CrossRef]
155. Shimomura, I.; Funahashi, T.; Takahashi, M.; Maeda, K.; Kotani, K.; Nakamura, T. Enhanced expression of PAI-1 in visceral fat: Possible contributor to vascular disease in obesity. *Nat. Med.* **1996**, *2*, 800–803. [CrossRef] [PubMed]
156. Alessi, M.C.; Peiretti, F.; Morange, P.; Henry, M.; Nalbone, G.; Juhan-Vague, I. Production of plasminogen activator inhibitor 1 by human adipose tissue: Possible link between visceral fat accumulation and vascular disease. *Diabetes* **1997**, *46*, 860–867. [CrossRef]
157. Darbandi, M.; Pasdar, Y.; Moradi, S.; Mohamed, H.J.J.; Hamzeh, B.; Salimi, Y. Discriminatory Capacity of Anthropometric Indices for Cardiovascular Disease in Adults: A Systematic Review and Meta-Analysis. *Prev. Chronic. Dis.* **2020**, *17*, E131. [CrossRef] [PubMed]
158. Morange, P.E.; Lijnen, H.R.; Alessi, M.C.; Kopp, F.; Collen, D.; Juhan-Vague, I. Influence of PAI-1 on adipose tissue growth and metabolic parameters in a murine model of diet-induced obesity. *Arterioscler. Thromb. Vasc. Biol.* **2000**, *20*, 1150–1154. [CrossRef]
159. Singh, P.; Peterson, T.E.; Barber, K.R.; Kuniyoshi, F.S.; Jensen, A.; Hoffmann, M. Leptin upregulates the expression of plasminogen activator inhibitor-1 in human vascular endothelial cells. *Biochem. Biophys. Res. Commun.* **2010**, *392*, 47–52. [CrossRef]
160. Félétou, M. Multiple Functions of the Endothelial Cells [Internet]. The Endothelium: Part 1: Multiple Functions of the Endothelial Cells—Focus on Endothelium-Derived Vasoactive Mediators. Morgan & Claypool Life Sciences. 2020. Available online: https://www.ncbi.nlm.nih.gov/books/NBK57148/ (accessed on 12 November 2020).
161. Hasan, S.S.; Fischer, A. The Endothelium: An Active Regulator of Lipid and Glucose Homeostasis. *Trends Cell Biol.* **2020**, *31*, 37–41. [CrossRef] [PubMed]
162. Kaur, R.; Kaur, M.; Singh, J. Endothelial dysfunction and platelet hyperactivity in type 2 diabetes mellitus: Molecular insights and therapeutic strategies. *Cardiovasc. Diabetol.* **2018**, *17*, 121. [CrossRef]
163. Pasquarelli-do-Nascimento, G.; Braz-de-Melo, H.A.; Faria, S.S.; Santos, I.O.; Kobinger, G.P.; Magalhães, K.G. Hypercoagulopathy and Adipose Tissue Exacerbated Inflammation May Explain Higher Mortality in COVID-19 Patients With Obesity. *Front. Endocrinol.* **2020**. Available online: https://www.ncbi.nlm.nih.gov/pmc/articles/PMC7399077/ (accessed on 11 November 2020). [CrossRef] [PubMed]
164. Hamming, I.; Timens, W.; Bulthuis, M.L.C.; Lely, A.T.; Navis, G.J.; Goor, H. Tissue distribution of ACE2 protein, the functional receptor for SARS coronavirus. A first step in understanding SARS pathogenesis. *J. Pathol.* **2004**, *203*, 631–637. [CrossRef]
165. Nägele, M.P.; Haubner, B.; Tanner, F.C.; Ruschitzka, F.; Flammer, A.J. Endothelial dysfunction in COVID-19: Current findings and therapeutic implications. *Atherosclerosis* **2020**, *314*, 58–62. [CrossRef]

166. Halder, M.; Petsophonsakul, P.; Akbulut, A.C.; Pavlic, A.; Bohan, F.; Anderson, E. Vitamin K: Double Bonds beyond Coagulation Insights into Differences between Vitamin K1 and K2 in Health and Disease. *Int. J. Mol. Sci.* **2019**, *20*, 896. [CrossRef]
167. Li, C.; Li, J.; He, F.; Li, K.; Li, X.; Zhang, Y. Matrix Gla protein regulates adipogenesis and is serum marker of visceral adiposity. *Adipocyte* **2020**, *9*, 68–76. [CrossRef]
168. Ravera, M.; Nickolas, T.; Plebani, M.; Iervasi, G.; Aghi, A.; Khairallah, P. Overweight—Obesity is associated with decreased vitamin K2 levels in hemodialysis patients. *Clin. Chem. Lab. Med.* **2020**, *59*, 581–589. [CrossRef] [PubMed]
169. Shea, M.K.; Booth, S.L.; Gundberg, C.M.; Peterson, J.W.; Waddell, C.; Dawson-Hughes, B. Adulthood obesity is positively associated with adipose tissue concentrations of vitamin K and inversely associated with Curr. Immunol. Rev.culating indicators of vitamin K status in men and women. *J. Nutr.* **2010**, *140*, 1029–1034. [CrossRef] [PubMed]
170. Wortsman, J.; Matsuoka, L.Y.; Chen, T.C.; Lu, Z.; Holick, M.F. Decreased bioavailability of vitamin D in obesity. *Am. J. Clin. Nutr.* **2000**, *72*, 690–693. [CrossRef]
171. Landrier, J.-F.; Marcotorchino, J.; Tourniaire, F. Lipophilic Micronutrients and Adipose Tissue Biology. *Nutrients* **2012**, *4*, 1622–1649. [CrossRef] [PubMed]
172. Dofferhoff, A.S.M.; Piscaer, I.; Schurgers, L.J.; Visser, M.P.J.; van den Ouweland, J.M.W.; de Jong, P.A. Reduced vitamin K status as a potentially modifiable risk factor of severe COVID-19. *Clin. Infect. Dis.* **2020**. [CrossRef]
173. Branca, F.; Lartey, A.; Oenema, S.; Aguayo, V.; Stordalen, G.A.; Richardson, R. Transforming the Food System to Fight Non-Communicable Diseases. *BMJ* **2019**. Available online: https://www.bmj.com/content/364/bmj.l296 (accessed on 19 November 2020). [CrossRef]
174. Bruins, M.J.; Van Dael, P.; Eggersdorfer, M. The Role of Nutrients in Reducing the Risk for Noncommunicable Diseases during Aging. *Nutrients* **2020**, *85*. Available online: https://www.ncbi.nlm.nih.gov/pmc/articles/PMC6356205/ (accessed on 19 November 2020). [CrossRef]
175. Rauber, F.; Da Costa Louzada, M.L.; Steele, E.M.; Millett, C.; Monteiro, C.A.; Levy, R.B. Ultra-Processed Food Consumption and Chronic Non-Communicable Diseases-Related Dietary Nutrient Profile in the UK (2008–2014). *Nutrients* **2018**, *10*, 587. [CrossRef]
176. Lee, I.-M.; Shiroma, E.J.; Lobelo, F.; Puska, P.; Blair, S.N.; Katzmarzyk, P.T. Effect of physical inactivity on major non-communicable diseases worldwide: An analysis of burden of disease and life expectancy. *Lancet* **2012**, *380*, 219–229. [CrossRef]
177. Sridhar, G.R.; Sanjana, N.S.N. Sleep, Curr. Immunol. Rev.cadian dysrhythmia, obesity and diabetes. *World J. Diabetes* **2016**, *7*, 515–522. [CrossRef] [PubMed]
178. Fricchione, G.L. The Challenge of Stress-Related Non-Communicable Diseases. *Med. Sci. Monit. Basic Res.* **2018**, *24*, 93–95. [CrossRef] [PubMed]
179. de Frel, D.L.; Atsma, D.E.; Pijl, H.; Seidell, J.C.; Leenen, P.J.M.; Dik, W.A. The Impact of Obesity and Lifestyle on the Immune System and Susceptibility to Infections Such as COVID-19. *Front. Nutr.* **2020**. Available online: https://www.frontiersin.org/articles/10.3389/fnut.2020.597600/full?&utm_source=Email_to_authors_&utm_medium=Email&utm_content=T1_11.5e1_author&utm_campaign=Email_publication&field=&journalName=Frontiers_in_Nutrition&id=597600 (accessed on 23 November 2020).

MDPI

St. Alban-Anlage 66

4052 Basel

Switzerland

Tel. +41 61 683 77 34

Fax +41 61 302 89 18

www.mdpi.com

Cells Editorial Office

E-mail: cells@mdpi.com

www.mdpi.com/journal/cells